Some Aspects of
Vacuum Ultraviolet Radiation Physics

Some Aspects of

VACUUM ULTRAVIOLET RADIATION PHYSICS

Edited by
Nicole Damany, Executive Editor,
Jacques Romand and Boris Vodar

Laboratoire des Interactions Moléculaires et des Hautes Pressions,
Centre National de la Recherche Scientifique,
Bellevue, 92 190 Meudon, France.

With 108 Figures, 4 Plates, and 28 Tables

AKADEMIE-VERLAG BERLIN 1974

Erschienen im Akademie-Verlag, 108 Berlin, Leipziger Straße 3—4

Lizenznummer: 202 · 100/437/74
Gesamtherstellung: VEB Druckhaus „Maxim Gorki", 74 Altenburg
Bestellnummer 761 759 3 (6054) · LSV 1165
Printed in GDR
EVP 68,—

Contents

Foreword

THE main content of this book is dated from January 1971 and gives the state of research until 1970.

Due to the long delays required for publication, the bibliography has been up dated by the authors until the middle of 1972, and some more recent references have been added on proofs.

N. DAMANY
J. ROMAND
B. VODAR

Contributors

C. Breton — Service de Recherche de la Fusion Contrôlée, Commissariat à l'Energie Atomique, 92260 Fontenay aux Roses, France.

H. Damany — Laboratoire des Interactions Moléculaires et des Hautes Pressions du CNRS, Bellevue, 92190, Meudon France.

N. Damany (Mrs) — Laboratoire des Interactions Moléculaires et des Hautes Pressions du CNRS, Bellevue, 92190 Meudon, France.

S. Leach — Laboratoire de Photophysique Moléculaire, Faculté des Sciences, 91405 Orsay, France.

B. A. Lombos — Sir George Williams University, Montreal 107, Quebec, Canada.

M. Pouey — Laboratoire des Interactions Moléculaires et des Hautes Pressions du CNRS, Bellevue, 92190 Meudon, France.

M. Priol — Laboratoire de Spectroscopie, Université de Rennes, 35031 Rennes-Cedex, France.

S. Robin (Mrs) — Laboratoire de Spectroscopie, Université de Rennes, 35031 Rennes-Cedex, France.

S. Robin — Laboratoire de Spectroscopie, Université de Rennes, 35031 Rennes-Cedex, France.

J. Romand — Laboratoire des Interactions Moléculaires et des Hautes Pressions du CNRS, Bellevue, 92190 Meudon, France.

J.-Y. Roncin — Laboratoire des Interactions Moléculaires et des Hautes Pressions du CNRS. Bellevue, 92199 Meudon, France.

J. L. Schwob — Laboratoire des Interactions Moléculaires et des Hautes Pressions du CNRS, Bellevue, 92190 Meudon, France.

G. Stephan — Laboratoire de Spectroscopie, Université de Rennes, 35031 Rennes-Cedex, France.

B. Vodar — Laboratoire des Interactions Moléculaires et des Hautes Pressions du CNRS, Bellevue, 92190 Meudon, France.

Acknowledgements

A PARTICULAR large debt of gratitude is due to Dr. Chandrasekharan, Maître de Recherche at the Laboratoire des Interactions Moléculaires et des Hautes Pressions, Vacuum ultraviolet Section, Bellevue (France) whose scientific competence in the subjects treated facilitated the important task of translation into English of several chapters of the present book.

We take pleasure in making acknowledgement of the help of some friends who have reread and criticized portions of the manuscript, particularly Dr. G. Tabisz (Winnipeg, Canada) and Dr. B. J. Orr (Sydney, Australia).

The co-operative effort necessary for the preparation of this book is due to the Centre National de la Recherche Scientifique, which supports the work of most of the authors.

We are also grateful to the editors of *Solid State Physics*, *Optica Acta*, *Journal of Optical Society*, *Applied Optics*, *Compte Rendus de l'Académie des Sciences*, *Physical Review*, *Physica Status Solidi*, *Journal of the Physical Society of Japan*, *Optics Communications*, *Journal of Chemical Physics*, *Astrophysical Journal*, *Journal of Molecular Spectroscopy*, *Canadian Journal of Physics*, *Physics Letters*, *Molecular Spectroscopy*, *International Journal of Mass Spectroscopy and Ion Physics*, *Review of Scientific Instruments*, *Proceeding of the Royal Society*, *Journal of Scientific Instruments* (*Journal of Physics*, E), *Revue d'Optique*, *Revue de Physique appliquée*, *Journal of Quantitative Spectroscopy, and Radiative Transfer*, the Academic Press Inc., North-Holland Publishing Co., and Wiley & Sons Inc. for permission to reproduce various figures, and also to individual authors for their permission: acknowledgements to authors are given in each case by reference under the figures.

We wish to thank Mrs. Giroir and Mrs. Roumier for great care in typing the manuscript.

N. DAMANY
J. ROMAND
B. VODAR

Introduction

J. Romand and B. Vodar
Laboratoire des Interactions Moléculaires et des Hautes Pressions du CNRS, Bellevue, 92190 Meudon, France

The vacuum ultraviolet region (v.u.v.) is commonly regarded as extending from about 2000 Å to a few angstroms, thus overlapping that of soft X-rays. The distinction resides essentially in the origin of the phenomena concerned; the far ultraviolet concerns optical spectra resulting from the transitions of the outer electrons of an atom or ion, while the soft X-rays arise from the transitions of the internal electrons. The extent of the overlap is rather important if one notes that on one side there are numerous examples of X-ray discontinuities situated above 100 Å (examples: Be (K) at 110 Å, Al (L) at 170 Å, Mg (L) at 250 Å, Ti ($M_{2,3}$) at 350 Å, etc.), while, on the other side, optical spectra from 2 to 20 Å are emitted by ions isoelectronic with hydrogen (O VIII, F IX, Ne X, Ti XXII) and helium (F VIII, Ne IX, Na X, Mg XI, Ti XXI, Fe XXV, Cu XXVIII), produced in plasmas at high temperature.

Schumann[1] was the first to penetrate into the v.u.v. in 1893 by discovering and overcoming the three obstacles to the extension of spectroscopy below 2000 Å: the opacity of quartz prisms and lenses, which he replaced by optical elements made of fluorite; the lack of transparency of air (due to oxygen) below 1850 Å; and the high absorption of the gelatine on photographic plates. Later, Lyman[2] took a fundamental step in introducing the concave grating as the dispersing element; this permitted him to measure wavelengths and to extend the spectra below the limit of transparency of fluorite (1250 Å) down to the region of 250 Å.

The various steps in the history of progression towards shorter wavelengths have been well described in the review by Tousey.[3] As early as 1927 the junction with X-rays (and even the overlapping of the spectra) had been made, on the one hand, by optical methods employing ruled gratings at grazing incidence,[4] and, on the other, by X-ray methods utilizing a fatty acid crystal of large lattice spacing.[5] To the same period belongs the work of Thibaud[6] in the soft X-ray area with a plane grating at grazing incidence by means of which he obtained, with a vacuum spark, spectra from about 4000 to 140 Å in a single exposure. At about the same time, Holweck[7] studied the v.u.v. range with non-dispersive techniques; at one time the name of "Holweck region" was given to the 600 to 10 Å range in which he chiefly worked.

The period from 1925 to 1942 saw the beginning of the study of molecular spectra, optical properties of solids, and photoelectric effects in the v.u.v. However, during this period, when much interest in physics was centred around the problems of atomic structure, emission spectroscopy led to the most striking results. A notable accomplishment was established by the work of Edlen:[8] from precise observations made in the laboratory on extreme u.v. spectra of highly ionized atoms, he proved conclusively that certain visible lines of the Solar Corona belong to spectra of ions (Fe XIV, for example), which implied a temperature of the order of 10^6 K for the Corona.

The book of Bomke[9] published in 1937 and the review made by Boyce[10] in 1941 give accounts of the situation in the v.u.v. spectroscopy up to the end of this first period, roughly 40 years after Schumann's discovery.

After 1945 progress was markedly accelerated for technical reasons as well as for physical motivations. Regarding the first, one may consider the following factors, which we believe were decisive in the development of v.u.v. instrumentation and experiment, and about which we give below some brief indications together with some very incomplete early references:

— The improvement in vacuum techniques.
— The improvements in grating-ruling techniques and—especially—the replication of concave gratings which made this spectral region, depending entirely on concave gratings, more easily accessible.
— The various new mountings using the concave grating, e.g. the purely rotational mounting monochromator of Seya[11] and the purely geometrical focusing of a grazing incidence monochromator.[12]
— The development of synthetic crystals: mainly lithium and calcium fluoride and also quartz of high purity; more recently, magnesium fluoride.
— Improvements in existing light sources and the discovery of new ones: (i) for line spectra the sliding and the triggered vacuum sparks,[13,14] representing practical improvements over the Millikan "hot spark"; (ii) for continuous spectra, progress was essential (as no good continuum source existed for the whole v.u.v. range) and is represented by the new rare gases discharges,[15,16] the vacuum-triggered high-intensity spark,[17,18] and the synchrotron radiation.[19]
— The appearance of new types of photomultipliers, mainly the solar-blind ones and the windowless type which is especially useful in the short-wavelength region.
— A great improvement lies in the development of appropriate coatings for mirrors and gratings[20,21] since the reflecting powers are usually low and highly dependent on surface contamination or oxidation (see below).
— The development of time resolution techniques, important in the study of plasmas, and of more sophisticated methods of measurement.

Details about the techniques may be found in two recent and important books, one by Zaidel and Schreider,[21] and the other by Samson,[22] the latter being devoted exclusively to experimental developments.*

* These books having both appeared in 1967, it may be noted that no book on v.u.v. had been published for the 30 years between 1937 and 1967.

A large variety of spectrographs and monochromators of good quality have now become available commercially. Nevertheless, much remains to be done in the production of apparatus for specialized purposes, e.g. absorption spectrophotometers: certain existing models have yet attained neither the degree of perfection nor the commercial availability of corresponding instruments in the other spectral regions—infrared, visible, and near u.v. This relative slowness in the progress of commercial instruments contrasts particularly with the rapid evolution of infrared instrumentation. This probably connects with the rather underdeveloped state of industrial application of v.u.v. spectroscopy.

Besides these stimuli of purely technical character, work in the v.u.v. has been encouraged for particular scientific reasons, notably the development of space astrophysics and the emergence of plasma research in the period following World War II.

The penetration of the opaque barrier formed by the terrestrial atmosphere with rockets and satellites throws open the entire v.u.v. region to instruments carried by such spacecrafts. In particular the spectra of ions of high-ionization stage thereby obtained, provide a rich source of information about celestial emitters; likewise the study of absorbed and scattered light is very important for the understanding of the earth's atmosphere and of interstellar matter. It is evident that for the interpretation (in particular for identification of spectra) these investigations must be completed by observations on laboratory "hot" sources. Thus a close relationship appears between this aspect of space research and the study of hot plasmas produced essentially by magnetic confinement in laboratory machines with a view to thermonuclear applications. These plasmas, of relatively low density but rather high temperature (10-500 eV), are excellent sources for the spectroscopist in the v.u.v. The recording of highly ionized spectra, the determination of transition probabilities, the study of collision effects (from line profiles), etc., are some presently flourishing fields of investigation. On the other hand, it is quite evident that the emitted spectra are closely dependent on the plasma parameters, and spectroscopy—especially in the v.u.v.—provides important diagnostic methods for the determination of certain of these parameters, such as the electronic temperature.

Some other areas of interest, though perhaps less spectacular, contribute as strongly to the development of v.u.v. research. Due to the large range of energy covered in this region (six to several hundred electron volts), v.u.v. radiation interacts strongly with matter; in fact, all atoms and molecules absorb totally or over part of this region. Two examples suffice to illustrate this remark.

In molecular physics the electronic spectra in the v.u.v. are due to transitions involving a large variety of excited states. In spite of the complexity of the spectra and of the multiplicity of the mechanisms (predissociation, non-radiative processes, breaking of bonds, photodissociation, etc.[23]), the v.u.v. spectra often constitute the best or a unique way of getting information about these states, particularly as to the geometry of the excited molecule, which sometimes is profoundly modified with respect to the ground state. As knowledge about the excited state is essential for the comprehension of photochemical or high-temperature reactions, careful study of these states by v.u.v. spectroscopy is an important field of research.

Another interesting field concerns transitions involving Rydberg states, which, indeed, may be observed only in the v.u.v. One may formally distinguish two types of state—*valence states* in which the excited molecular orbitals are correlated with atomic orbitals of the constituent atoms whose principal quantum numbers are all the same as those of the ground state, and the other, *Rydberg states*, in which one of the quantum numbers is greater than in the ground state. In a Rydberg transition the excited electron goes far from the core; structure may generally be observed for the lower terms of the series. The various Rydberg states corresponding to increasing values of the quantum number n of the excited electron give rise to a series of bands, closely analogous to series in atomic spectra, and this is represented by a Rydberg formula; they converge towards a short-wavelength limit, corresponding to the removal of the electron, and hence to the formation of a molecular ion followed by a photoionization continuum. There are several Rydberg series, ionization potentials, and photoionization continua, and generally autoionization takes place.

The spectroscopic investigation of the important phenomena connected with the ionization of molecules necessitates observations in the v.u.v., and many investigations have had such motivation. In particular, much attention has been paid to the dependence of ionization potentials on the nature of bonds, the presence of substituents, etc.[24, 25] These potentials could be determined by several methods which allow comparison and checking of different data, extrapolation of line series to the limit, direct measurements of photoionization current, and also ionization under energetically controlled electron impacts. The association of mass spectrometry permits at the same time the characterization of ions and radicals formed during the process. However, these methods are faulty when autoionization takes place. Therefore one prefers now the modern photoelectron spectroscopy method which consists in provoking ionization by means of highly energetic photons and investigating the energy spectrum of the ejected photoelectrons. This method, which gives information about the structure of the ions and also affords analytic applications, is presently undergoing a large development as evidenced by the appearance of commercially available instruments.

A second example of development of v.u.v. spectroscopy is concerned with solid-state studies. The electronic properties of solids may be characterized by the complex dielectric constant which is connected with the optical constants by classical formulae. These data may be interpreted in terms of various mechanisms of excitation and allow construction of the band diagrams of solids. The main types of transitions are: interband transitions (the simplest case being that of direct transitions keeping the **k** vector unchanged), excitonic transitions resulting in the creation of an electron-hole pair by excitation of an electron from the valence band, and plasmons corresponding to a collective oscillation of electrons, analogous to those present in gaseous plasmas.

For determination of these optical constants in the v.u.v., transmittance studies can be made only in the rare regions of transparency, and reflectance data are generally used. It is important to recognize that these optical constants are strongly sensitive to the structure of the investigated material (single crystal, polycrystalline or amorphous sample, evaporated film, etc.) and also to the surface roughness and cleanliness (pollution, adsorption, oxidation, etc.). A classical example is that of aluminium, which was believed to be a poor reflector in the v.u.v., while in fact only the reflectance of the superficial aluminium oxide film was being measured. Great care is taken now to eliminate all the parasitic factors and to study the properties of a material as well defined as possible, by using ultra-high vacuum techniques, cleavage of crystals under vacuum, and control of structure by other methods (X-ray diffraction). However, the v.u.v. is certainly a very favourable region for the study of surface phenomena, such as adsorption, etc., a field in which some important developments may be foreseen in the coming years.

Astrophysics, plasma physics, molecular physics, and solid-state physics are the research areas which have been mentioned in this brief review as utilizing largely v.u.v. radiation and as having contributed to the development of vacuum spectroscopy. It is noticeable that these wide fields (along with instrumentation) formed the topics of the first international conference on v.u.v. radiation physics which took place in 1962,[26] whereas the second conference, in 1968, was devoted uniquely to the radiation interactions with solids in the v.u.v.

Beside the broad themes and relatively well-developed fields outlined above, some other aspects of v.u.v. research, which are presently less developed, deserve a short mention. Among them photochemistry certainly holds the first rank: the more detailed studies of strongly excited molecular states, metastable states, the species formed under irradiation, could provide precious information on the chemical reaction mechanism and lead to important applications; we may expect an increasing and rapid development of these researches in the near future. A closely connected problem concerns the study of molecular interactions in dense media (dense gases or molecules trapped in solid matrices). Its importance and specificity in that region lie in the fact that, due to the high values of photon energies, one can study perturbations of excited states which, in thermodynamic equilibrium, do exist only under extreme temperature conditions, generally not easily attainable. Analytical applications should also be mentioned, and some preliminary work has already shown the possibilities offered by absorption spectra,[27] where the high cross-sections in the v.u.v.—particularly for the Rydberg lines—is beneficial, and also by emission spectra, where the sensitivity of lines due to highly ionized atoms is profitably exploited.[28] Nevertheless, in spite of the recent progress in development of the v.u.v., provided now with elaborate instruments and techniques, the access of this region is still rather difficult to some modern advances, in particular those relying on lasers, since, in spite of some recent promising results,[29,30] short wavelength u.v. lasers are not yet readily available. Thus we should note that, whereas the holographic gratings, as presently made, can have great advantages in the v.u.v. (e.g. because of the possibility of very close rulings), some of the aberration corrections for concave gratings which they promise seem not yet fully applicable to this spectral range because such corrections are obtained by using, in the production process, a laser light, of much longer wavelength. Nevertheless, the possibility of correcting certain aberrations by using non-linear rulings, earlier discussed by Zernicke and now feasible by holography, might bring some innovations in the future to vacuum spectroscopy, which relies almost entirely on the use of concave gratings. Indeed, the still poor reflectances, making impractical the use of multiple reflection devices, have prevented development of interferometric methods; as a partial consequence, our knowledge of hyperfine structure in this spectral range is almost non-existent.

Until now, matters of technique and instrumentation have been preponderant in the books on v.u.v. spectroscopy. The purpose of the present work is, on the contrary, to give to the reader, whether physicist or graduate student, some information on the present state of research in the v.u.v. in relation to large areas of physics; it does not pretend to be complete, and the questions concerning astrophysics as well as those relating to the very short-wavelength region (less than 100 Å—region of soft X-ray spectroscopy) have been left aside, as these subjects could only have been touched on in the restricted scope of this book. Therefore a limited number of well-defined fields have been described.

The first part, devoted to the optical properties of solids, is divided into three chapters treating, respectively, the specific methods of determination of optical constants in v.u.v., the properties of metals, and those of ionic insulators.

In the second part, dealing with molecular spectroscopy, four chapters describe the spectra of diatomic and simple polyatomic molecules, those of paraffins, those of condensed phases, and, finally photoelectron spectroscopy.

The third part is concerned with some aspects of emission spectroscopy in the v.u.v. in relation to laboratory plasmas.

Although the technical aspects of v.u.v. spectroscopy have been largely omitted, the fourth part of this book deals with image formation by concave gratings, spectrophotometry, and diffusion by surfaces, with emphasis on some refinements which may be brought into the instrumental optics in this short-wavelength region.

References

1. Schumann's papers are listed in Lyman's book (see ref. 2).
2. Lyman, T., *The Spectroscopy of the Extreme Ultraviolet*, 2nd ed., Longmans, London, 1928.
3. Tousey, R., *Appl. Opt.* **1**, 679 (1962).
4. Osgood, T. H., *Phys. Rev.* **30**, 567 (1927).
5. Dauvillier, A., *J. Phys. Radium* **8**, 1 (1927).
6. Thibaud, J., *J. Phys. Radium* **8**, 13, 447 (1927).
7. Holweck, F., *De la lumière aux rayons X*, Presses Universitaires de France, Paris, 1927.
8. Edlen, B., *Z. Astrophys.* **20**, 30 (1942).
9. Bomke, H., *Vakuumspektroskopie*, Barth, Leipzig (1937).

10. Boyce, J. C., *Rev. Mod. Phys.* **13**, 1 (1941).
11. Seya, M., *Sci. Light, Tokyo* **2**, 8 (1952).
12. Salle, M., and Vodar, B., *Compt. Rend.* **230**, 380 (1950).
13. Vodar, B., and Astoin, N., *Nature* **166**, 1029 (1950).
14. Balloffet, G., *Ann. Phys.* **5**, 1243 (1960).
15. Tanaka, Y., and Zelikoff, M., *J. Opt. Soc. Am.*, **44**, 254 (1954).
16. Tanaka, Y., *J. Opt. Soc. Am.* **45**, 710 (1955).
17. Balloffet, G., Romand, J., and Vodar, B., *Compt. Rend.* **252**, 4139 (1961).
18. Damany, H., Roncin, J.-Y., and Damany-Astoin, N., *Appl. Opt.* **5**, 297 (1966).
19. Hartmann, P. L., and Tomboulian, D. H., *Phys. Rev.* **91**, 1577 (1953).
20. Hass, G., Hunter, W. R., and Tousey, R., *J. Opt. Soc. Am.* **46**, 1009 (1956).
21. Zaidel, A. H., and Schreider, E. R., *Spectroscopia Vakuumnovo Ultraviolet*, Naouka, Moscow, 1967. English translation: *Vacuum Ultraviolet Spectroscopy*, Ann-Arbor, Humphrey Service, 1970.
22. Samson, J. A. R., *Techniques of Vacuum Ultraviolet Spectroscopy*, Wiley, New York, 1967.
23. Ramsay, D. A., Electronic spectra of polyatomic molecules and the configurations of molecules in excited electronic states, in *Determination of Organic Structures by Physical Methods*, vol. 2 (eds. Nachod, F. C., and Philipps, W. D.), Academic Press, New York, 1962.
24. Walsh, A. D., *J. Chem. Soc.* 2260 (1953).
25. Turner, D. W., Far and vacuum ultraviolet spectroscopy, in *Determination of Organic Structures by Physical Methods*; see ref. 23.
26. Proceedings first international conference on vacuum ultraviolet radiation physics, *J. Quant. Spectrosc Radiative Transfer* **2**, 313 (1962).
27. Kaye, W., *Appl. Spectrosc.* **15**, 89 (1961), and *Anal. Chem.* **34**, 287 (1962).
28. Romand, J., and Berneron, R., *Proceedings of XIIIth Colloquium Spectroscopicum Internationale, Ottawa, 1967*, Hilger, London, 1968, p. 434.
29. Waynant, R. W., *Phys. Rev. Letters* **28**, 533 (1972).
30. Koehler, H. A., Ferderber, L. J., Redhead, D. L., and Ebert, P. J., *Appl. Phys. Lett.* **21**, 198 (1972).

PART I

OPTICAL PROPERTIES OF SOLIDS IN THE VACUUM ULTRAVIOLET

General Principles

S. Robin (Mrs)
Laboratoire de Spectroscopie, Université de Rennes, 35031 Rennes-Cedex, France

The aim of this part is to summarize the principal work on the optical properties of solids in the far u.v. and to show that this spectral region is capable of providing important contributions to solid-state physics based upon a theory in full development. Indeed, the great experimental advances of recent years have provided measurements of more refinement for comparison with theoretical predictions and leading to expectation of further improvement of theory. We are obliged to limit ourselves generally to the research carried out only in the last 10 years. The reader should also consult the following review articles on the optical properties of solids and their interpretations by Madden,[1] Fan,[2] and Phillips,[3] and the latest book by Greenaway and Harbeke.[4]

The optical properties of solids are studied in the region 6-50 eV chiefly through measurements of reflectance or transmittance. As all materials are strongly absorbing in this region, the latter measurements are made on very thin films produced by evaporation, which may have very different properties from those of a bulk substance. The cleanliness of the surfaces being decisive, the best experimental conditions should be sought, e.g. evaporated films in ultra-high vacuum (10^{-9} torr) or crystals cleaved *in situ*. Unfortunately, such conditions apply only in the most recent work. Finally, we exclude discussion of other techniques such as photoelectric effect, electroabsorption, etc., but mention them only as an indispensable complement for the proposed interpretation.

In Chapter 1, various formulae are first summarized which are needed for a phenomenological basis, and then the methods of determination of the optical constants in the far u.v. are dealt with. The study of optical properties of solids is divided into two parts: metals, and ionic insulators, each corresponding to a chapter. However, it seemed useful to outline briefly the process of obtaining the directly exploitable physical quantities from the optical constants and describing the various solid-state phenomena (excitons, interband transitions, and plasmons).

The real and imaginary parts of the relative dielectric constant, $\hat{\varepsilon} = \varepsilon_1 - i\varepsilon_2$, are related to the optical constants n and k through the equations

$$\varepsilon_1 = n^2 - k^2,$$

$$\varepsilon_2 = 2nk.$$

The quantum theory of interband transitions gives

$$\varepsilon_2(\omega) = \frac{e^2h^2}{4\pi m\varepsilon_0} \Sigma_{ij} \frac{1}{\Omega} \int_S \frac{f_{ij}(\mathbf{k})}{E_{ij}\ |\nabla E_{ij}|}\, dS_{\mathbf{k}}\, |f(E_i) - f(E_j)|, \qquad E_{ij} = \hbar\omega,$$

in the case of a direct dipole electronic transition from the state of the valence band $E_j(\mathbf{k})$ associated with the wave vector $\mathbf{k}$, to the corresponding state of the conduction band $E_i(\mathbf{k})$ (neglecting the relaxation term). $f(E)$ is the distribution function, $dS_{\mathbf{k}}$ a surface element of the equal energy surface S in $\mathbf{k}$-space, $E_i(\mathbf{k}) - E_j(\mathbf{k}) = \hbar\omega$, and Ω the volume of the Brillouin zone.

$$f_{ij}(\mathbf{k}) = \frac{2}{3m} \frac{|<i, k|\, p\, |j, k>|^2}{E_{ij}}$$

is the oscillator strength.

ε_2 will give information about the interband density of states.

$$N(E) = \frac{dn}{dE} = \frac{I}{\Omega} \int_S \frac{dS_{\mathbf{k}}}{|\nabla_{\mathbf{k}} E_{ij}|}$$

provided that the other factors are considered as constants. The singularities of ε_2 correspond, then, to the critical points of the joint density of states (van Hove singularities); the latter occur at the points of the Brillouin zone where

$$\hbar\omega_{ij}(\mathbf{k}) = E_i(\mathbf{k}) - E_j(\mathbf{k})$$

is stationary, or

$$\nabla_{\mathbf{k}}\omega_{ij} = 0.$$

The singularities occur very often at points of high symmetry of the Brillouin zone (e.g. Γ, X, L, points for a face-centred cubic system, Fig. 1). The four types of singularities (this number is limited by considerations of symmetry[6]) named M_0, M_1, M_2, M_3 are represented in Fig. 2. It is to be noted that a single critical point does not produce a narrow peak, but only an absorption edge: a peak is due to two singularities M_1 and M_2 accidentally having nearly the same energy.

The simplest case is that of permitted electric-dipole direct transitions, the bottom of the conduction band and the top of the valence band being on the same vertical in the $E(\mathbf{k})$ diagram. However, it is sometimes possible to observe either indirect transitions involving a phonon, or non-direct transitions, for which the conservation of the wave vector is no longer an important rule. The study of the energy distribution in the photoemission of electrons by noble metals[7] has afforded supplementary evidence for these non-direct transitions.

Secondly, a photon creates a positive hole by removing an electron from the valence band, and this hole attracts the excited electron: the electron-hole pair bounded by Coulomb attraction is called an exciton.[8] Thereby, the minimum excitation energy of the crystal becomes smaller than the forbidden band-gap for semiconductors and insulators; the formation of excitons is not limited to the absorption threshold; it is to be noted that the excitonic lines or bands are characterized by resonance peaks in the dielectric constant.

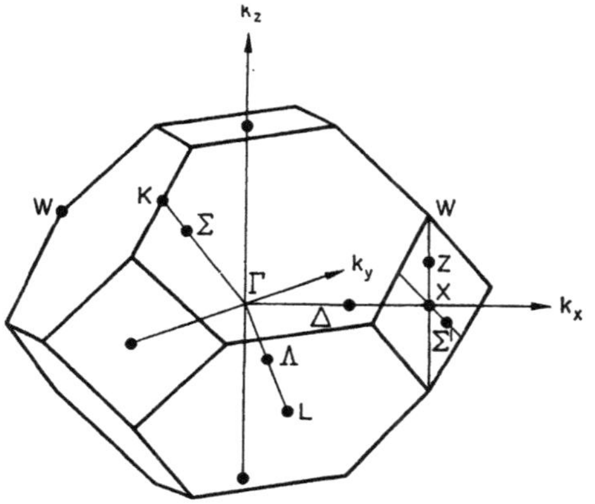

FIG. 1.
Brillouin zone for a f.c.c. crystal.

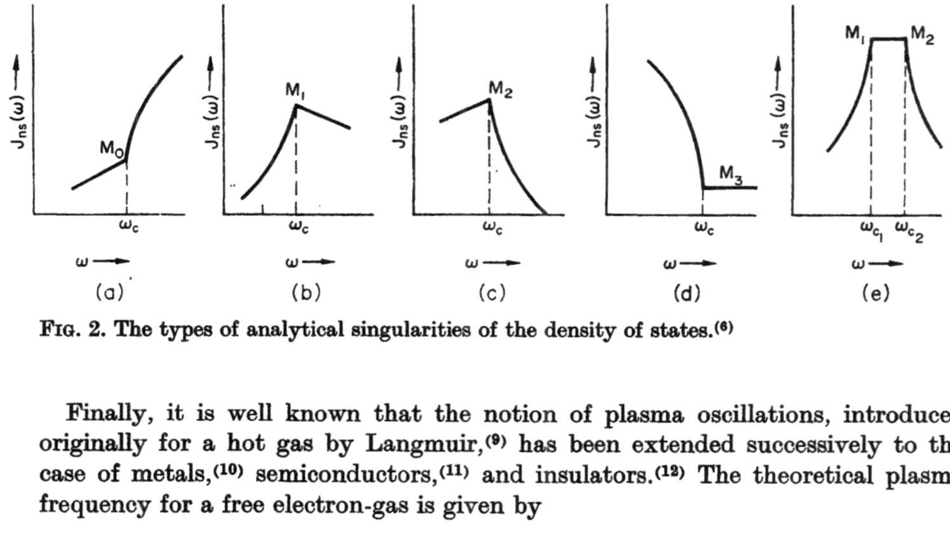

FIG. 2. The types of analytical singularities of the density of states.[6]

Finally, it is well known that the notion of plasma oscillations, introduced originally for a hot gas by Langmuir,[9] has been extended successively to the case of metals,[10] semiconductors,[11] and insulators.[12] The theoretical plasma frequency for a free electron-gas is given by

$$\omega_p^2 = \frac{N_e e^2}{m\varepsilon_0},$$

where N_e is the density of the free electrons of charge e and mass m. It has been shown in some recent reviews that evidence for the plasmons may be provided by optical and characteristic energy loss measurements. Indeed, according to the random phase approximation (RPA),[16] in the case of isotropic solids the transverse and longitudinal dielectric constants could be made equal. Then, from the optical constants n and k, one can calculate the energy loss function

$$\operatorname{Im}\left(\frac{\varepsilon_2}{\varepsilon_1^2+\varepsilon_2^2}\right),$$

which is proportional to the probability that the incident electron gives up an energy $\hbar\omega$ to the electrons of the material.[17] The spectrum of Im (ε^{-1}) will therefore be identical with the characteristic loss function and permits distinction between plasmons and interband transitions. A plasmon is characterized by a peak of Im (ε^{-1}) corresponding to $\varepsilon_1 = 0$ and $\varepsilon_2 \ll 1$. The resonance manifests itself either as a discrete sharp line or as a broadened line which may be displaced due to an interband transition (the displacement being always away from the position of the interband transition[14]). An interband transition corresponds to discontinuities in the curves of Im (ε^{-1}) and ε_2, which characterize respectively the longitudinal and transverse resonances. A so-called "surface" plasmon may be associated with a "volume" plasmon of frequency ω_p, the frequency ω_s of the former being given by[18]

$$\omega_s = \frac{\omega_p}{\sqrt{2}}$$

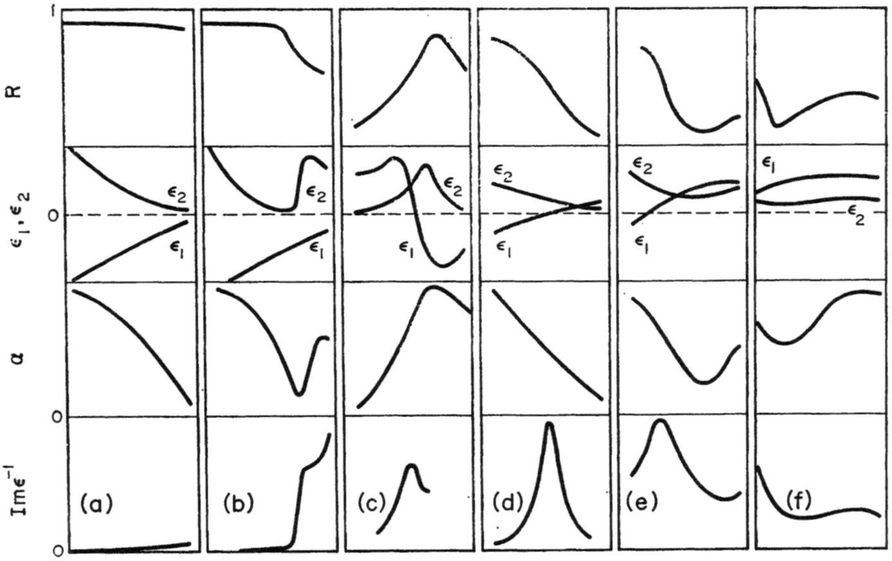

FIG. 3. Typical behaviour of optical constants associated with various excitation processes in solids.[20] Metals: (a) free-electron range, (b) low energy interband transition. Insulators, semiconductors, and metals: (c) interband transition, (d) plasma region, (e) plasma region modified by interband transition, (f) excitation of deeper bands above plasma frequency.

on the assumption that the solid is placed in vacuum and has a "clean" surface—in particular free from oxide coating. These plasmons are characterized by a peak of Im $\left(\frac{1}{1+\varepsilon}\right)$ and a decrease in the reflectance depending on the roughness of the surface.[19]

The differing behaviour of the reflectance R, the absorption coefficient $\alpha = \frac{4\pi k}{\lambda}$, ε_1, ε_2, and Im (ε^{-1}), depending on the excitation process and the type of solid (metal, insulator, or semiconductor), are assembled in Fig. 3, taken from Ehrenreich.[20]

Finally, two sum rules can be written from the reciprocity relation between ε_1 and ε_2[21] for a finite area of integration. The first of these permits calculation of the number n_{eff} of electrons per atom or molecule participating in the optical properties within the region 0 to ω_0:

$$\int_0^{\omega_0} \omega' \varepsilon_2(\omega')\, d\omega' = \int_0^{\omega_0} \omega' \,\text{Im}\,(\varepsilon^{-1})\, d\omega' = \frac{\pi\omega_p^2}{2} = \frac{\pi}{2}\frac{Ne^2}{m\varepsilon_0} n_{\text{eff}}$$

The curve of $n_{\text{eff}}(\omega)$ gives information about the dispersion of the oscillator strengths.

The second rule gives the effective dielectric constant

$$\varepsilon_{s,\text{eff}} = 1 + \frac{2}{\pi}\int_0^{\omega_0} \frac{\varepsilon_2(\omega')}{\omega'}\, d\omega'.$$

The variations of $\varepsilon_{s,\text{eff}}$ are connected with the principal interband transitions responsible for the value of the dielectric constant. In general, the saturation value should coincide with the electronic part of the static dielectric constant. On the other hand, a small difference $\delta\varepsilon_{1,s}$ may indicate, as, for example, in the case of certain semiconductors, the d-band contribution to the dielectric constant.

References

1. Madden, R. P., *Physics of Thin Films*, I, Academic Press, New York, 1963.
2. Fan, H. Y., *Handbuch der Physik*, Springer-Verlag, Berlin, 25/2a, 157 (1967)
3. Phillips, J. C., *Solid St. Phys.*, **18**, 55 (1966).
4. Greenaway, D. L., and Harbeke, G., *Optical Properties and Band structure of Semiconductors*, Pergamon Press, Oxford, 1968.
5. van Hove, L., *Phys. Rev.* **89**, 1189 (1953).
6. Phillips, J. C., *Phys. Rev.* **104**, 1263 (1956).
7. Berglund, C. N., and Spicer, W. E., *Phys. Rev.* **136** A, 1030 (1965).
8. Knox, R. S., *Theory of Excitons*, Academic Press, New York, 1963.
9. Langmuir, I., *Proc. Natn. Acad. Sci. U.S.A.* **14**, 627 (1928).
10. Bohm, D., and Pines, D., *Phys. Rev.* **85**, 338 (1952).

11. Ehrenreich, H., and Cohen, M., *Phys. Rev.* **115**, 786 (1959).
12. Miyakawa, T., *J. Phys. Soc. Japan* **24**, 768 (1968).
13. Pradal, F., Gout, C., and Fabre, D., *J. de Phys.* **6**, 372 (1965).
14. Gauthé, B., *Pertes d'énergie caractéristiques des électrons dans les solides*, Gauthier Villars, Paris, 1968.
15. Raether, H., *Springer Tracts in Modern Physics*, Springer-Verlag, Berlin, 1964, vol. 38, p. 84.
16. Pines, D., *Elementary Excitations in Solids*, W. A. Benjamin, New York, 1963.
17. Fröhlich, H., and Pelzer, H., *Proc. Phys. Soc.* A, **68**, 525 (1955).
18. Ritchie, R. H., *Phys. Rev.* **106**, 274 (1957).
19. Steinmann, W., *Phys. Status Solidi* **28**, 437 (1968).
20. Ehrenreich, H., *The Optical Properties of Solids*, Rendiconti Scuola Internationale "E. Fermi" Corso XXXIV, Academic Press, New York, 1966.
21. Stern, F., *Solid St. Phys.* **15**, 299 (1963).

CHAPTER 1

Phenomenological description of optical properties of solids and methods of determination of optical constants in the vacuum ultraviolet

H. Damany

Laboratoire des Interactions Moléculaires et des Hautes Pressions du CNRS, Bellevue, 92190 Meudon, France

Contents

As an introduction to Part I, this chapter has two aims. Firstly, it introduces the optical constants on a phenomenological basis and condenses essential formulae such as reflectances, dispersion relations, and sum rules. Secondly, it surveys the methods of determination of the optical constants with emphasis on the specific problems of the vacuum ultraviolet (v.u.v.).

PHENOMENOLOGICAL DESCRIPTION OF THE OPTICAL PROPERTIES OF SOLIDS

1. Notation

Electric field **E**.
Magnetic field **B**.
Current density **J**.
Electric polarization **P**.
Conductivity tensor σ_{ij}.
Polarizability tensor α_{ij}.
Dielectric tensor ε_{ij}.
System of units, m.k.s.a.
ε_0 = dielectric constant of vacuum = 8.854×10^{-12} F-m^{-1}.

The convention for the time dependence of a monochromatic wave is

$$\mathbf{E} = \mathbf{E}_0(\mathbf{r})e^{i\omega t}. \tag{1}$$

As a consequence, there is a minus sign between the real and imaginary parts of ε and $\hat{n}$ and no minus sign before Im (ε^{-1}).

2. Introduction of the optical constants

The propagation of electromagnetic waves in a medium obeys Maxwell's equations, where the properties of the medium are characterized by the fields **J** and **P**, the magnetic properties playing no part at the frequencies of the v.u.v. ($\nu > 10^{15}$ s^{-1}). We suppose that the material has no optical activity and the field strengths are not too high, so that at each point the following linear relations are valid:

$$J_i = \sigma_{ij}E_j, \tag{2}$$

$$P_i = \alpha_{ij}E_j. \tag{3}$$

For a monochromatic wave[1] and an isotropic material, eqs. (2) and (3) become

$$\mathbf{J} = \sigma(\omega)\mathbf{E}_0(\mathbf{r})e^{i\omega t}, \tag{4}$$

$$\mathbf{P} = \alpha(\omega)\mathbf{E}_0(\mathbf{r})e^{i\omega t}, \tag{5}$$

where $\sigma(\omega)$ and $\alpha(\omega)$ are scalar functions of the frequency, and describe two different physical properties of the material. But, as they are associated, through

Maxwell's equations, by the traditional electric displacement

$$\mathbf{D} = \varepsilon_0 \mathbf{E} + \mathbf{P} - i(\sigma/\omega)\mathbf{E} = \varepsilon_a \mathbf{E}, \tag{6}$$

another complex function, namely the absolute dielectric constant

$$\varepsilon_a = \varepsilon_{a1} - i\varepsilon_{a2}, \tag{7}$$

is introduced.

The imaginary part ε_{a2} represents absorption, whatever process this may involve, either damping of free charges [σ term in eqn. (6)] or damping of the dipole orientation (expressed by a complex polarizability $\alpha = \alpha_1 - i\alpha_2$). The distinction between the two processes would lead to

$$\varepsilon_{a1} = \varepsilon_0 + \alpha_1, \qquad \varepsilon_{a2} = \alpha_2 + \sigma/\omega. \tag{8}$$

Instead of the absolute dielectric constant ε_a (unit Farad-meter^{-1}), the dimensionless relative dielectric constant is introduced as

$$\varepsilon = \varepsilon_a/\varepsilon_0 = \varepsilon_1 - i\varepsilon_2. \tag{9}$$

Consider now a monochromatic plane wave

$$\mathbf{E} = \mathbf{E}_0 \exp i(\omega t - \mathbf{k} \cdot \mathbf{r}), \tag{10}$$

where $\mathbf{E}_0$ is a constant vector. Maxwell's equations yield the relation[1]

$$\mathbf{k} \cdot \mathbf{k} = \omega^2 \varepsilon / c^2 = 4\pi^2 \hat{n}^2/\lambda^2, \tag{11}$$

where, by analogy with the case of no absorption, a complex refractive index is introduced

$$\hat{n} = n - ik, \tag{12}$$

where n is the refractive index and k the absorption index related to the absorption coefficient K of Beer's law by

$$K = 4\pi k/\lambda. \tag{13}$$

From eqns. (9) and (11),

$$\varepsilon_1 = n^2 - k^2, \qquad \varepsilon_2 = 2nk. \tag{14}$$

The name "optical constants" refers either to the pair ε_1, ε_2 or to the pair n, k. They are not true constants but functions of the frequency. The reflectances being generally expressed as functions of n and k, it is n and k which are deduced first from the experimental data.

For an anisotropic material, without optical activity, the dielectric tensor is shown to be symmetrical[2] and may be written, in the system of principal dielectric axes under the simple form*

$$\varepsilon_{ij} = \begin{vmatrix} \hat{n}_1^2 & 0 & 0 \\ 0 & \hat{n}_2^2 & 0 \\ 0 & 0 & \hat{n}_3^2 \end{vmatrix}. \tag{15}$$

The $\hat{n}_j$, principal refractive index, are complex numbers of the form $n_j - ik_j$. Such a material is then characterized by six optical constants.

3. Dispersion relations and sum rules

The optical constants ε_1 and ε_2 are not really two independent functions of ω. This is a direct consequence of the causality principle which requires that the function $\varepsilon(\omega)$ satisfies certain criteria, particularly that no singularity exists in the lower half-plane of the complex frequency $\omega = \omega' + i\omega''$;[3] then it may be proved that ε_1 and ε_2 are related by the integral relations (known as Kramers-Kronig relations[4])

$$\varepsilon_1(\omega) - 1 = \frac{1}{\pi} \fint_{-\infty}^{+\infty} \frac{\varepsilon_2(x)}{x - \omega}\, dx, \tag{16}$$

$$\varepsilon_2(\omega) = \frac{-1}{\pi} \fint_{-\infty}^{+\infty} \frac{\varepsilon_1(x) - 1}{x - \omega}\, dx, \tag{17}$$

where $\fint$ denotes the Cauchy principal value of the integral.

For a material having a direct current conductivity σ_0 these relations are modified; in eqn. (16) $\varepsilon_2(x)$ has to be replaced by $\varepsilon_2(x) - \sigma_0/(\varepsilon_0 x)$, and eqn. (17) becomes

$$\varepsilon_2(\omega) = \frac{-1}{\pi} \fint_{-\infty}^{+\infty} \frac{\varepsilon_1(x) - 1}{x - \omega}\, dx + \frac{\sigma_0}{\varepsilon_0 \omega}. \tag{18}$$

Taking into account the fact that $\varepsilon_1(\omega)$ must be an even function and $\varepsilon_2(\omega)$ an odd function of ω,[3] we write eqn. (16) and (17) as

$$\varepsilon_1(\omega) - 1 = \frac{2}{\pi} \fint_{0}^{\infty} \frac{x\varepsilon_2(x)}{x^2 - \omega^2}\, dx, \tag{19}$$

$$\varepsilon_2(\omega) = -\frac{2\omega}{\pi} \fint_{0}^{\infty} \frac{\varepsilon_1(x) - 1}{x^2 - \omega^2}\, dx, \tag{20}$$

* With the restriction that the absorption axes and optical symmetry axes coincide.

or, as $\int_0^\infty \frac{dx}{x^2-\omega^2} = 0$, under the symmetrical form,

$$\varepsilon_1(\omega) - 1 = \frac{2}{\pi} \int_0^\infty \frac{x\varepsilon_2(x) - \omega\varepsilon_2(\omega)}{x^2 - \omega^2}\, dx, \tag{21}$$

$$\varepsilon_2(\omega) = -\frac{2\omega}{\pi} \int_0^\infty \frac{\varepsilon_1(x) - \varepsilon_1(\omega)}{x^2 - \omega^2}\, dx, \tag{22}$$

where the singularity at $x = \omega$ is removed.

The basic relations (16) and (17) are not specific to the optical constants. In other fields of physics,[5] e.g. in the theory of electrical circuits,[6] similar relations relate the real and imaginary parts of every function of frequency associating two physical quantities linearly.

Let us consider, for instance, the amplitude reflection coefficient for normal incidence,*

$$r = \sqrt{R}e^{-i\theta}, \tag{23}$$

where R is the reflectance and θ the phase change at reflection. Then,

$$\ln r = \frac{1}{2} \ln R - i\theta. \tag{24}$$

The real and imaginary parts of $\ln r$ are connected by relations similar to (19)-(22), and[7]

$$\theta(\omega) = \frac{\omega}{\pi} \int_0^\infty \frac{\ln [R(\omega)/R(x)]}{x^2 - \omega^2}\, dx. \tag{25}$$

Integrating by parts yields

$$2\pi\theta(\omega) = \int_0^\infty \ln \left| \frac{x-\omega}{x+\omega} \right| \frac{d \ln R(x)}{dx}\, dx. \tag{26}$$

This form shows the local character of the Kramers-Kronig transforms.[8] Equation (25) or (26) is very important: it provides the means of getting $\theta(\omega)$ from measurements of R, and hence, of deducing n and k:

$$n = (1 - R)/(1 + R - 2\sqrt{R} \cos \theta),$$
$$k = 2\sqrt{R} \sin \theta/(1 + R - 2\sqrt{R} \cos \theta). \tag{27}$$

* The minus sign in the exponent is in accordance with the convention adopted in § 1.

We give some other useful relations deduced in a similar way.[9] For example, the characteristic energy losses of electrons in solids lead to the consideration of $\varepsilon^{-1} = (\varepsilon_1 + i\varepsilon_2)/(\varepsilon_1^2 + \varepsilon_2^2)$ and of the loss function

$$\mathrm{Im}\,(\varepsilon^{-1}) = \varepsilon_2/(\varepsilon_1^2 + \varepsilon_2^2). \tag{28}$$

It is shown that Re (ε^{-1}) and Im (ε^{-1}) are connected by the relations

$$\mathrm{Re}\,[\varepsilon(\omega)^{-1}] - 1 = \frac{-2}{\pi} \int_0^\infty \frac{x\,\mathrm{Im}\,[\varepsilon^{-1}(x)]}{x^2 - \omega^2}\,dx, \tag{29}$$

$$\mathrm{Im}\,[\varepsilon(\omega)^{-1}] = \frac{2\omega}{\pi} \int_0^\infty \frac{\mathrm{Re}\,[\varepsilon^{-1}(x)] - 1}{x^2 - \omega^2}\,dx. \tag{30}$$

Finally, n, k, and the absorption coefficient K are connected by:

$$n(\omega) - 1 = \frac{2}{\pi} \int_0^\infty \frac{x k(x)}{x^2 - \omega^2}\,dx, \tag{31}$$

$$k(\omega) = \frac{-2\omega}{\pi} \int_0^\infty \frac{n(x) - 1}{x^2 - \omega^2}\,dx, \tag{32}$$

$$n(\omega) - 1 = \frac{1}{2\pi^2} \int_0^\infty \frac{K(\lambda)\,d\lambda}{1 - \lambda^2/\lambda_\omega^2}. \tag{33}$$

SUM RULES

From the dispersion relations are deduced the following relations called sum rules. Some of them are direct consequence, others need special assumptions.

If, in eqn. (33), we put $\omega = 0$, $\lambda_\omega = \infty$, $n_s = n(0)$, n_s, being the static refractive index,

$$n_s - 1 = [1/(2\pi^2)] \int_0^\infty K(\lambda)\,d\lambda. \tag{34}$$

In the same way, from eqn. (19),

$$\varepsilon_{1s} - 1 = n_s^2 - 1 = (2/\pi) \int_0^\infty [\varepsilon_2(x)/x]\,dx, \tag{35}$$

valid for a dielectric; for a metal, we must change $\varepsilon_2(x)$ into $\varepsilon_2(x) - \sigma_0/(\varepsilon_0 x)$ in the integral.

The following relations are deduced with the help of the additional hypotheses: (a) ω is much higher than the frequency of the last absorption band, (b) the free electron approximation is valid at high frequencies. Then eqn. (19) is written

$$\varepsilon_1(\omega) - 1 = -\omega_p^2/\omega^2 = -[2/(\pi\omega^2)] \int_0^\infty x\varepsilon_2(x)\, dx,$$

where

$$\omega_p^2 = N_e e^2/(m\varepsilon_0). \tag{36}$$

N_e is the density of free electrons with charge e and mass m, and we have then the sum rule

$$\int_0^\infty x\varepsilon_2(x)\, dx = \frac{\pi}{2}\, \omega_p^2. \tag{37}$$

With the same hypotheses, eqn. (30) yields

$$\int_0^\infty \mathrm{Im}\, [\varepsilon^{-1}(x)] x\, dx = \frac{\pi}{2}\, \omega_p^2. \tag{38}$$

If the frequency range in eqn. (37) is limited to $(0, \omega_0)$, we may write eqn. (8) as,

$$\int_0^{\omega_0} x\varepsilon_2(x)\, dx = \frac{\pi}{2} \frac{e^2}{m\varepsilon_0} N n_e, \tag{39}$$

where N is the atom density and n_e the number of valence electrons per atom contributing to the optical constants in the frequency interval $(0, \omega_0)$.

4. Generalized Fresnel equations for reflectances

Before dealing with the methods of determination of the optical constants, we give the reflectance formulae for oblique incidence of light on an absorbing medium.

4.1. ISOTROPIC MEDIUM

Consider a monochromatic plane wave

$$\mathbf{E}_1 = \mathbf{E}_{01} \exp i(\omega t - \mathbf{k}_1 \cdot \mathbf{r})$$

incident at an angle θ_1 from a transparent medium of refractive index n_1 (real) on the boundary of an absorbing medium of refractive index $\hat{n}_2 = n_2 - ik_2$;

the axes x, y, z are such that (x, y) is the plane of incidence and the boundary lies at $x = 0$. The incident wave gives rise to a reflected wave

$$\mathbf{E}_1' = \mathbf{E}_{01}' \exp i(\omega t - \mathbf{k}_1' \cdot \mathbf{r})$$

and a transmitted wave

$$\mathbf{E}_2 = \mathbf{E}_{02} \exp i(\omega t - \mathbf{k}_2 \cdot \mathbf{r}),$$

with similar expressions for $\mathbf{B}_j$.

The wave vectors $\mathbf{k}_j$ are such that $\mathbf{k}_j \cdot \mathbf{k}_j = 4\pi^2 n_j^2/\lambda^2$ and, on account of the phase boundary conditions, they lie in the plane $z = 0$ and have the same component along y. Their components along x and y are

$$\left.\begin{aligned} \mathbf{k}_1 &= \left(\frac{2\pi}{\lambda} n_1 \cos\theta_1, \quad \frac{2\pi}{\lambda} n_1 \sin\theta_1\right), \\ \mathbf{k}_1' &= \left(-\frac{2\pi}{\lambda} n_1 \cos\theta_1, \quad \frac{2\pi}{\lambda} n_1 \sin\theta_1\right), \\ \mathbf{k}_2 &= \left(\frac{2\pi}{\lambda} (p - iq), \quad \frac{2\pi}{\lambda} n_1 \sin\theta_1\right). \end{aligned}\right\} \tag{40}$$

Equations (40) comprise the laws of reflection and refraction; k_{2x} is complex, due to the attenuation of the transmitted wave.*

It is easily shown that

$$p^2 - q^2 = n_2^2 - k_2^2 - n_1^2 \sin^2\theta_1, \qquad pq = n_2 k_2, \tag{41}$$

or

$$\left.\begin{aligned} 2p^2 &= [(n_2^2 - k_2^2 - n_1^2 \sin^2\theta_1)^2 + 4n_2^2k_2^2]^{1/2} + (n_2^2 - k_2^2 - n_1^2 \sin^2\theta_1), \\ 2q^2 &= [(n_2^2 - k_2^2 - n_1^2 \sin^2\theta_1)^2 + 4n_2^2k_2^2]^{1/2} - (n_2^2 - k_2^2 - n_1^2 \sin^2\theta_1). \end{aligned}\right\} \tag{42}$$

The other boundary conditions for the amplitudes yield the Fresnel coefficients as functions of the k_j's components.

Electric field perpendicular to the plane of incidence (s-polarization):

$$r_s = \frac{E_{01}'}{E_{01}} = \frac{k_{1x} - k_{2x}}{k_{1x} + k_{2x}} = \sqrt{R_s} e^{-i\theta_s}, \qquad t_s = \frac{E_{02}}{E_{01}} = \frac{2k_{1x}}{k_{1x} + k_{2x}}. \tag{43}$$

Electric field parallel to the plane of incidence (p-polarization):

$$\left.\begin{aligned} r_p = \bar{r}_s &= \frac{B_{01}'}{B_{01}} = \frac{\hat{n}_2^2 k_{1x} - n_1^2 k_{2x}}{\hat{n}_2^2 k_{1x} + n_1^2 k_{2x}} = \frac{k_{1y}^2 - k_{1x}k_{2x}}{k_{1y}^2 + k_{1x}k_{2x}} \frac{k_{1x} - k_{2x}}{k_{1x} + k_{2x}} \\ &= \sqrt{R_p} e^{-i\theta_p}, \\ \bar{t}_s &= \frac{B_{02}}{B_{01}} = \frac{2\hat{n}_2^2 k_{1x}}{\hat{n}_2^2 k_{1x} + n_1^2 k_{2x}}. \end{aligned}\right\} \tag{44}$$

* The attenuation is of the form $\exp[-4\pi q x/\lambda]$, involving only the distance along the normal to the surface.

It is worthy of note that eqns. (43) and (44) are also valid for the interface between two absorbing media: k_{1y} has to be replaced by the component along y of the **k**-vector of the first medium (supposed transparent).

From eqns. (40)-(44) the reflectances and phase changes are easily deduced. Leaving apart the transmittances which are seldom used,

$$R_s = \frac{(p - n_1 \cos \theta_1)^2 + q^2}{(p + n_1 \cos \theta_1)^2 + q^2}, \tag{45}$$

$$R_p = R_s \frac{(p - n_1 \sin \theta_1 \tan \theta_1)^2 + q^2}{(p + n_1 \sin \theta_1 \tan \theta_1)^2 + q^2}, \tag{46}$$

and the phase difference is given by

$$\tan \Delta = \tan (\theta_s - \theta_p) = \frac{2qn_1 \sin \theta_1 \tan \theta_1}{p^2 + q^2 - n_1^2 \sin^2 \theta_1 \tan^2 \theta_1}. \tag{47}$$

For normal incidence,

$$r = \frac{n_1 - \hat{n}_2}{n_1 + \hat{n}_2} = \sqrt{R}\, e^{-i\theta}, \qquad R = \frac{(n_1 - n_2)^2 + k_2^2}{(n_1 + n_2)^2 + k_2^2},$$

$$\tan \theta = \frac{-2n_1 k_2}{n_1^2 - n_2^2 - k_2^2}. \tag{48}$$

So, for an absorbing isotropic medium, the reflectances are expressed in a simple form as functions of p, q, but this simplicity is only apparent since the relations (42) between p, q and the optical constants are very intricate.

4.2. ANISOTROPIC MEDIUM

The case of an anisotropic absorbing medium is much more complicated. In the general case, to one incident plane wave there correspond two transmitted plane waves, and if the incident wave is linearly polarized, the reflected one is elliptically polarized. Nevertheless, there exist particular positions of the principal axes with respect to the plane of incidence for which equations are available.

A typical and somewhat simpler case deals with an uniaxial crystal cut either parallel or perpendicular to the optic axis; only two indices, $\hat{n}_0$ and $\hat{n}_e$, are involved and, for particular positions of the optic axis, the Fresnel coefficient is a combination of both indices sufficiently simple for calculation from the measured reflectances. These rather specific cases will be treated along with the methods of determination of the optical constants.

METHODS OF DETERMINATION OF OPTICAL CONSTANTS IN THE VACUUM ULTRAVIOLET

As will appear in the following chapters dealing with the various experimental investigations, study of the optical properties of solids in the v.u.v. has evolved considerably during the past 10 years, with important technical progress and parallel development of the theory. Experimental progress has accompanied greater care in the choice of materials, their purity, and concern with crystalline structure, surface cleanliness, and—importantly—the search for methods yielding absolute values of the optical constants so that the experimental data could be confronted with theory.

In this section we will investigate the methods suitable for the determination of the optical constants in the v.u.v. and we shall have the opportunity of showing the increasing part played by polarimetry in spite of the considerable difficulties encountered in this spectral range. Most materials are strongly absorbing in the v.u.v., so that reflectance methods are the best suited for bulk materials, and the larger part of the text will be devoted to reflectance measurements with a discrimination between the use of unpolarized light and the use of polarized light. We shall suppose the reflecting surface to be ideally plane and clean, qualities which may be approached by the present techniques (cleavage under ultra-high vacuum). Needless to say, along with the structure of the material the surface condition may affect strongly the results in the v.u.v.[8] These factors must be specified by the experimenter in order to attach full meaning to the optical constants deduced from the measurements, and we exclude detailed discussion of this question.

1. Reflectance measurements with unpolarized light

1.1. NORMAL INCIDENCE—KRAMERS-KRONIG ANALYSIS

Strictly normal incidence is not feasible, but, happily, the shapes of R_p and R_s curves versus incidence angle θ are such that, in most cases, an incidence angle up to 20° introduces no significant difference. Yet an incidence angle as small as possible is preferable, and a special mounting devised by Hunter[10] allows a 4° angle to be attained; under such conditions, $R_p = R_s = R$, and a partial polarization of the incident light has no influence.

The optical constants are then deduced from the measurements of R, following the method of Robinson and Price.[11] The method is based on formulae (25)-(27); if R is known over the complete frequency range from zero to infinity, θ and, hence, n and k may be calculated. Practically, R is measured only in a limited spectral range, and the restriction of the range of integration introduces an error in θ and, hence, in the optical constants. Bowlden and Wilmshurst[12] have investigated this effect for absorption bands in the infrared and shown that a suitable choice of the integration range can minimize the error. But this conclusion, which results from the local character of the Kramers-Kronig (K.-K.) transforms[8] cannot be applied to the v.u.v. It is necessary to use data available in the literature for the optical constants of the material at low frequencies (see, for instance, refs. 13 and 14), and, for the high-frequency side, where no data exists, a correction is needed.

Various extrapolation procedures have been used; they assume a simple behaviour of $R(\omega)$ at high frequencies: for instance, ω_1 being the upper experimental limit and R_1 the corresponding measured value of $R: R = R_1(\omega_1/\omega)^4$ (free electron approximation),[15] or $R = R_1(\omega_1/\omega)^p$,[16] or $R = R_1 \exp B(\omega_1 - \omega)$. For discussion, see refs. 8, 9, and 17. As pointed out by Roessler,[17] the assumption of a smooth behaviour of $R(\omega)$ is not always justified, and he proposed a more general method which may be summarized in this way: (a, b) being the experimental frequency range, eqn. (26) may be written, with the help of the mean-value theorem for integrals,

$$\theta(\omega) = \frac{1}{2\pi} \ln \left| \frac{\omega - a}{\omega + a} \right| \ln R(\xi) + \frac{1}{2\pi} \int_a^b \ln \left| \frac{x - \omega}{x + \omega} \right| \frac{d \ln R(x)}{dx} \, dx$$

$$+ \frac{1}{2\pi} \ln \left| \frac{b - \omega}{b + \omega} \right| \ln R(\eta) \tag{49}$$

with ξ in the interval $(0, a)$ and η in the interval (b, ∞) being determined by the condition $\theta = 0$ below the absorption edge. The essential assumption is that the dependence of ξ and η on ω may be neglected. As shown by Roessler, this assumption is not unreasonable, and the method has proved useful for analysing reflectance spectra.

A somewhat different procedure has been proposed by MacRae *et al.*;[18, 19] it consists in writing θ as $\theta_1 + \Delta\theta$, where $\Delta\theta$ is the contribution of the high frequencies where no experimental data are available, and interpolating $\Delta\theta$ by a smooth curve, provided the frequency interval be flanked by two frequency intervals where n and k, and hence $\Delta\theta$, have been determined by other methods.

Let us note that the confrontation of the results of a K.-K. analysis with those obtained by another method (e.g. a "two-angle" method) generally shows a good agreement.[13, 20–22]

For the sake of completeness, we indicate here that the K.-K. analysis can also be performed on reflection data obtained under oblique incidence[23, 24] and also on transmission data (e.g. thin films).[25]

Finally, another method, also using normal incidence reflectance data but different from the K.-K. analysis, has been given recently by Verleur.[26] He takes a dispersion relation of the form

$$\varepsilon = \varepsilon_1 - i\varepsilon_2 = \varepsilon_\infty + \sum_{i=1}^{N} \{s_i/[1 - (\omega^2/\omega_i^2) + i\Gamma(\omega/\omega_i)\} \tag{50}$$

and looks for a set of parameters ε_∞, s_i, ω_i, Γ_i to get a good agreement over the whole spectral region observed between R_{exp} and $R_{\text{calc}} = \left| \frac{\varepsilon^{1/2} - 1}{\varepsilon^{1/2} + 1} \right|^2$. An initial guess of the set of parameters is necessary; then, an automatic-fitting procedure, with a computer program, yields the final set of parameters giving the best fit. The method is illustrated by the determination of the optical constants of silicon between 0 and 10 eV and shows a good agreement with the K.-K. analysis on the same data. No extrapolation outside the spectral range is required. The method applies also to transmittance data.

1.2. TWO-ANGLE METHOD

n and k may be determined, from the measurements of $R = {}^1/_2\,(R_s + R_p)$ at two angles of incidence, by a graphical method first described by Tousey.[27] For each angle, a set of curves of constant R in the plane (n, k) is needed. Such sets have been published in the literature[27] and reviewed recently by Holl.[28] Equivalent forms are curves of constant n and constant k in the plane R_{θ_1}, R_{θ_2}, where θ_1 and θ_2 are the two angles of incidence choosen.[29] The accuracy on n and k can be increased if more than two angles are employed along with a data-reduction method with the help of a computer.[30, 31]

The method is valid only if the light is totally depolarized. As gratings give always some percentage of polarization, it is necessary either to add a depolarizer (e.g. a thick birefringent plate) or to measure R for two orthogonal positions of the plane of incidence and take the arithmetic mean.[29] Otherwise a polarization parameter should be introduced and evaluated as discussed by Tousey.[27]

2. Production of partially or totally polarized light in the vacuum ultraviolet

Before dealing with polarimetric methods, we must examine the ways of polarizing v.u.v. radiation.

2.1. PARTIAL POLARIZATION BY GRATINGS

At present all spectrometers for the v.u.v. use gratings and, generally, the exit beam is partially plane polarized; this polarization is defined either by the ratio $g = I_\perp/I_\parallel$, where $I_\perp$ and $I_\parallel$ are the intensities of the components vibrating perpendicular and parallel to the plane of incidence respectively, or by the degree of polarization $P = (I_\perp - I_\parallel)/(I_\perp + I_\parallel)$. P varies with the wavelength, and its magnitude depends on the type of mounting of the grating. For normal incidence mountings, P is generally small.[32–34] On the other hand, P may attain a large value for grazing incidence mountings[35–37] and for Seya-Namioka mounting,[38, 39] as illustrated in Fig. 1. As seen from these examples, P may be of the order of 20-40% in some spectral range. P is dependent on the type of grating, the blaze angle, and the coating,[38, 40] and, furthermore, it may change with time.[38] Thus the use of this partial polarization for the measurements of R_p and R_s requires either the simultaneous measurement of P or, at least, a frequent control of its possible evolution.

The methods of determination of P are the following:

Consider P as an adjustable parameter in the expressions of oblique incidence reflectances and look for the P value giving self-consistent results for the optical constants of the material under study.[27, 32, 37]

Use an analyser of known properties; it may be a mirror whose optical constants are already known[41] or are simultaneously determined by the methods of § 2;[39]

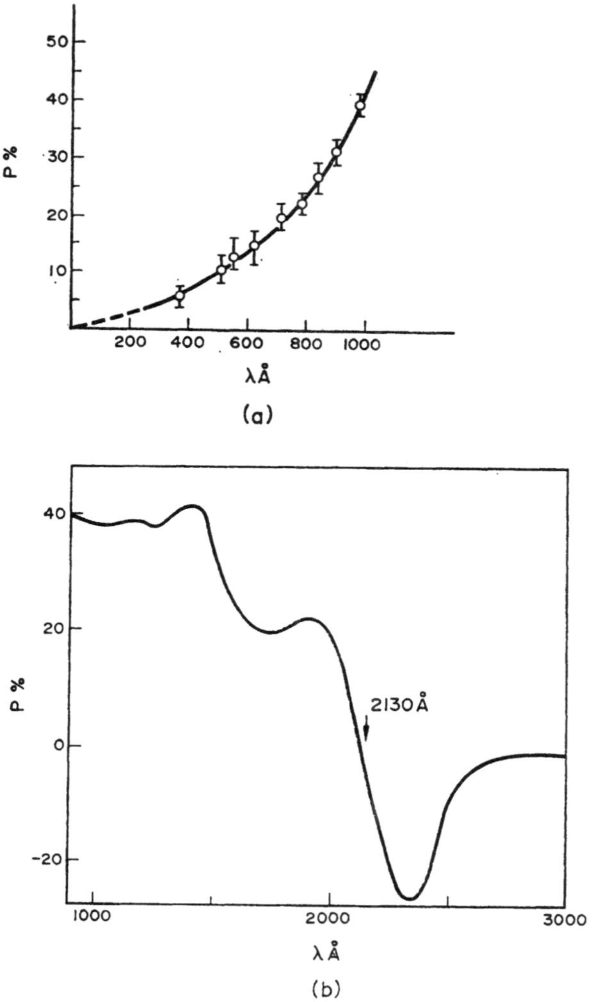

FIG. 1. Degree of polarization of monochromatic light. (a) Grazing incidence (82°), unblazed glass grating, radius of curvature 3 m, 600 lines per mm;[36] (b) Seya-Namioka mounting, aluminium grating coated with MgF_2, radius of curvature of 0.498 mm, 1200 lines per mm, blazed for 1230 Å (in the Seya-Namioka mounting).[39]

it may also be a three-mirror polarizer,[40] or a pile-of-plates polarizer,[40] or a Brewster angle polarizer.

A more general method[38, 42] rests on the Abelès relation $R_p = R_s^2$[43] for the incidence angle 45°. The beam emergent from the monochromator falls on a plane reflector at 45° angle of incidence. The plane of incidence of this mirror may be perpendicular with (H position) or parallel (V position) to the exit slit. Let R_H and R_V be the corresponding reflectances. If it is supposed that the polarization vector of the partially plane-polarized light is parallel or perpendicular to the exit slit—which seems true in most cases—P is given by

$$P = \frac{R_H - R_V}{R_H + R_V} = \frac{1 + [1 + 4(R_H + R_V)]^{1/2}}{3 - [1 + 4(R_H + R_V)]^{1/2}}. \tag{51}$$

This method is valid provided the Abelès relation be applicable, i.e. the reflecting mirror has a clean surface; it is essential to choose a material without thin surface films (oxide, for example). Gold and silica appear well suited for this purpose in the 500-2000 Å range.[38]

The polarization given by a grating can also be measured along with its efficiency[40, 44] with the help of a polarizer (pile-of-plates or triple-reflectance polarizer) as illustrated by Fig. 2.

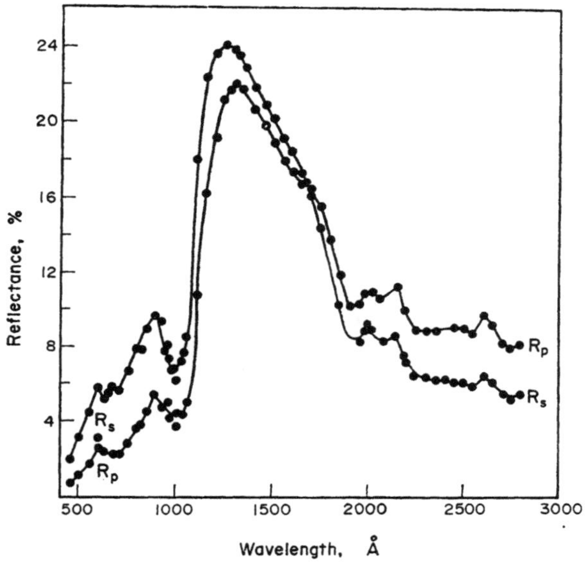

Fig. 2. Reflectance of aluminized grating blazed at 1500 Å, overcoated with MgF_2.[40]

2.2. SOURCE OF HIGHLY POLARIZED LIGHT

This is the specific case when synchrotron radiation is used as a source for the extreme ultraviolet (x.u.v.). In addition to the 180 MeV NBS synchrotron,[45] several of these machines are used for x.u.v. spectroscopy, e.g. the Deutsches Elektronen-Synchrotron DESY at Hamburg and the Tokyo 750 MeV Synchrotron. The instrumental aspects of synchrotron x.u.v. spectroscopy have been reviewed recently by Madden *et al.*[46] One important feature of the synchrotron radiation is that it is highly polarized in the plane of the synchrotron. The degree of polarization is, generally, higher than 80%,[46–48] and this source is well adapted to polarimetric measurements.[49]

2.3. POLARIZERS FOR THE VACUUM ULTRAVIOLET

In the v.u.v. there is a limited choice of polarizers and, as a result of the optical properties of the materials in this region, their "transmittance" is rather low. The principles involved are the same as in other spectral regions, namely linear dichroism, birefringence, and polarization by reflection or by transmission.

Dichroic polarizers

Just one example is found in the literature: a thin, calcite plate cut parallel to the optic axis;[50] between 1650 and 1900 Å the extraordinary ray alone is transmitted with a maximum at 1730 Å. The use of this type of polarizer is evidently limited by its narrow spectral range.

Birefringent polarizers

They are prisms of the type Wollaston, Rochon, or Senarmont. On Fig. 3 are shown the various configurations of the optic axes and of the transmitted beams. In the v.u.v. no cement can be used for assembling the two parts of a prism and, if optical contacting is not feasible, the angle θ must be inferior to the critical angle of the first part, and, thus, the deviation α is limited.

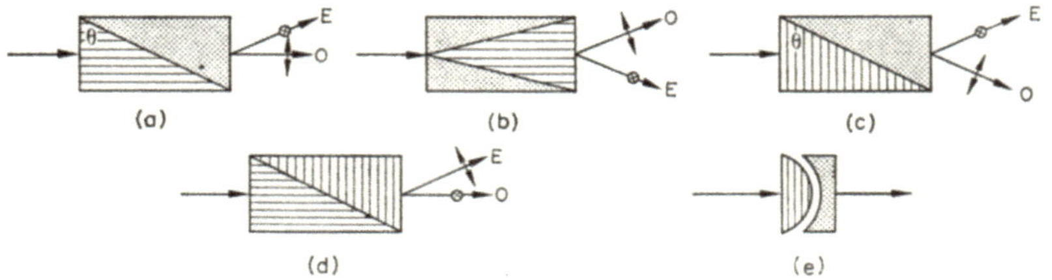

FIG. 3. Schematic views of birefringent polarizers. (a) Rochon prism; (b) double Rochon prism; (c) Wollaston prism; (d) Senarmont prism; (e) birefringent lenses.

Johnson[51] used successfully a MgF_2 Wollaston prism down to 1200 Å, but the two parts were not optically contacted; the angular separation 2α was 1.12° in the range 1400-2100 Å.

Compared to the Wollaston prism, the Rochon and Senarmont prisms have half an angular separation, but one beam, namely the ordinary one, is undeviated. The angular separation is given, to a good approximation, by $\alpha = \Delta n \tan\theta$, where $\Delta n = n_e - n_0$ is the birefringence of the material. The choice of the latter is limited to quartz and magnesium fluoride. Data on the birefringence of both materials (and sapphire) have been published recently,[52] and may be used for the determination of θ in order to get a given α.

The good transmission of synthetic quartz down to 1500 Å and the ease of polishing allow the production of optically contacted Rochon prisms. Such a prism has been employed down to 1500 Å for crystal optics studies and linear dichroism measurements[53] (see below); the dimensions are 10 by 10 by 28 mm and α(rad) increases from 3.65×10^{-2} at 2000 Å to 5.44×10^{-2} at 1550 Å.

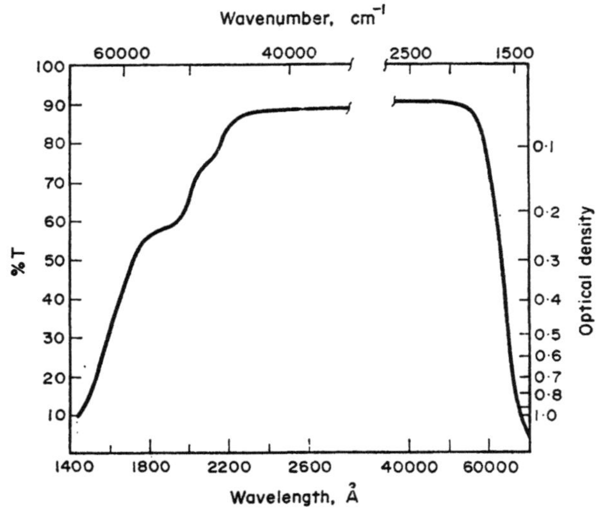

Fig. 4. Transmission characteristic of a MgF_2 single Rochon prism of dimensions 12 by 12 by 50 mm.[55]

Magnesium fluoride has a good transmission down to 1200 Å (cut off at ~1150 Å) and appears quite promising. However, optical contacting is rather difficult to achieve and optically contacted prisms have been reported only recently: double Rochon and double Senarmont (dimensions ~8 by 8 by 16 mm),[54] single Rochon (12 by 12 by 50 mm).[55] Figure 4 shows the transmission characteristic of the latter prism. Such prisms can be used down to 1300 Å, but below this the rapid decrease of Δn,[52a] and consequently of the angular deviation α, makes them inefficient as polarizers.

Apart from birefringent prisms, Figure 3 (e) shows another mounting tried in the v.u.v.[52b] It consists of two MgF_2 lenses respectively plano-concave and plano-convex with crossed optic axes: such a pair acts as a convergent lens of focal length $F_V = R/\Delta n$ (R radius of curvature) for one polarization and as a divergent lens with $F_H = -F_V$ for the other polarization. A diaphragm of suitable size and shape is placed in the focal plane F_V. Due to the magnitude of Δn, a sufficient convergence is achieved with three pairs of lenses; with $R = 9.9$ mm $F_{tot} = 235$ mm at 1608 Å, and the degree of polarization obtained is more than 98%. The drawback is that surface losses reduce the transmission.

Pile-of-plates polarizers

Although this type of polarizer implies reflection phenomena, we choose to treat it separately for it is a transmission polarizer. Its use is then limited to the Schumann region. We do not enter into details of the theory or the design of this polarizer, much used in the infrared.[56, 57] The degree of polarization and the transmittance of the favoured component are given by

$$P = \frac{1 - \cos^{4m}(\theta_i - \theta_r)}{1 + \cos^{4m}(\theta_i - \theta_r)}, \qquad T_p = \left[1 - \frac{\tan^2(\theta_i - \theta_r)}{\tan^2(\theta_i + \theta_r)}\right]^{2m}, \tag{52}$$

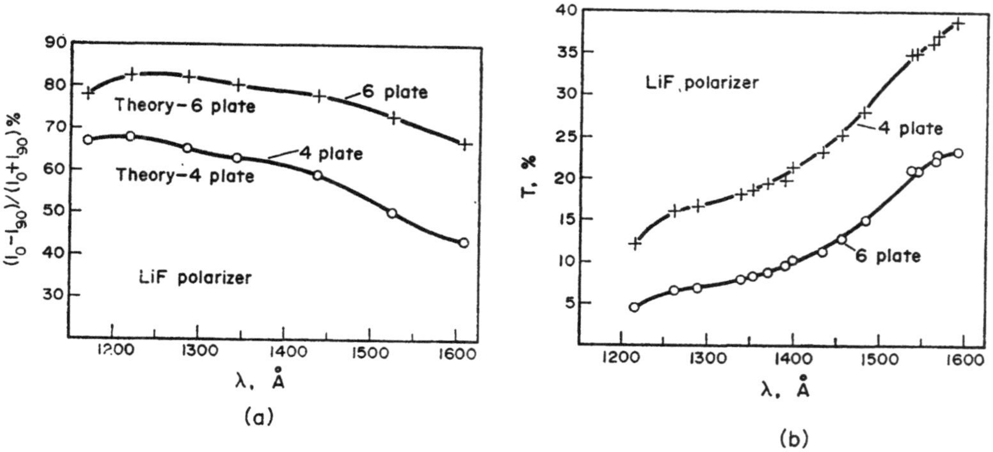

Fig. 5. Spectral dependence of the properties of LiF pile-of-plates polarizers. (a) Polarizance; (b) transmittance, i.e. ratio of the intensities of the polarized transmitted beam to the unpolarized incident beam.[59]

where θ_i and θ_r are the angles of incidence and of refraction respectively, and m the number of plates. The material may be calcium fluoride[58] or lithium fluoride,[59] the latter being preferable since plates can be easily cleaved and surface losses due to polishing are reduced. Figure 5 shows the degree of polarization and transmittance obtained with four and six plates.[59] The lower limit in wavelength is approximately 1200 Å. Note that this polarizer is commercially available in the United States:[60] a typical set is described by Heath: it consists of eight plates with thicknesses 0.25-0.4 mm, so that the total thickness of LiF is 2.5 mm for an angle of incidence of 60°.[61]

Reflection-type polarizers

Single reflection. A single reflection at Brewster or pseudo-Brewster angle θ_B has proved advantageous in the Schumann region, either with transparent (LiF, CaF_2, MgF_2) or with slightly absorbing ($k < 0.1$) materials (glass, quartz, sapphire, biotite[62]). However, the efficiency of such a polarizer depends also on the reflecting power $R_S(\theta_B)$.[63] Figure 6 presents the variation with λ of P and $R_S(\theta_B)$ for some materials;[64] it is to be noted that sapphire combines a high P with a good R_S in the spectral range 700-2000 Å. From the known optical constants,

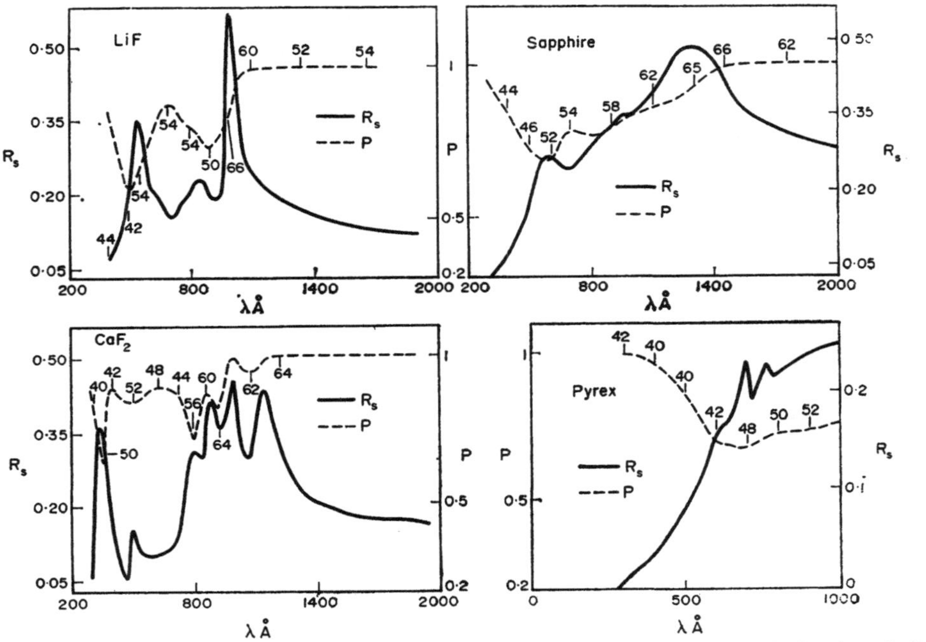

FIG. 6. Efficiency of several materials as polarizers at pseudo-Brewster angle θ_B (θ_B is indicated along with the curve P vs. λ).[64]

Sasaki and Fukutani have studied the polarizing properties at the fixed angle of incidence 60° of various materials as LiF, MgF_2, Si, Ge, and Au,[65] and their conclusions are in agreement with those of Hunter[66] that, at shorter wavelengths ($\lambda < 1000$ Å), gold has the most uniform polarization characteristics. Other materials, such as ZnS and anodically formed films of Al_2O_3, have also been studied along with multilayer unbacked metal films (both by reflection and by transmission).[66] Finally, Abelès[67] has investigated theoretically the possibilities of a thin film on a bulk substrate, namely the combinations oxide/metal, ionic crystal/metal and metal/ionic crystal; he concludes that a high polarization can be achieved, but in a narrow spectral range and with a small angular aperture.*

* The use, as a polarizing mirror, of a highly anisotropic material is worthy of consideration, calcite, for instance.[39,68]

Let us notice that the problem of angular aperture may be partly solved if, instead of a plane mirror, a toroidal mirror is used. The radii of curvature are chosen so that the stigmatic condition is satisfied for an angle of incidence close to the angle of maximum polarization. Thus the angle of incidence is stationary, even with a divergent beam, and the degree of polarization may be high.(69)

Triple reflection.(70, 71) In order to avoid deviation of the beam, the triple reflection arrangement of Figure 7 (b) is very convenient and, with gold mirrors, it is possible to obtain 95% polarization with approximately 15% transmission over a wide spectral range as illustrated by Figure 7.(71)

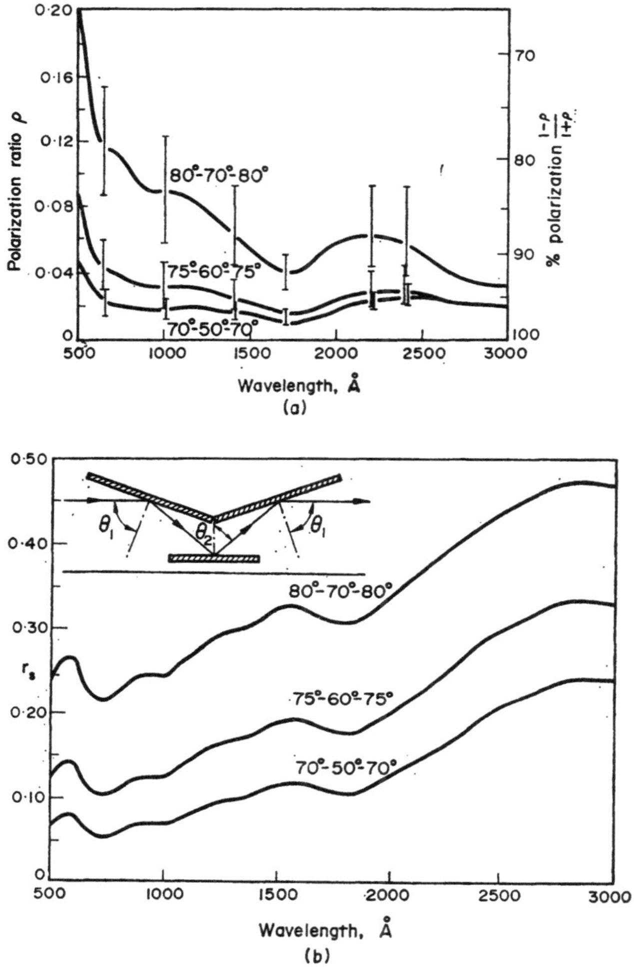

Fig. 7. Efficiency of triple-reflection polarizers of different geometry (gold mirrors). (a) Polarization ratios; (b) effective reflectance.(71)

Four reflection. Such an arrangement has been adopted by Rosenbaum *et al.*[48] The four gold mirrors are at an angle of incidence of 60° and have their normals in one plane; this offers the additional advantage of not inverting the beam—a condition required in some applications.

Transmission-type polarizers

The transmission through thin unbacked films is a possibility for the x.u.v. which has been investigated theoretically by Hunter[66] who considered Al and the combinations Al_2O_3-Al-Al_2O_3 and Au-Al-Au for the spectral region 300-800 Å. The drawbacks are the same as for reflecting multilayers, and attempts to verify performance are impeded by various experimental difficulties.

Circular polarization

Besides plane polarization, polarimetry includes many other important problems such as modulation of polarization for differential effect measurements, phase-angle measurements, and ellipsometry, but the present state of v.u.v. techniques seems to prohibit most of these types of measurements. Nevertheless, the important problem of circular polarization, connected with phase retarders, has found several recent approaches.

Plane-polarized light can be converted into circularly polarized by reflection on a mirror whose plane of incidence makes an angle of 45° with the direction of the vibration provided the light is incident at the principal angle θ_p, and $R_s(\theta_p)$, and $R_p(\theta_p)$ are nearly equal. The angle θ_p, defined as giving $\Delta = \delta_s - \delta_p = 90°$, can be calculated analytically from the optical constants.[28, 63]

As θ_p is wavelength-dependent, this kind of retarder is only well adapted for use in a narrow spectral range, and has been reported by McIlrath[72] for the study of the Lyman a line of hydrogen; the mirror consists of an evaporated aluminium film oxidized in air for which $\theta_p = 52° \pm 2°$ and $R_p(\theta_p) = R_s(\theta_p) = 0.50 \pm 0.05$.

Piezobirefringence provides another possibility, and Metcalf and Baird[73] produced a retarder by straining an LiF crystal and obtained light 95% circularly polarized, with a transmission of 5% at 1216 Å for the complete polarizer train (LiF reflecting polarizer + LiF crystal).

In the Schumann region, other methods may be considered, e.g. double total reflection prism (Fresnel rhomb) (though the state of polish of CaF_2 or LiF surfaces affects the total reflection), or an achromatic quarter-wave plate by associating several birefrigent materials[52a] though thicknesses are small and the tolerance of flatness and of thickness are very stringent.

3. Reflectance measurements with polarized light

3.1. ISOTROPIC MATERIALS, OBLIQUE INCIDENCE, TWO OR MORE ANGLES OF INCIDENCE

We consider here the case where the back surface of the sample plays no part (bulk material or strongly absorbing thin films), so that the reflection phenomena are governed by eqns. (41) to (46). The problem consists in getting R_s and R_p

from intensity measurements and this implies rotation of either the polarizer or of the whole reflectometer; these technical aspects will be discussed below in § 5.

As regards theory, the analysis made by Humphreys-Owen[74, 75] is valid, but the choice of method is determined by the particular conditions pertaining to the vacuum ultraviolet, especially the accuracy of reflectance measurements—not so high in this spectral range—and the use of partial polarization. Thus methods which entail the determination of one of the three special angles along with a reflectance or a reflectance ratio are hardly feasible. For instance, the method[76] developed by Potter for the infrared or the visible, namely the determination of the optical constants from the angle θ_B at which $P = (R_p - R_s)/(R_p + R_s)$ maximizes and the value P_{max} of this maximum may be used only as a check once the optical constants have been determined by another method.[35] Practically, only the methods of class I[74] need be considered:

Method A R_p and R_s at one angle of incidence.
Method B R_p at two angles of incidence.
Method C R_s at two angles of incidence.
Method D R_p/R_s at two angles of incidence.

The first problem is how to determine R_p and R_s when the light incident on the sample is partially polarized with a degree of polarization $V = (I_V - I_H)/(I_V + I_H)$. Suppose we measure R_H and R_V when the plane of incidence is respectively horizontal and vertical. Then[36]

$$R_s = \frac{R_H - R_V + V(R_H + R_V)}{2V}, \qquad R_p = \frac{R_V - R_H + V(R_H + R_V)}{2V}. \quad (53)$$

So, provided V has been measured (§ 2.1), R_s and R_p are easily determined from the measurements of R_H and R_V.

Method A. R_p and R_s at one angle of incidence θ. From eqns. (41), (45), and (46), we deduce[36]

$$\left.\begin{aligned} 2p &= \frac{(R_s - R_p)(1 - R_s)\cos\theta(\cos^2\theta - \sin^2\theta\tan^2\theta)}{(R_p + R_s)(R_s - 1) + 2R_s(1 - R_p)\cos^2\theta}, \\ q^2 &= 2p(1 + R_s)\cos\theta/(1 - R_s) - p^2 - \cos^2\theta, \end{aligned}\right\} \quad (54)$$

and

$$\left.\begin{aligned} n^2 &= \frac{1}{2}(p^2 - q^2 + \sin^2\theta) + \left[\frac{1}{4}(p^2 - q^2 + \sin^2\theta)^2 + p^2q^2\right]^{1/2}, \\ k &= pq/n. \end{aligned}\right\} \quad (55)$$

It appears that the accuracy can be improved if the angle θ is high, e.g., 60°.[36]

Methods B, C, and D. In contrast with method A, these methods do not allow an analytical determination of n and k. The procedure to be employed is similar to that of § 1.2, namely a graphical solution with the help of constant R_s or R_p

curves in the plane (n, k) for various angles (the sets of curves are available in the literature[28]) or an equivalent computer analysis.[49] As far as we know, method D has not been employed in the v.u.v., and the choice remains between C and B. From the shape of the constant R_s and R_p curves, Hunter[77] concludes that the method B gives the best precision except in the case where both $1 - n$ and k are small, as discussed by Feuerbacher.[49] In other respects, the accuracy of methods A, B, and C has been discussed by Uzan[78] with a few numerical examples. Furthermore, the accuracy of the determination of n and k can be improved if more than two angles are used.[49]

A slightly different method has been described by Graves and Lenham for isotropic materials,[79] and, although not applied to the v.u.v., is worthy of consideration. This is a "three-angle" method consisting in the measurement of the ratios $R_p(\theta_1)/R_p(\theta_2)$ and $R_p(\theta_3)/R_p(\theta_2)$ and the use of the constant value curves of ε_1 and ε_2 in the plane $R_p(\theta_1)/R_p(\theta_2)$ vs. $R_p(\theta_3)/R_p(\theta_2)$. Its use requires completely polarized incident light, which may prove restrictive in the v.u.v.

3.2. ANISOTROPIC MATERIALS

Due to the connection with band structures, the study of optical anisotropy effects in crystals deserves special attention, and an increasing number of papers are devoted to such effects. Let us consider first the simpler and important case of uniaxial crystals which includes hexagonal, tetragonal, and trigonal classes.

Uniaxial crystals: normal incidence and Kramers-Kronig analysis

First method. Polarized light, one crystal sample cut parallel to the optic axis c. The method consists in measuring R_0 for E perpendicular to c and R_e for E parallel to c; a subsequent use of the K.-K. analysis yields n_0 and k_0 from R_0 and n_e, and k_e from R_e.[80, 81]

Second method. Unpolarized light, two crystal samples one cut parallel to the optic axis, the other cut perpendicular to the optic axis. Measurements performed on the first sample yield $R_1 = (R_0 + R_e)/2$ and on the second sample $R_2 = R_0$,[82] whence R_0 and R_e are readily deduced. The question which naturally arises is how to ascertain whether the surface conditions are identical, even if the same polishing process has been applied to both samples.

Uniaxial crystals: oblique incidence

The treatment of weakly absorbing anisotropic media as found, for instance, in Born and Wolf,[2] is of limited use in the v.u.v., and, until recently, the only approach to the case of strongly absorbing media was that of Drude.[83] If the optic axis is arbitrarily oriented with respect to the incidence and surface planes, the formulae for the reflectance amplitudes can be calculated[84] but are very complicated. However, the intensity ratios can be derived easily from them in the four cases summarized below.

Let θ_1 be the angle of incidence and n_1 the refractive index (real) of the medium of incidence, r the amplitude reflectance ratio.

Case I. Optic axis c normal to the plane of incidence.

$$r_s^I = \frac{n_1 \cos\theta_1 - (\hat{n}_e^2 - n_1^2 \sin^2\theta_1)^{1/2}}{n_1 \cos\theta_1 + (\hat{n}_e^2 - n_1^2 \sin^2\theta_1)^{1/2}} = r_s(\hat{n}_e) \tag{56}$$

$$r_p^I = \frac{\hat{n}_0^2 \cos\theta_1 - n_1(\hat{n}_0^2 - n_1^2 \sin^2\theta_1)^{1/2}}{\hat{n}_0^2 \cos\theta_1 + n_1(\hat{n}_0^2 - n_1^2 \sin^2\theta_1)^{1/2}} = r_p(\hat{n}_0). \tag{57}$$

Case II. Optic axis c parallel to the plane of incidence. α = angle between the optic axis and the normal to the surface.

$$r_s^{II} = \frac{n_1 \cos\theta_1 - (\hat{n}_0^2 - n_1^2 \sin^2\theta_1)^{1/2}}{n_1 \cos\theta_1 + (\hat{n}_0^2 - n_1^2 \sin^2\theta_1)^{1/2}} = r_s(\hat{n}_0), \tag{58}$$

$$r_p^{II} = \frac{\hat{n}_0\hat{n}_e \cos\theta_1 - n_1(\hat{n}_e^2 \cos^2\alpha + \hat{n}_0^2 \sin^2\alpha - n_1^2 \sin^2\theta_1)^{1/2}}{\hat{n}_0\hat{n}_e \cos\theta_1 + n_1(\hat{n}_e^2 \cos^2\alpha + \hat{n}_0^2 \sin^2\alpha - n_1^2 \sin^2\theta_1)^{1/2}}, \tag{59}$$

the latter formulae covering the cases III and IV.

Case III. Optic axis c perpendicular to the surface. $\alpha = 0$.

Case IV. Optic axis c parallel to the plane of incidence and to the surface. $\alpha = 90°$.

The general formulae of case II include, thus, case III, which has been treated recently by Mosteller and Wooten.[85] The notation $r(\hat{n}_j)$, with $j = o$ or e, indicates that the formula is the same as for an isotropic material with complex refractive index $\hat{n}_j$. From eqns. (56)-(58) the calculation of reflectances is identical to the isotropic case by putting

$$p_j - iq_j = (\hat{n}_j^2 - n_1^2 \sin^2\theta_1)^{1/2}, \qquad J = o, e. \tag{60}$$

For eqn. (59) it is necessary to introduce

$$A = n_0 n_e - k_0 k_e, \qquad B = n_0 k_e + n_e k_0 \tag{61}$$

and

$$p - iq = (\hat{n}_0^2 \sin^2\alpha + \hat{n}_e^2 \cos^2\alpha - n_1^2 \sin^2\theta_1)^{1/2} \tag{62}$$

so that

$$R_p^{II} = \frac{(A\cos\theta_1 - p)^2 + (B\cos\theta_1 - q)^2}{(A\cos\theta_1 + p)^2 + (B\cos\theta_1 + q)^2}. \tag{63}$$

Note that:

$\alpha = 0$ (case III) gives $p = p_e,\ q = q_e$.

$\alpha = 90°$ (case IV) gives $p = p_0,\ q = q_0$.

Since it is inconvenient to employ two samples, the more advantageous method rests on measurements on a sample cut parallel to the optic axis. By rotation of the polarizer, or of the sample in its plane, it is possible to measure $R_s(n_e, k_e, \theta_1)$, $R_p(n_0, k_0, \theta_1)$, $R_s(n_0, k_0, \theta_1)$, and $R_p^{IV}(n_0, k_0, n_e, k_e, \theta_1)$ at any angle of incidence θ_1 (cases I and IV). By measurements at two or more angles of incidence, and following the same procedure as for isotropic materials, n_0 and k_0 can be determined. Independently, n_e and k_e can be deduced from $R_s(n_e, k_e, \theta_1)$ and $R_s(n_e, k_e, \theta_2)$, but, as discussed by Engelsrath and Loewenstein,(86) the accuracy is poor. Yet, Uzan(78) has shown that the case IV they had discarded is quite valuable: by introducing the previously determined values of n_0, k_0 in the expression for R_p^{IV}, n_e and k_e can be obtained by intersection of constant R_p^{IV} curves in the plane n_e, k_e. As these curves are portions of circles,(84) an analytical solution is also feasible.(39) Nevertheless, the discussion of accuracy(78) emphasizes the need of a sufficiently high degree of polarization and precise determination of this when partially polarized light is used. Finally, in the case of a material like graphite which, for mechanical reasons, can only be cut perpendicular to the optic axis, case III needs consideration. Provided the measurements are performed at a large number of angles and analysed by computer, precise results for the optical constants can be obtained.(107)

Biaxial crystals: Orthorhombic class only

Drude(83) has given expressions for the ratios r_p/r_s for six orientations characterized by the principal axes lying respectively in and normal to the plane of incidence. Those equations are not very convenient and, following the same proce-

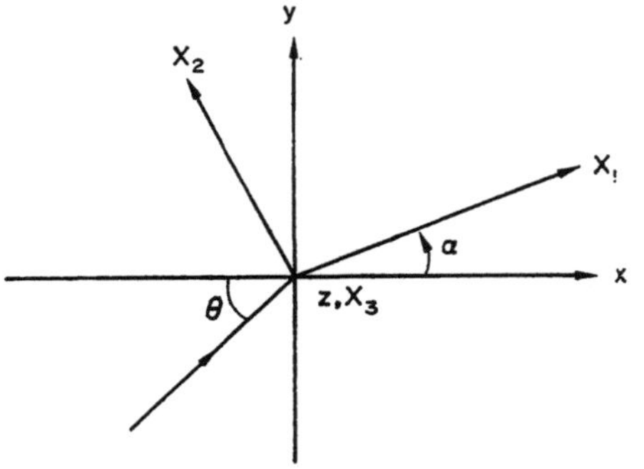

FIG. 8. Respective positions of the coordinate axes x, y, z, and of the principal axes X_1, X_2, X_3 of the crystal; x, y is the plane of incidence, the interface between the incidence medium, and the crystal is the plane $x = 0$.

dure as for uniaxial crystals, we may write directly r_s and r_p for the more general case of Fig. 8. X_1, X_2, X_3 are the principal axes corresponding respectively to the indices $\hat{n}_1$, $\hat{n}_2$, $\hat{n}_3$, n is the refractive index of the incidence medium, θ the angle of incidence, X_1 and X_2 lie in the plane of incidence x, y, and α is the angle between the normal x to the boundary and the X_1 axis. Then

$$r_s = \frac{n \cos\theta - (\hat{n}_3 - n^2 \sin^2\theta)^{1/2}}{n \cos\theta + (\hat{n}_3^2 - n^2 \sin^2\theta)^{1/2}} = r_s(\hat{n}_3), \tag{64}$$

$$r_p = \frac{\hat{n}_1\hat{n}_2 \cos\theta - n(\hat{n}_1^2 \cos^2\alpha + \hat{n}_2^2 \sin^2\alpha - n^2 \sin^2\theta)^{1/2}}{\hat{n}_1\hat{n}_2 \cos\theta + n(\hat{n}_1^2 \cos^2\alpha + \hat{n}_2^2 \sin^2\alpha - n^2 \sin^2\theta)^{1/2}}. \tag{65}$$

These equations include the cases $\alpha = 0$ and $\alpha = 90°$ considered by Drude, and the four other cases are deduced by permutation of the indices.

The only attempt to use Drude's equations is by Jacobsen,[32a] who established a computer program, but no experimental work on biaxial crystals in the v.u.v. has so far been reported.

4. Other methods

We present here briefly other methods which are of more limited use in the v.u.v.

4.1. MODIFIED CRITICAL-ANGLE METHOD

This method, developed by Hunter,[88] can be applied to materials having a refractive index n less than unity and a small k. It implies the measurement of the reflectance vs. angle of incidence, then the determination of the angle θ_c for which the slope of the reflectance curve is a maximum. If k is zero, n is given by $n = \sin \theta_c$; for non-zero values of k but less than 0.2, a slight correction to this formula is necessary. The method is suitable for metals above the plasma frequency.[89]

4.2. INTERFERENCE IN THIN PLATES OR FILMS

The observation of interference effects with thin plates is a classical way of determining the index of refraction of a transparent material. This observation may be performed as follows.

At oblique incidence: by varying the angle of incidence at a given wavelength, maxima and minima of reflectance are observed, conforming with the equation

$$N\lambda = P\lambda + 2T(n^2 - \sin^2 \theta)^{1/2}, \tag{66}$$

where T is the thickness and $P\lambda$ represents the phase change at reflection. This method is complementary to the modified critical angle method. Discussion of the determination of N and P will be found in ref. 89.

At normal incidence: by varying the wavelength, a channelled spectrum is observed in transmission through a thin transparent unbacked plate; a continuous spectrum or a spectrophotometer is required. The maxima of transmission are given by $2nT = N\lambda$, and the method has been applied in the v.u.v. to crystals which can be cleaved[90] or polished[97] to a very small thickness.

At normal incidence for a film on an absorbing substrate: if the thickness of the film varies linearly along the large dimension of the supporting plate, a channelled spectrum is observed in reflection by scanning the film with the beam of light at a given wavelength. The variation of thickness is obtained by evaporation through a rotating mask.[92] From the oscillations of R, n and k of the film can be deduced[93] provided k be small (< 0.1).

4.3. THIN DIELECTRIC FILM ON A METAL SURFACE

The method consists in measuring the normal incidence reflectance of the metal alone, then of the metal coated with a thin transparent dielectric film of known refractive index and thickness.(94) In the v.u.v., the dielectric may be LiF.(95)

4.4. MEASUREMENTS OF DIFFERENTIAL EFFECTS

Under this heading we include effects such as birefringence, linear or circular dichroism, electroreflectance, and so on. We consider here only those which have been measured in the v.u.v.

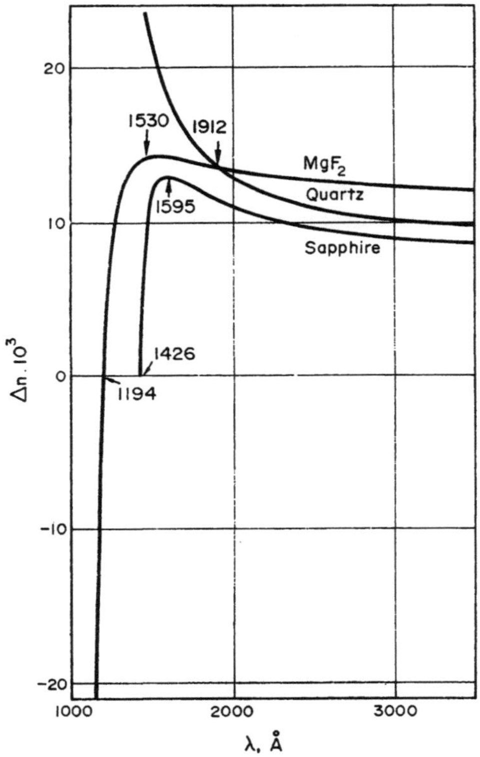

FIG. 9. Dispersion of the birefringence of sapphire, magnesium fluoride, and quartz. $\Delta n = -(n_0 - n_e)$ for sapphire and $\Delta n = (n_0 - n_e)$ for the other two crystals.(52)

The birefringence already quoted in § 2.3 is measured in a classical way by channelled spectra obtained with a birefringent plate placed between crossed polarizers.(52, 96) Figure 9 shows the dispersion of birefringence of quartz, sapphire, and magnesium fluoride in the v.u.v.(52)

Linear dichroism may be measured, either by the method described by Jaffé *et al.*,(97) consisting in associating a polarizer and a multiwave plate, or by modulation of the polarization with a rotating polarizer such as a Rochon prism.(53)

It is to be noted that such methods give the difference of two quantities, for instance two refractive indices, with a much higher accuracy than either of them measured separately.

Due to the problem of retarders, measurement of circular dichroism has not yet been extended to the v.u.v., and the present limit is approximately 1900 Å, either with commercial apparatus or with the method reported by Holzwarth.[98]

Finally, an application of the rotating polarizer has been the measurement of the rotatory power of quartz down to 1480 Å.[99]

5. Cause of errors in reflectance measurements and design of reflectometers for the vacuum ultraviolet

We have seen in § 3 that reflectance measurements are of prime importance in the v.u.v. The technical aspects of methods of measurement have been omitted in the preceding sections. Since brief treatment of the experimental methods is not possible, depending as they do on a number of special techniques of vacuum spectroscopy, excellently surveyed by Samson,[100] the following summary is restricted to papers containing practical designs and sources of error.

The surge of interest in the optical properties of solids in the v.u.v. accounts for the rather large number of v.u.v. reflectometers described in the literature over the past 10 years.[10, 101–5] Without entering into the details of the various designs, for which the reader is referred to the original papers, we may say that the trends are towards combining coupled mirror-detector movements (by an angle-doubling device) with rotation of the whole system about the axis of incident light (in polarimetric experiments) and using ultra-high vacuum conditions.[105]

As for errors and their consequences, the most comprehensive study can be found in two papers by Hunter,[77, 106] where a thorough discussion of the errors and the ways to eliminate or reduce them is presented followed by a description of a reflectometer meeting various essential requirements.[106]

References

1. Feynman, R. P., Leighton, R. B., and Sands, M., *The Feynman Lectures of Physics*, Addison-Wesley, Reading, Massachusetts, 1964, vol. 2, ch. 33.
2. Born, M., and Wolf, E., *Principles of Optics*, Pergamon Press, London, 1965, p. 665.
3. Landau, L. D., and Lifshitz, E. M., *Electrodynamics of Continuous Media*, Pergamon Press, Oxford, 1963, p. 256.
4. Kronig, R. de L., *J. Opt. Soc. Am.* **12**, 547 (1926).
5. Landau, L. D., and Lifshitz, E. M., *Statistical Physics*, Pergamon Press, London, 1958.
6. Bode, H. W., *Network Analysis and Feedback Amplifier Design*, van Nostrand, New York, 1945.
7. Toll, J. S., *Phys. Rev.* **104**, 1760 (1956).
8. Greenaway, D. L., and Harbeke, G., *Optical Properties and Band Structures of Semiconductors*, Pergamon Press, Oxford, 1968.
9. Stern, F., *Solid St. Phys.* **15**, 299 (1963).

10. MADDEN, R. P., and CANFIELD, L. R., *J. Opt. Soc. Am.* **51**, 838 (1961).
11. ROBINSON, T. S., and PRICE, W. C., *Proc. Phys. Soc. London* B, **66**, 969 (1953).
12. BOWLDEN, H. J., and WILMSHURST, J. K., *J. Opt. Soc. Am.* **53**, 1073 (1963).
13. PRIOL, M., SEIGNAC, A., and ROBIN, Mme S., *Compt. Rend.* **258**, 5398 (1964).
14. WHITE, J. J., and STRALEY, J. W., *J. Opt. Soc. Am.* **58**, 759 (1968).
15. CARDONA, M., and GREENAWAY, D. L., *Phys. Rev.* **133**, A, 1685 (1964).
16. PHILIPP, H. R., and TAFT, E. A., *Phys. Rev.* **113**, 1002 (1959).
17. ROESSLER, D. M., *Br. J. Appl. Phys.* **16**, 1119 (1965); **17**, 1313 (1966).
18. MACRAE, R. A., ARAKAWA, E. T., and WILLIAMS, M. W., *Phys. Rev.* **162**, 615 (1967).
19. ARAKAWA, E. T., and WILLIAMS, M. W., *J. Phys. Chem. Solids* **29**, 735 (1968).
20. PRIOL, M., LARVOR, M., and ROBIN, S., *Compt. Rend.* **259**, 3983 (1964).
21. HINSON, D. C., and STEVENSON, J. R., *Phys. Rev.* **159**, 711 (1967).
22. LEMONNIER, J. C., STÉPHAN, G., and ROBIN, S., *Compt. Rend.* **262**B, 355 (1966).
23. ROESSLER, D. M., *Br. J. Appl. Phys.* **16**, 1359 (1965).
24. BERREMAN, D. W., *Appl. Opt.* **6**, 1519 (1967).
25. KOZIMA, K., SUËTAKA, W., and SCHATZ, P. N., *J. Opt. Soc. Am.* **56**, 181 (1966).
26. VERLEUR, H. W., *J. Opt. Soc. Am.* **58**, 1356 (1968).
27. TOUSEY, R., *J. Opt. Soc. Am.* **29**, 235 (1939).
28. HOLL, H. B., *J. Opt. Soc. Am.* **57**, 683 (1967).
29. SASAKI, T., FUKUTANI, H., and ISHIGURO, K., *Japan. J. Appl. Phys.* **4**, suppl. I, 527 (1965).
30. JUENKER, D. W., *J. Opt. Soc. Am.* **55**, 295 (1965).
31. JUENKER, D. W., LEBLANC, L. J., and MARTIN, C. R., *J. Opt. Soc. Am.* **58**, 164 (1968).
32. TOUSEY, R., *Phys. Rev.* **50**, 1057 (1936).
33. ROBIN-SALOMOND, S., *Rev. Opt.* **33**, 193 (1954).
34. HINSON, D. C., *J. Opt. Soc. Am.* **56**, 408 (1966).
35. UZAN, E., DAMANY, H., and ROMAND, J., *Compt. Rend.* **260**, 5735 (1965).
36. STEPHAN, G., LEMONNIER, J. C., and ROBIN, S., *Compt. Rend.* **260**, 6819 (1965); *J. Opt. Soc. Am.* **57**, 486 (1967).
37. COLE, T. T., and OPPENHEIMER, F., *Appl. Opt.* **1**, 709 (1962).
38. RABINOVITCH, K., CANFIELD, L. R., and MADDEN, R. P., *Appl. Opt.* **4**, 1005 (1965).
39. LE CALVEZ, Y., STEPHAN, G., and ROBIN, S., *Optica Acta* **15**, 583 (1968).
40. HANSON, W. F., and ARAKAWA, E. T., *J. Opt. Soc. Am.* **56**, 124 (1966).
41. SASAKI, T., and ISHIGURO, K., *Japan. J. Appl. Phys.* **2**, 289 (1963).
42. DAMANY, H., *Rev. Opt.* **44**, 1 (1965).
43. ABELÈS, F., *Compt. Rend.* **230**, 1942 (1950).
44. HAMMER, D. C., ARAKAWA, E. T., and BIRKHOFF, R. D., *Appl. Opt.* **3**, 79 (1964).
45. CODLING, K., and MADDEN, R. P., *J. Appl. Phys.* **36**, 380 (1965).
46. MADDEN, R. P., EDERER, D. L., and CODLING, K., *Appl. Opt.* **6**, 31 (1967).
47. BATHOW, G., FREYTAG, E., and HAENSEL, R., *J. Appl. Phys.* **37**, 3449 (1966).
48. ROSENBAUM, G., FEUERBACHER, B., GODWIN, R. P., and SKIBOWSKI, M., *Appl. Opt.* **7**, 1917 (1968).
49. FEUERBACHER, B., SKIBOWSKI, M., GODWIN, R. P., and SASAKI, T., *J. Opt. Soc. Am.* **58**, 1434 (1968).
50. SCHELLMAN, J., CHANDRASEKHARAN, V., DAMANY, H., and ROMAND, J., *Compt. Rend.* **260**, 117 (1965).
51. JOHNSON, W. C., JR., *Rev. Sci. Inst.* **35**, 1375 (1964).
52. CHANDRASEKHARAN, V., and DAMANY, H., (a) *Appl. Opt.* **7**, 939 (1968); **8**, 671 (1969); (b) unpublished.
53. BRAHMS, J., PILET, J., DAMANY, H., and CHANDRASEKHARAN, V., *Proc. Natn. Acad. Sci. U.S.A.* **60**, 1130 (1968).
54. STEINMETZ, D. L., PHILLIPS, W. G., WIRICK, M., and FORBES, F. F., *Appl. Opt.* **6**, 1001 (1967).
55. MORRIS, G. C., and ABRAMSON, A. S., *Appl. Opt.* **8**, 1249 (1969).

56. SHURCLIFF, W. A., *Polarized Light*, Harward Univ. Press, Cambridge, Massachusetts, 1962.
57. WEINBERG, J. L., *Appl. Opt.* **3**, 1057 (1964).
58. SCHELLMAN, J., CHANDRASEKHARAN, V., and DAMANY, H., *Compt. Rend.* **259**, 4560 (1964).
59. WALKER, W. C., *Appl. Opt.* **3**, 1457 (1964).
60. Crystal-Solid State Division of the Harshaw Chemical Co.
61. HEATH, D. F., *Appl. Opt.* **7**, 455 (1968).
62. ROBIN, M. B., KUEBLER, N. A., and YOH-HAN PAO, *Rev. Scient. Instrum.* **37**, 922 (1966).
63. DAMANY, H., *Optica Acta* **12**, 95 (1965).
64. STEPHAN, G., LEMONNIER, J. C., LE CALVEZ, Y., and ROBIN S., *Compt. Rend.* **262**B, 1272 (1966).
65. SASAKI, T., and FUKUTANI, H., *Japan. J. Appl. Phys.* **3**, 125 (1964).
66. HUNTER, W. R., *Japan. J. Appl. Phys.* **4**, suppl. I, 520 (1965).
67. ABELÈS, F., *Japan. J. Appl. Phys.* **4**, suppl. I, 517 (1965).
68. CHANDRASEKHARAN, V., and DAMANY, H., *Compt. Rend.* **265**B, 1474 (1967).
69. UZAN, E., CHANDRASEKHARAN, V., and DAMANY, H., *Optics Commun.* **1**, 221 (1969).
70. HAMM, R. N., MACRAE, R. A., and ARAKAWA, E. T., *J. Opt. Soc. Am.* **55**, 1460 (1965).
71. HORTON, V. G., ARAKAWA, E. T., HAMM, R. N., and WILLIAMS, M. W., *Appl. Opt.* **8**, 667 (1969).
72. MCILRATH, T. J., *J. Opt. Soc. Am.* **58**, 506 (1968).
73. METCALF, H., and BAIRD, J. C., *Appl. Opt.* **5**, 1407 (1966).
74. HUMPHREYS-OWEN, S. P. F., *Proc. Phys. Soc. London* **77**, 949 (1961).
75. HEAVENS, O. S., *Physics of Thin Films*, (eds. HASS, G., and THUN, R. E.), Academic Press, New York, 1964, vol. 2, p. 193.
76. POTTER, R. F., *J. Opt. Soc. Am.* **54**, 904 (1964); *Optical Properties of Solids* (eds. NUDELMAN, S., and MITRA, S. S.), Plenum Press, New York, 1969, ch. 16.
77. HUNTER, W. R., *J. Opt. Soc. Am.* **55**, 1197 (1965).
78. UZAN, E., *Optica Acta* **15**, 237 (1968).
79. GRAVES, R. H. W., and LENHAM, A. P., *J. Opt. Soc. Am.* **58**, 884 (1968).
80. GREENAWAY, D. L., and HARBEKE, G., *J. Phys. Chem. Solids* **26**, 1585 (1965).
81. CARDONA, M., and HARBEKE, G., *Phys. Rev.* **137**A, 1467 (1965).
82. LEIGA, A. G., *J. Opt. Soc. Am.* **58**, 880 (1968).
83. DRUDE, P., *Wied. Ann. Phys.* **32**, 585 (1887).
84. DAMANY, H., and UZAN, E., *Optica Acta* **17**, 131 (1970).
85. MOSTELLER, L. P., and WOOTEN, F., *J. Opt. Soc. Am.* **58**, 511 (1968).
86. ENGELSRATH, A., and LOEWENSTEIN, E. V., *Appl. Opt.* **5**, 565 (1966).
87. JACOBSEN, R. T., *J. Opt. Soc. Am.* **54**, 1170 (1964).
88. HUNTER, W. R., *J. Opt. Soc. Am.* **54**, 15 (1964).
89. SUTHERLAND, J. C., ARAKAWA, E. T., and HAMM, R. N., *J. Opt. Soc. Am.* **57**, 645 (1967).
90. CHANDRASEKHARAN, V., and DAMANY, H., *Compt. Rend.* **262**, 1268 (1966).
91. CHANDRASEKHARAN, V., and DAMANY, H., *Appl. Opt.* **7**, 687 (1968).
92. FABRE, D., and DAMANY, H., *J. Phys.* **22**, 138A (1961).
93. FABRE, D., ROMAND, J., and VODAR, B., *Optica Acta* **9**, 73 (1962); FABRE, D., *Rev. Opt.* **43**, 394 and 504 (1964).
94. VINCENT-GEISSE, J., and LECOMTE, J., *Appl. Opt.* **1**, 575 (1962).
95. SEIGNAC, A., PRIOL, M., and ROBIN, S., *Compt. Rend.* **258**, 4948 (1964).
96. SERVANT, R., *Ann. Phys.* **12**, 424 (1939).
97. JAFFE, J. H., JAFFE, H., and ROZENHECK, K., *Rev. Scient. Instrum.* **38**, 935 (1967).
98. HOLZWARTH, G., *Rev. Scient. Instrum.* **36**, 59 (1965).
99. CHANDRASEKHARAN, V., and DAMANY, H., *Appl. Opt.* **8**, 1506 (1969).
100. SAMSON, J. A. R., *Techniques of Vacuum Ultraviolet Spectroscopy*, Wiley, New York, 1967.
101. SMITH, A., *J. Opt. Soc. Am.* **50**, 863 (1960).
102. CHUBB, C., and HENSLEY, E., *Rev. Scient. Instrum.* **36**, 547 (1965).
103. PRIOL, M., DAUDÉ, A., and ROBIN, S., *Le Vide* **123**, 244 (1966).

104. VEHSE, R. C., SUTHERLAND, J. C., and ARAKAWA, E. T., *Rev. Scient. Instrum.* **39**, 268 (1968).
105. FEUERBACHER, B., GODWIN, R. P., and SKIBOWSKI, M., *Rev. Scient. Instrum.* **40**, 305 (1969).
106. HUNTER, W. R., *Appl. Opt.* **6**, 2140 (1967).
107. GREENAWAY, D. L., HARBEKE, G., BASSANI, F., and TOSATTI, E., *Phys. Rev.* **178**, 1340 (1969).

Additional bibliography

AHRENKIEL, R. K., Modified Kramers-Kronig analysis of optical spectra, *J. Opt. Soc. Am.* **61**, 1651 (1971); **62**, 1009 (1972).

ARMSTRONG, K. R., and BELL, R. J., Generalized principal angle of incidence and critical angle, *J. Opt. Soc. Am.* **60**, 701 (1970).

BALDINI, G., and RIGALDI, L., Ultra-violet optical constants of thin films determined by reflectance measurements, *J. Opt. Soc. Am.* **60**, 495 (1970).

BERMAN, M., KERCHNER, H. R., and ERGUN, S., Determination of the optical properties of absorbing uniaxial crystals from reflectance at oblique incidence, *J. Opt. Soc. Am.* **60**, 646 (1970).

BURGE, R. E., DAVIDSON, A. T., DRAPER, J. C., FIELD, G. R., and MURPHY, E., New focusing reflectometer for measuring optical constants of thin films, *Appl. Opt.* **10**, 342 (1971).

CAZAUX, J., Sur l'emploi des matériaux très anisotropes comme polariseurs pour l'ultraviolet extrême, *Nouvelle Rev. Opt.* **2**, 397 (1971).

CHANDRASEKHARAN, V., and DAMANY, H., Birefringent lens polarizer for the vacuum ultraviolet, *Appl. Opt.* **10**, 681 (1971).

CORVAN, J. J., ARAKAWA, E. T., and PAINTER, L. R., Time studies of the polarization due to gratings, *Appl. Opt.* **8**, 1734 (1969).

FEDOROV, F. I., and FILIPPOV, V. V., Reflection at a boundary between an uniaxial crystal and an isotropic medium, *Zh. prikl. Spektroskopiya* **9**, 1031 (1968) [*J. Appl. Spectrosc.* **9**, 1380 (1968)].

FIELD, G. R., and MURPHY, E., Method of using the reflectance ratios of different angles of incidence for the determination of optical constants, *Appl. Opt.* **10**, 1402 (1971).

GRAVES, R. H. W., Ellipsometry using imperfect polarizers, *Appl. Opt.* **10**, 2679 (1971).

GRAVES, R. H. W., Generalized principal angle of incidence, *J. Opt. Soc. Am.* **61**, 274 (1971).

HUEN, T., IRANI, G. B., and WOOTEN, F., Scanning ultra-high vacuum reflectometer, *Appl. Opt.* **10**, 552 (1971).

KOLB, D. M., Determination of the optical constants of solids by reflectance-ratio measurements at non-normal incidence, *J. Opt. Soc. Am.* **62**, 599 (1972).

MATSUI, A., and WALKER, W. C., Polarization of three vacuum ultraviolet monochromators measured with a biotite polarizer, *J. Opt. Soc. Am.* **60**, 64 (1970).

NESTELL, J. E., JR., and CHRISTY, R. W., Derivation of optical constants of metals from thin-film measurements at oblique incidence, *Appl. Opt.* **11**, 643 (1972).

NYTYANANDA, R., and RAMASESHAN, S., Dispersion of the birefringence of quartz, *J. Opt. Soc. Am.* **60**, 1531 (1970).

POTTER, R. F., Pseudo-Brewster angle and the optical constants, *Appl. Opt.* **9**, 1717 (1970).

SCHLEDERMANN, M., and SKIBOWSKI, M., Determination of the ellipcity of light and of optical constants by use two reflection polarizers, *Appl. Opt.* **10**, 321 (1971).

SCHMIDT, E., Simple method for the determination of optical constants of absorbing materials, *Appl. Opt.* **8**, 1905 (1969).

THETFORD, A., BATES, B., MACDONALD, J., On the design of Fabry-Perot reflectors for the vacuum ultraviolet, *Appl. Opt.* **9**, 35 (1970).

VASILINA, Z. S., and ROMANYAK, N. A., Reflection spectra in the v.u.v. under large angles of incidence, *Optika i Spektrosk.* **32**, 188 (1972) [*Optics Spectrosc.* **32**, 95 (1972)].

WEBBER, S. E., and SCHARBER, S. R., JR, A method of determining the optical constants of semi-transparent films, *Appl. Opt.* **10**, 338 (1971).

CHAPTER 2

Optical properties of metals

M. Priol and S. Robin
Laboratoire de Spectroscopie, Université de Rennes, 35031 Rennes-Cedex, France

Contents

Until 1960 the unreliability of measurements of the optical properties of metals in the far u.v. inhibited attempts at reasonable interpretation. Spectacular advances have since occurred, both experimentally, e.g. through the availability of clean surfaces and the use of detectors of better sensitivity, and in the application of high-speed computers for calculations of band structures. In turn, the results have enabled the theorists to construct more detailed formulations. We recognize, particularly, the importance of the electron-electron Coloumb-interactions, screening effects, and plasma oscillations, easily predicted by the RPA method.

We present different possible interpretations of the optical properties of metals by grouping them according to similarity of behaviour; firstly, those having free electrons (alkali metals, Al, Mg, Be), then divalent metals, rare earths, transition metals, and various others; finally, examples of alloys and other compounds having a metallic behaviour. Other types of experiment will be mentioned which afford supplementary information, e.g. the photoelectric effect which gives the absolute energy of the electronic levels and electron-energy loss experiments which concern especially volume and surface plasmons.

1. Metals obeying the free-electron theory

The metals obeying this theory[1] are classified in two groups: (a) the alkali metals, and (b) Al, Mg, and Be. All are characterized by a very high reflectance up to the plasma frequency.

According to the Maxwell equations and the free-electron theory, the dielectric constant $\hat{\varepsilon}(\omega)$ may be written

$$\hat{\varepsilon}(\omega) = 1 - \frac{\omega_p^2}{\omega(\omega - i/\tau_c)},$$

where ω_p is the plasma frequency (angular) and τ_c the relaxation time.

Separating real and imaginary parts,

$$\varepsilon_1 = 1 - \frac{\omega_p^2}{\omega^2 + 1/\tau_c^2}, \qquad \varepsilon_2 = \frac{\omega_p^2}{\omega\tau_c(\omega^2 + 1/\tau_c^2)}.$$

In the case where $\omega\tau_c \gg 1$,

$$\varepsilon_1 = 1 - \frac{\omega_p^2}{\omega^2}, \qquad \varepsilon_2 = 0.$$

$$\text{If } \omega_p > \omega \begin{cases} n = 0, \\ k = \left(\dfrac{\omega_p^2}{\omega_2} - 1\right)^{1/2}. \end{cases} \qquad \text{If } \omega_p < \omega \begin{cases} n = \left(1 - \dfrac{\omega_p^2}{\omega^2}\right)^{1/2}, \\ k = 0. \end{cases}$$

This shows that for frequencies higher than ω_p the solid becomes transparent.

1.1. ALKALI METALS

There are some optical measurements on the alkali metals in the near and far u.v. regions. The most extensive work has been carried out by Ives and Briggs,[2] Fleischmann *et al.*,[3–9] Sutherland *et al.*,[10, 11] and recently by Mayer *et al.*[12–14] up to 4 eV under ultra-high vacuum conditions.

Sodium and potassium have received great attention because their plasma frequencies lie in a spectral region which has long been easily accessible, and they obey the free-electron theory closely (the Fermi surface being practically spherical); their critical frequencies correspond to relatively small energies (3.7 eV for K) and are slightly displaced by the influence of interband transitions. Cohen[15] has shown that ε_1 can be put into the form

$$\varepsilon_1 = A - \frac{B}{\omega^2},$$

where A includes the total polarizability and B is given by

$$B = \frac{N_c e^2}{m_{\text{eff}} \varepsilon_0} = \omega_a^2$$

where N_c is the number density of conduction electrons which, for alkali elements, is equal to the number of atoms per unit volume, and m_{eff} is the effective mass of these electrons. The preceding equation then becomes

$$\varepsilon_1 = 1 + \alpha_a - \frac{\omega_a^2}{\omega^2} \quad \text{with} \quad \alpha_a = \frac{2e^2}{m^2 \hbar \varepsilon_0} \sum_{n,l} \frac{2}{(2\pi)^3} \int d^3k \, \frac{|P_{\text{ln}}|^2}{\omega_{\text{ln}}^3}.$$

α_a may be taken to be a constant only for photon energies less than the interband transition threshold (e.g. less than 2.2 eV for sodium). For energies between 2.2 and 30.8 eV (the threshold of excitation for inner shells of the ion), Sutherland and Arakawa[10, 11] employ an identical relation,

$$\varepsilon_1 = A' - \frac{B'}{\omega^2},$$

where A' is a term including the polarizability of the internal electrons, and B' is given by the relation

$$B' = \frac{N_c e^2}{m_{\text{opt}} \varepsilon_0} = \omega_b^2.$$

Finally,

$$\varepsilon_1 = 1 + \alpha_b - (\lambda^2/\lambda_b^2).$$

The curve of ε_1 versus λ^2 is then a straight line from which the optical mass and the plasma frequency may be deduced.

Figure 1 represents the variation of ε_2 for sodium as a function of energy.[2, 10–14] The first maximum between 1.2 and 2.2 eV cannot be explained by a one-electron theory (see refs. 34 and 37); the second maximum corresponds to the threshold of the interband transitions; a third maximum appears between 6 and 7 eV, which may be attributed to an interband transition or, after Esposito *et al.*,[16]

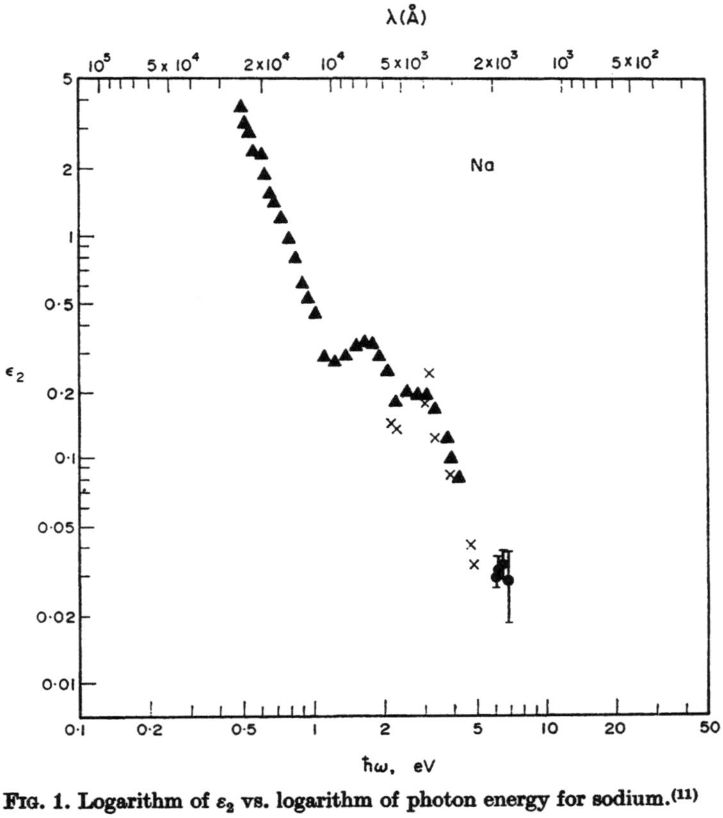

FIG. 1. Logarithm of ε_2 vs. logarithm of photon energy for sodium.[11]

to a plasmon contributing to an absorption process, since the relaxation time is sufficiently large (indeed, this maximum ought to be at 8.8 eV). According to the latter hypothesis, the other alkali metals should likewise exhibit maxima at about 5.8 eV for potassium, 4.5 eV for rubidium, and 4.25 eV for caesium: unfortunately, no experimental evidence of these exists at present.

Finally, the dielectric constant may be put in the form

$$\hat{\varepsilon} = 1 + \frac{Ne^2/m\varepsilon_0}{-\omega^2 + i\omega\gamma} + \sum_{(k)} \frac{N_k e^2/m\varepsilon_0}{\omega_k^2 - \omega^2 + i\omega\gamma_k},$$

where the indices k refer to the bound electrons. When ω_p occurs in the u.v. region, the damping constant is negligible compared to ω, and, as we are remote from the resonance frequencies ω_k of the bound electrons, we can also neglect γ_k. Then $\varepsilon \simeq 0$ and the equation reduces to

$$1 = \frac{Ne^2/m\varepsilon_0}{\omega_p^2} + \sum_{(k)} \frac{N_k e^2/m\varepsilon_0}{\omega_p^2 - \omega_k^2}.$$

The last term appears as a correction to the theoretical frequency, the value thus obtained being closer to the experimental one than ω_p. The plasmon energies determined from optical measurements are compared with characteristic energy losses of electrons[17–26] in Table 1.

TABLE 1. VOLUME PLASMON AND SURFACE PLASMON ENERGIES FOR ALKALI METALS (eV).

Metal	Theoretical values ω_p	Optical experiments Im (ε^{-1})	Characteristic energy losses of electrons	
			Volume plasmons	Surface plasmons
Li	8.02	6[18]	7.12[25]	4.2[25]
Na	5.95	5.9[18]	5.72[25]	2.85[25]
K	4.29	3.7[2]	3.72[25]	2.63[25]
Rb	3.86	3.4[18]	3.41[25]	2.46[25]
Cs	3.44	2.86[25]	2.9[25]	2.1[25]

Also, some authors[27–31] have recently studied radiative and non-radiative processes—with or without **k** conservation—in thin metal films near their plasma resonances. For example, a metal illuminated with light polarized parallel to the plane of incidence at the plasma frequency, re-emits radiation in a direction coincident with that of neither the incident nor the reflected radiation.

Again, Mayer and El Naby[18] mention some anomalies which may be compared to those found in the far u.v. for other metals. An example is the anomaly in the function $\omega\varepsilon_2$ for sodium at 0.5 eV, which is attributed either to an indirect transition[32] or to antiferromagnetic properties of the alkalis.[33] Cohen and Phillips[34–37] offer the possibility of a resonance between a pair of particles—one in the ground and the other in the excited state.

1.2. ALUMINIUM, MAGNESIUM, AND BERYLLIUM

The critical wavelength being situated at 830, 1198, and 630 Å for Al, Mg, and Be respectively, these metals should be very good reflectors in the v.u.v. up to these limits, with a transmission zone on the higher energy side. However, they oxidize easily, and growth of a thin oxide film causes a very rapid diminution of reflectance. Consequently the most reliable results are those recently obtained with films produced by fast evaporation or in ultra-high vacuum. In particular, aluminium has been much studied in reflection[38–47] and in transmission[43, 48–54] because of its intrinsic theoretical and practical importance. For magnesium, principally transmittance (51, 53, 55–57) and recently some reflectance measurements in ultra-high vacuum(46, 58) have been reported. For beryllium, the references on reflectance and absorption are 51, 53, and 59. Figure 2 represents the most recent reflectance values for these metals. The values for beryllium were determined in ordinary vacuum ($\sim 10^{-5}$ torr) which doubtless explains their wide disparity with the free-electron theory.

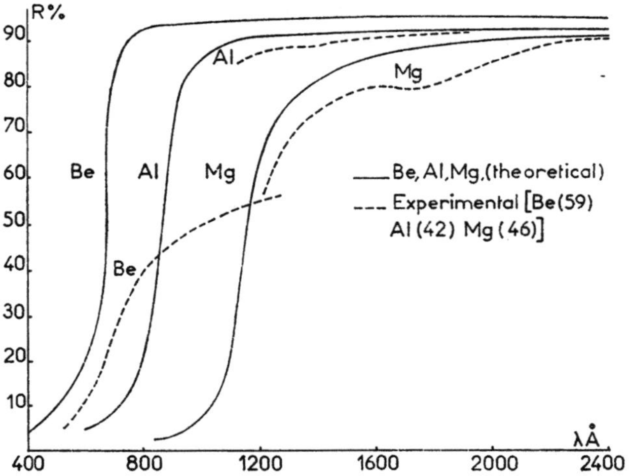

FIG. 2. Reflectance of Al, Mg, and Be vs. wavelength.(42,46)

The imaginary part of the dielectric constant ε_2(exp) may be represented as the sum of the contribution of the free electrons $\varepsilon_2^{(f)}$ and the bound electrons $\delta\varepsilon_2^{(b)}$:(60)

$$\varepsilon_2\,(\text{exp}) = \varepsilon_2^{(f)} + \delta\varepsilon_2^{(b)}.$$

$\varepsilon_2^{(f)}$ may be calculated by the classical Drude's equation

$$\varepsilon_2^{(f)} = \frac{\hbar}{\tau_e E}\,\frac{E_p^2}{E^2 + (\hbar/\tau_e)^2},$$

where E_p is the energy of the volume plasmon for the free electrons. Thus one may evaluate the contributions $\delta\varepsilon_2^{(b)}$ due to transitions at very low energies (1.5 eV for aluminium for example), and $\Delta\delta\varepsilon_2^{(b)}$ due to transitions at very high energies. Thus $\Delta\delta\varepsilon_2^{(b)}$ exhibits a maximum at 10.0 eV for aluminium and 7.4 eV for magnesium. Williams *et al.*(45) assign the peak at 10.0 eV to two unresolved transitions $X_1 \to X_5$ at 7.9 eV and $X_4 \to X_1$ at 12.0 eV. Further, the optical results and the values of the characteristic energy losses for the plasmons(61–76) are collected in Table 2.2; the experimental evidence for the surface plasmons in aluminium and magnesium is clear, but for beryllium optical measurements do not exist.

TABLE 2. PLASMON ENERGIES FROM OPTICAL SPECTRA AND CHARACTERISTIC ENERGY LOSSES OF ELECTRONS (eV)

Metal	Surface plasmon			Volume plasmon		
	Theoretical	Optical experimental	Energy loss experimental	Theoretical	Optical experimental	Energy loss experimental
Al	10.4	10.7[45]	10.3[65]	14.8	15.2[40]	15.3[65]
Mg	7.5	7.8[46]	7.1[65]	10.9	10.6[46]	10.6[65]
Be	13.5			19.1	18.4[59]	18.9[71]

A comparative study of characteristic energy losses for aluminium in both solid and liquid states[76] shows that the losses due to plasmons vary with the temperature, and also that, at the melting point, there is an abrupt displacement corresponding to the change of density.

Finally, Steinmann *et al.*[77–79] observe with p-polarized light a maximum in reflectance and absorption of thin films of aluminium as well as a maximum photoelectric efficiency at about the plasma frequency.

2. Divalent metals

2.1. ALKALINE EARTH METALS

Optical measurements on these materials are scarce in the far u.v. The reflectance of thin films of barium have been recently measured in ultra-high vacuum.[80] The results confirm the existence of a volume plasmon at 7.6 eV, in rough agreement with the theoretical frequency (6.7 eV) and the characteristic energy losses (7.1 eV);[25] the latter method shows up also a surface plasmon at 3.2 eV.

Optical studies on thin strontium films produced in ultra-high vacuum are limited to about 5 eV;[81] the characteristic energy loss spectrum[25, 82] presents two peaks, one at 4.0 eV and the other at 8.1 eV. These are attributed either to two volume plasmons or to one volume and one surface plasmon.

For calcium there are no optical measurements in the far u.v.; the energy loss spectrum[18, 20, 67, 83] indicates the existence of volume or surface plasmons at 1.8, 4.1, 6.8, and 8.7 eV. It is relevant that the outermost shells of these group II elements contain two weakly bound electrons, and double electron transitions have been observed in the v.u.v. absorption spectra of the corresponding vapours.[84, 85]

2.2. CADMIUM AND ZINC

Cadmium and zinc, forming hexagonal close-packed systems, have band structure closely approaching that of the free-electron theory as shown by the reflectance curve for a zinc crystal in ultra-high vacuum.[86] The rest of the results for zinc[87, 88] and cadmium[87, 89, 90] have been obtained in ordinary vacuum. A plasma oscillation appears at 8.55 eV for cadmium and at 9.6 eV for zinc. In zinc it is difficult to distinguish the volume and surface plasmons.[87] Study of these metals becomes complicated for energies above the plasma frequencies. For zinc, band-structure calculations[91] indicate that the d states lie about 11 eV below the Fermi level, so that transitions from $3d$ states become possible in this region. For cadmium, Jelinek *et al.*[90] find a maximum of Im (ε^{-1}) at 1240 and 950 Å, the first peak being due to the threshold excitation of the $N_{4,5}$ shells.

Finally, characteristic energy loss results for zinc[92–94] and for cadmium[67, 95, 96] are in agreement with the optical results and with the emission studies of Herickhoff.[97]

3. Rare earths

From lanthanum to lutecium, the elements of the rare-earth group form a well-defined class. Lanthanum having no $4f$ electrons is trivalent and approaches the transition metals in its properties. Europium and ytterbium have no $5d$ electrons, but, instead, two $6s$ valence electrons: they are divalent, are the only rare earths crystallizing in the cubic system, and their properties resemble those of the alkaline earths in certain respects. For the other elements of the group, the $4f$ electrons increase in number with z and are responsible for the magnetic properties of rare earths. The band structures of gadolinium and of tamarium show that the bands nearest to the Fermi energy may be derived from the atomic states $5d$ and $6s$ and have a high density of states. A discontinuity in the reflectance spectrum would be attributed to transitions between the $4f$ levels and the conduction bands above the Fermi levels (about 7 eV for gadolinium[99]). Theoretical calculation[98] indicates similar Fermi surfaces for Gd, Dy, Er, Lu, and Yb.

Gadolinium is practically the only case studied in the v.u.v.[99] Its reflectance spectrum exhibits two structures—one at 3.7 and the other near 8 eV; the latter is in agreement with the calculations of Dimmock *et al.*[100] which predict a transition at about 7 eV, between the $4f$ levels and the conduction band. Berglund *et al.*[32, 101] and Spicer[102] derived the conditions for the occurrence of non-direct transitions; Blodgett *et al.*[99] have obtained the energy distribution curve from 3.4 to 11.4 eV and have also shown that non-direct transitions prevail. Their results are in agreement with the calculated curve of joint density of states;[103] however, even on this curve no structure appears below −4 eV from the Fermi level. The maximum of the density of states at −5.8 eV observed by Blodgett *et al.*[99] could be due to f states which are neglected in the calculated joint density of states. The Curie point of gadolinium being about 16 °C, preliminary studies at low temperatures[99, 104] have been attempted to investigate the effect of transition from paramagnetism to ferromagnetism, but this has so far eluded observation in the v.u.v.

4. Transition metals

References on the optical properties in the v.u.v. of these metals, which have been widely studied are given in Table 3.[105–51, 239] The transition metals are characterized by partially filled *d* bands, responsible for their magnetic properties. They do not obey the free-electron model[143] (no measurement, even in the far infrared[152–5] has provided evidence of free-electron behaviour). We elaborate below the results for the three elements Ni, Pd, and Pt.

TABLE 3. OPTICAL AND PHOTOELECTRIC MEASUREMENTS ON TRANSITION METALS

Ti	107, 126, 151
V	126, 136, 151
Cr	107, 109, 114, 145, 151
Mn	107, 112, 151
Fe	107, 129, 132, 151
Co	130, 146, 147, 151
Ni	107, 108, 117, 118, 124, 125, 129, 130, 132, 133, 134, 140, 141, 142, 143, 149, 150, 151
Zr	107, 110, 126, 136
Nb	126, 136
Mo	107, 121, 126, 135
Rh	105, 106, 111, 123, 136, 144
Pd	107, 108, 127, 128, 130, 131, 136, 139, 143, 147
Ta	110, 121, 239
W	110, 120, 121, 122, 239
Re	121, 239
Ir	136, 137, 138
Pt	106, 107, 109, 113, 115, 119, 123, 124, 128, 148, 239

4.1. NICKEL

In nickel the structure observed in the ε_2 curve[143] between 4.5 and 14.0 eV is associated with interband transitions, in agreement with energy loss spectra.[63, 156–9] A surface plasmon is found at 7.0 eV and a volume plasmon at 10.5 eV. A maximum at 68.5 eV is attributed to the transition $M_{II,III}$.[160] Recently[150] the −4.5 eV peak on the density of states curve of electrolytically polished nickel as well as the −4.0 and −5.4 eV peaks for caesium-covered nickel, have been attributed to interband transitions and to a surface plasmon of 4-4.5 eV. It is to be noted that the joint density of states curves for the three ferromagnetic metals Fe, Co, and Ni[147] have a great similarity despite the fact that they crystallize in different symmetries. The peaks at −0.3, −2.2, and −4.6 eV of the joint density of states curve for nickel almost coincide with those of iron and cobalt which could not be explained by a rigid band model. According to this scheme, the curves for iron and cobalt could be deduced from the curve for nickel by a simple translation of the Fermi level, to take account of diminution in the atomic number: this places the Fermi level of iron 2 eV below that of nickel.

The peak, near −5 eV, which appears on these curves, seems connected with ferromagnetism, since it is missing on the palladium curve.

In general the calculated curve of the joint density of states differs in a certain number of respects from the optical density of states, the many-body effect being possibly present in the latter (e.g. the −4.6 eV peak in nickel according to Phillips).[161] It might be considered that for these metals the density of states obtained by X-ray spectroscopy or by ion neutralization[162] is different from that obtained from photoemission work. Mott[163] remarks that the resonance effects will not be detected by the first two methods. Cuthill *et al.*[164] have proposed the hypothesis that the peak at −4.6 eV is due to simultaneous excitation of two electrons by a single photon, making it possible to interpret the optical and photoelectric results as due to direct transitions. Finally, Nesbet and Grant[165] have suggested non-direct transitions in terms of second-order processes: the electrons excited by a phonon are subjected to a well-defined energy loss in the metal, equal to 5 eV for nickel, a value in agreement with a peak in the electron energy-loss spectrum.[159]

4.2. PALLADIUM

The ε_2 curve of palladium (paramagnetic) obtained from reflectance measurements on evaporated films[131, 143, 147] exhibits several features; according to Vehse *et al.*[143] these are at 3.5 eV, between 8 and 16 eV, and at 20.5 eV, with surface and volume plasmons at 7.3 and 7.6 eV. The results of Yu *et al.*[147] obtained with layers deposited by evaporation in ultra-high vacuum, are in good agreement with those of Robin[131] up to 5.5 eV, but, at higher energy, their ε_2 values are less than that of the latter author.

From photoemission measurements[147] the non-direct transitions seem to predominate; nevertheless, these measurements provide the first evidence of weak direct transitions in the *d*-band region. A plasmon is observed at 7.5 eV. There is general agreement with the results obtained from characteristic energy losses of electrons.[94]

4.3. PLATINUM

The reflectance of platinum has often been measured in the v.u.v.[124, 128, 148] because it is relatively high (16% at about 12 eV) and is very stable in time, even when exposed to air, in contrast to the reflectance of aluminium. The ε_2 curve exhibits two features at 6.5 and 11 eV, in agreement with the characteristic energy-loss spectrum.[67, 166]

5. Noble metals

Noble metals are among those which have been especially studied in u.v., but the various authors have given different interpretations in explanation of their results. More precise measurements, taking into consideration variations with parameters like temperature, piezoreflectance, etc., supplemented by refined theoretical study (band scheme calculated with improved choice of crystal potential, many-body resonances, etc.) are still needed for better understanding of the phenomena observed.

5.1. COPPER

The important references on reflectance and absorption are 60, 107, 108, 143, 167, 169, and 175. The reflectance data for electrolytically polished bulk copper are shown in Fig. 3.[60] Up to 2.1 eV, the threshold of interband transitions, the

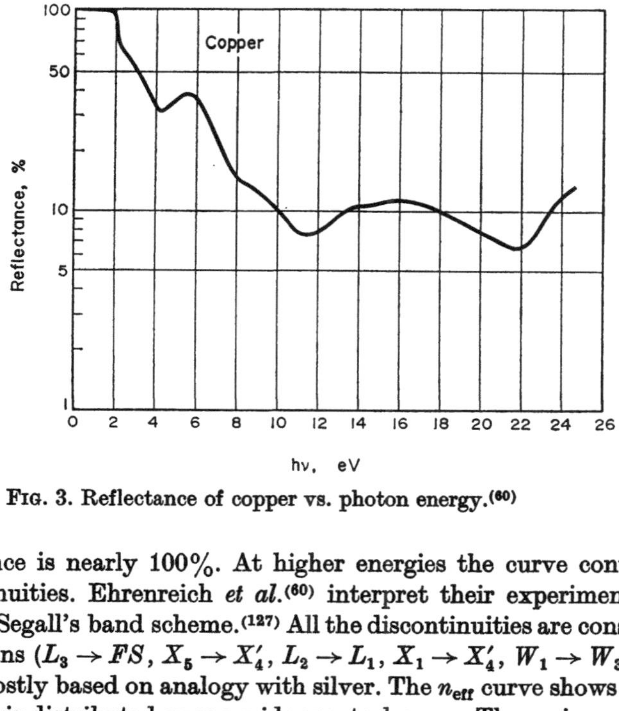

FIG. 3. Reflectance of copper vs. photon energy.[60]

reflectance is nearly 100%. At higher energies the curve contains a number of discontinuities. Ehrenreich *et al.*[60] interpret their experimental results on the basis of Segall's band scheme.[127] All the discontinuities are considered to be direct transitions ($L_3 \rightarrow FS$, $X_5 \rightarrow X_4'$, $L_2 \rightarrow L_1$, $X_1 \rightarrow X_4'$, $W_1 \rightarrow W_3$), the assignments being mostly based on analogy with silver. The n_{eff} curve shows that the oscillator strength is distributed over a wide spectral range. The main uncertainty concerns the position of the *d* band with respect to the Fermi level. The results of photoemission measurements of caesiated copper films are interpreted by Berglund and Spicer[32, 101] taking account of non-direct transitions (the rule for the conservation of **k** not being strict for transitions involving the *d* band). Cooper *et al.*[170] believe that the direct transitions also play an important role in Cu, Ag, and Au. However more precise recent measurements on photoemission[176] from copper films prepared in ultra-high vacuum seem to indicate that these direct transitions are too weak to be evident from such experiments. The major part of the structure is attributed to non-direct transitions except for the direct transition at 4.2 eV ($L_2' \rightarrow L_1$). Further, the optical density of states curve calculated from these measurements is in agreement with that calculated by Snow[177] and with that obtained by X-ray emission or ion neutralization spectroscopy.[162] Finally, the peak at 7.5 eV in Im (ε^{-1}) seems to be due to a volume plasmon displaced towards lower energy. The optical results are also in good agreement with the electron energy-loss spectrum.[67, 75, 178, 179]

5.2. SILVER

Silver has also been thoroughly studied during the last decade both in reflection[38, 60, 131, 168, 170, 172, 181–3] and in absorption.[175] Figure 4 displays the reflectance from 3 to 50 eV for films evaporated *in situ*.[131] Figure 5 presents three ε_2 curves evaluated from (a) optical,[131] (b) photoelectric[176] measurements, and (c)

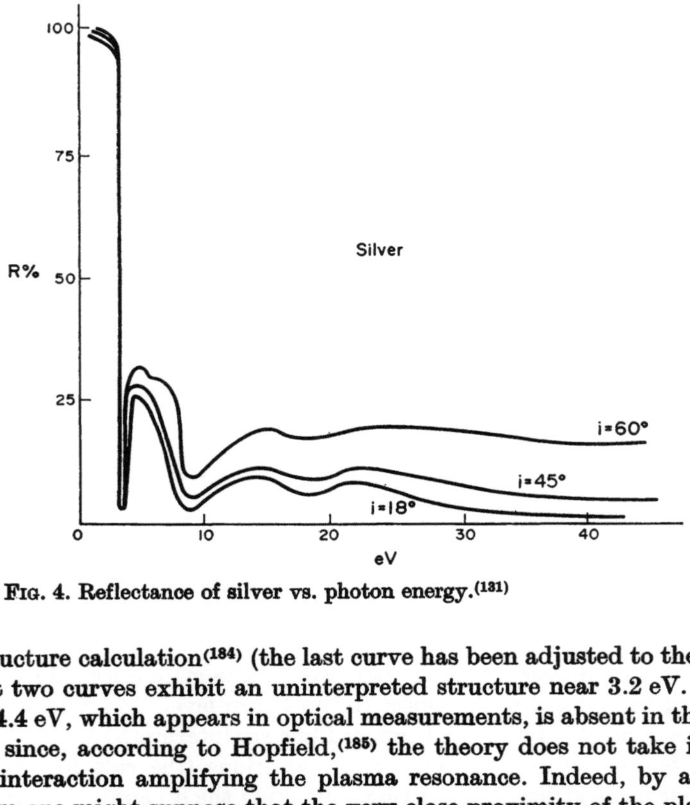

Fig. 4. Reflectance of silver vs. photon energy.[131]

band-structure calculation[184] (the last curve has been adjusted to the other two). The first two curves exhibit an uninterpreted structure near 3.2 eV. The narrow peak at 4.4 eV, which appears in optical measurements, is absent in the calculated ε_2 curve since, according to Hopfield,[185] the theory does not take into account phonon interaction amplifying the plasma resonance. Indeed, by analogy with potassium one might suppose that the very close proximity of the plasma oscillation qualitatively affects the ε_2 curve. There is a good agreement between the structure of ref. 184 and the shoulders at 5.4 and 6.5 eV observed in ref. 131. The correspondence is less clear in ref. 176, perhaps because very weak direct transitions are involved which cannot appear in the photoelectric effect. The shoulder at 5.4 eV is probably due to a transition from the top of the d band at the point X towards X_4'.[170] Daniels[203] has recently given an ε_2 curve calculated from electron energy loss values, in good agreement with the optical curve.[131] As in the case of copper, there is a predominance of non-direct transitions; no direct transition is observed from the d band but the transition $L_2' \to L_1$ between the states of the s and p bands is direct. L_2' and L_1 are 0.3 eV below and 3.7 eV above the Fermi level respectively. The position of the point X_4' is 1.8 eV above as in the case of copper. As for sodium and potassium, the very strong absorption by electrons having an energy near that of the Fermi level could not be explained by direct interband transitions.[36]

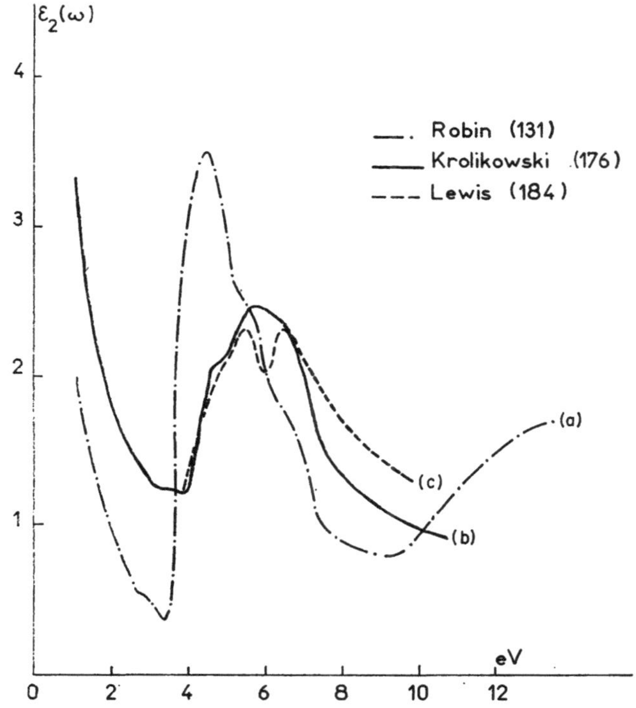

FIG. 5. Comparison of calculated ε_2 for silver with ε_2 obtained from reflectivity measurements. (a) obtained from reflectivity measurements.[131] (b) calculated from the optical density of states.[176] (c) calculated from the band structure.[184]

The region between 3.5 and 4.5 eV is very complex and has been studied extensively.[178, 186–206] The theoretical plasma oscillation at 9.2 eV is found actually at 3.8 eV because of the neighbouring interband transition threshold. The peak of Im (ε^{-1}) obtained optically at 3.8 eV could be due to unresolved volume and surface plasmons according to the different energy loss spectra.[178, 203]

To test the predictions of the theory given by Ferrell in 1968, experiments were undertaken recently to distinguish between the two types of plasmons—radiative and non-radiative.[78] The volume plasmon at 3.8 eV, excited by electrons of high energy, re-emits radiation with a peak situated at the plasmon energy whose intensity is, as predicted, a function of the thickness and the angle of emission. Finally, also undere xcitation by a continuous spectrum of p-polarized light, radiation with a strong peak at the plasma frequency is emitted (without conservation of the **k** vector). Stern(207) has given an interpretation of the phenomenon based on surface defects, which seems to be confirmed by Schreiber's experiments.(202)

5.3. GOLD

Chiefly because of the great stability of its films, gold has been studied in reflection(38, 169, 170, 172, 180, 208–10) and in absorption(175, 211, 212) by various authors. The region which corresponds to free-electron theory is limited to 2 eV. The results obtained for ε_2 are widely divergent in the v.u.v.

Photoelectric measurements(176) indicate that non-direct transitions are predominant; there probably exists no direct transition from the d bands (the peak at 3.1 eV cannot be attributed to a direct transition). From the energy-loss spectrum, Lynch and Swan(166) indicate unresolved volume and surface plasmons at 3.3 eV. The peak at 62.9 eV corresponds to the ionization energy of the O_3 level.

6. Other metals

We classify the rest of the metals according to groups III, IV, and V of the Periodic Table.

6.1. GROUP III

Gallium

Though, to our knowledge, there is no experimental work in the v.u.v. region, the energy-loss spectrum(76, 213, 214) for the liquid and solid states indicates a volume plasmon at 13.75 eV.

Thallium

The reflectance measurements of Jelineck *et al.*(90) show a plasmon at 10.25 eV in agreement with the energy-loss value.(213)

Indium

This has been studied widely(38, 49, 182, 217), chiefly because of its transmission characteristics in the extreme u.v. (maximum of transmission at 770 Å with an abrupt decrease at 730 Å caused by the absorption threshold of $N_{4,5}$ at 744 Å(53, 160)). Recent optical measurements demonstrate a volume plasmon at 11.6 eV (theoretical value of 12.6 eV) and a surface plasmon at 7.7 eV, both in agreement with the energy-loss spectrum.(63, 93, 97, 217, 218)

6.2. GROUP IV

Tin

There are reflection, transmission, and photoelectric measurements.(38, 219–21) They indicate a volume plasmon at 13.8 eV (theoretical energy of 14.3 eV) in agreement with electron-loss values.[67, 222, 223] A maximum of the function $\mathrm{Im}\left(\frac{1}{1+\varepsilon}\right)$ at 9.3 eV corresponds to a surface plasmon. Absorption measurements extend to 200 eV.[175]

Lead

Since the early results of Sabine[107] very few measurements have been made; the most recent are on films evaporated *in situ*.[145] The reflectance values are higher than the earlier ones, and a minimum is observed at 2400 Å. Im (ε^{-1}) presents a maximum at 12.7 eV, while the theoretical plasmon energy, taking four electrons per atom, is situated at 13.4 eV. The shoulder at 10 eV and the peak at 5 eV are probably due to interband transitions.[224]

6.3. GROUP V

There are two semi-metals in this group—antimony and bismuth.

Antimony

Work in the v.u.v. is scarce.[107, 116, 225–7] The reflectance of evaporated films between 4 and 40 eV[226, 227] exhibits a fall at 15 eV corresponding to the transmission threshold. Im (ε^{-1}) presents a maximum at 14.8 eV[226] or 16 eV[227] attributed to a volume plasmon (theoretical value 15.1 eV). The peak at 32 eV corresponds to the $N_{4,5}$ transition.[160, 228]

Bismuth

Transmittance measurements[38, 54, 175, 229] have chiefly been made, revealing a threshold at 18 eV. Toots and Marton[227] measured the reflectance between 10 and 26 eV and found a broad peak in Im (ε^{-1}) near 14 eV. The electron energy-loss spectrum[67] indicates a volume plasmon at 14.7 eV, a surface plasmon at 10 eV, and also an interband transition at 11.5 eV.[76] The threshold $O_{4,5}$ is situated at 25 eV.[160]

7. Alloys and metallic compounds

7.1. ALLOYS

The dependence of the optical properties of alloys on their composition has been studied extensively in the v.u.v. and a theory has been developed by various authors.[230, 231]

As an example, Fig. 6 represents the loss function Im (ε^{-1}) (up to 10 eV) for Cu-Zn compounds of different compositions.[232] Plasma resonances are apparent at about 2.7 and 7 eV. The main additional structure is an interband transition in the neighbourhood of 2.70 eV, which decreases with increasing temperature and with zinc concentration.

Alloys of noble metals,[233] noble metals with polyvalent metals,[180, 234] polyvalent metals only,[235] and noble metals with transition metals,[236] have been investigated. In the recent work of Spicer[207] on Cu-Ni alloys, the *d* band of copper has been clearly localized both by optical and photoemission experiments. Evidence for new states in the virtual state model is also presented.

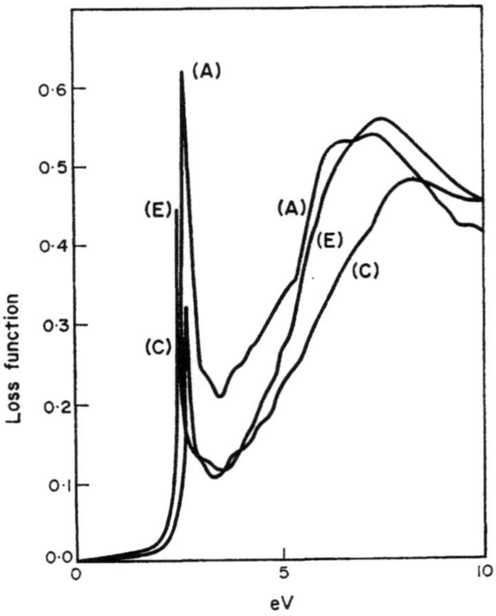

FIG. 6. Loss function for three β brass alloys at low temperature.[232] (A) 56% Cu, 44% Zn; (C) 52% Cu, 48% Zn; (E) 48% Cu, 52% Zn.

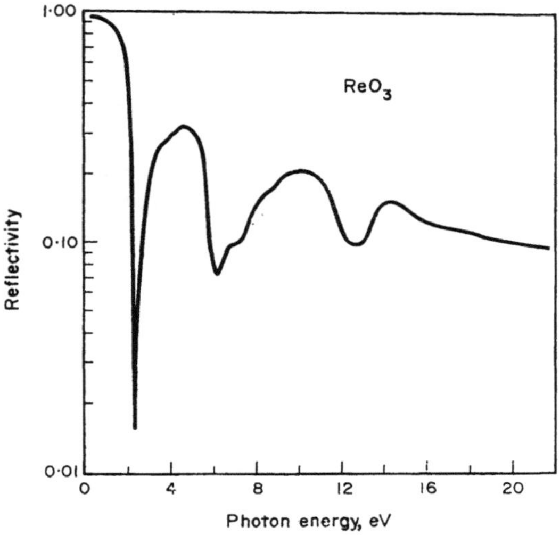

FIG. 7. Reflectance of ReO_3 vs. photon energy.[237]

7.2. METALLIC COMPOUNDS

Certain metallic compounds have optical properties approaching those of metals; this is the case for some oxides of transition metals. Study of these substances affords information on the *d* bands of the solid and of the transition metals also.

ReO_3 is a typical compound having a high conductivity at room temperature. From the optical point of view[237] it behaves as a metal following Drude's model up to 2.3 eV, as can be seen from the reflectivity curve (Fig. 7). For higher energies interband transitions are predominant. The function Im (ε^{-1}) shows a plasmon at 2.3 eV, which can be compared with a similar peak in silver.

Another example is TiC,[238] which has a plasma energy at 11 eV (the theoretical value of 20.3 eV is displaced in energy by the presence of interband transitions).

References

1. Drude, P., *Theory of Optics*, Longmans, London, 1902.
2. Ives, H. E., and Briggs, H. B., *J. Opt. Soc. Am.* **26**, 238 (1936); **27**, 181 and 395 (1937).
3. Fleischmann, R., *Z. Phys.* **174**, 104 (1963).
4. Junge, W., and Fleischmann, R., *Z. Phys.* **174**, 116 (1963).
5. Brünig, M., and Fleischmann, R., *Z. Phys.* **174**, 121 (1963).
6. Schopper, H., *Z. Phys.* **174**, 125 (1963).
7. Deichsel, H., *Z. Phys.* **174**, 136 (1963).
8. Müller, H., *Z. Phys.* **174**, 147 (1963).
9. Paris, D., *Z. Phys.* **174**, 180 (1963).
10. Sutherland, J. C., Arakawa, E. T., and Hamm, R. N., *J. Opt. Soc. Am.* **57**, 645 (1967).
11. Sutherland, J. C., Hamm, R. N., and Arakawa, E. T., *J. Opt. Soc. Am.* **59**, 1581 (1969).
12. El Naby, M. H., *Z. Phys.* **174**, 269 (1963).
13. Mayer, H., and El Naby, M. H., *Z. Phys.* **174**, 280 (1963).
14. Mayer, H., and Hietel, B., *Optical Properties and Electronic Structure of Metals and alloys*, North-Holland, Amsterdam, 1966, p. 47.
15. Cohen, M. H., *Phil. Mag.* **3**, 762 (1958).
16. Esposito, R. J., Muldawer, L., and Blomfield, P. E., *Phys. Rev.* **168**, 764 (1968).
17. Leder, L. B., *Phys. Rev.* **103**, 1721 (1956).
18. Wood, R. W., *Physical Optics*, MacMillan, New York (1936), p. 560.
19. Fellenzer, H., *Z. Phys.* **165**, 419 (1961).
20. Robins, J. L., and Best, P. E., *Proc. Phys. Soc. London* **79**, 110 (1962).
21. Swan, J. B., *Phys. Rev.* **135** A, 1467 (1964).
22. Hölzl, J., *Kurzn. Wiss. Göttingen* **8**, 39 (1965).
23. Kunz, C., *Phys. Letters* **15**, 312 (1965).
24. Sueoka, O., *J. Phys. Soc. Japan* **20**, 2249 (1965).
25. Kunz, C., *Z. Phys.* **196**, 311 (1966).
26. Kloos, T., *Z. Phys.* **210**, 303 (1968).
27. Brambring, J., and Raether, H., *Z. Naturforsch.* **21** A, 1527 (1966).
28. Bösenberg, J., and Raether, H., *Phys. Rev. Letters* **18**, 397 (1967).
29. Bösenberg, J., *Phys. Letters* **26** A, 74 (1967).
30. Brambring, J., *Z. Phys.* **200**, 186 (1967).
31. Bösenberg, J., *Z. Phys.* **218**, 282 (1969).
32. Berglund, C. N., and Spicer, W. E., *Phys. Rev.* **136** A, 1030 (1964).
33. Overhauser, A. W., *Phys. Rev. Letters* **13**, 190 (1964).
34. Cohen, M. H., and Phillips, J. C., *Phys. Rev. Letters* **12**, 662 (1964).
35. Cohen, M. H., *Phys. Rev. Letters* **12**, 664 (1964).
36. Phillips, J. C., *Phys. Rev.* **137** A, 1835 (1965).
37. Phillips, J. C., *Solid St. Phys.* **18**, 55 (1966).
38. Walker, W. C., Rustgi, O. P., and Weissler, G. L., *J. Opt. Soc. Am.* **49**, 471 (1959).
39. Madden, R. P., Canfield, L. R., and Hass, G., *J. Opt. Soc. Am.* **53**, 620 (1963).
40. Ehrenreich, H., Philipp, H. R., and Segall, B., *Phys. Rev.* **132**, 1918 (1963).

41. Rustgi, O. P., *J. Opt. Soc. Am.* **55**, 630 (1965).
42. Daudé, A., Priol, M., and Robin, S., *Compt. Rend.* **263**, 1178 (1966).
43. Ditchburn, R. W., and Freeman, G. H. C., *Proc. Roy. Soc. London* A, **294**, 20 (1966).
44. Vehse, R. C., Arakawa, E. T., and Stanford, J. L., *J. Opt. Soc. Am.* **57**, 551 (1967).
45. Williams, M. W., Arakawa, E. T., and Emerson, L. C., *Surf. Sci.* **6**, 127 (1967).
46. Daudé, A., Priol, M., Robin, S., and Robin, S., *Proceedings of the 2nd Colloquium on Thin Films, Budapest, 1967*, p. 287.
47. Feuerbacher, B. P., and Steinmann, W., *Optics Commun.* **1**, 81 (1969).
48. Hunter, W. R., *J. Phys.* **25**, 154 (1964).
49. Hunter, W. R., *J. Opt. Soc. Am.* **54**, 208 (1964).
50. Tomboulian, D. H., and Pell, E. M., *Phys. Rev.* **83**, 1196 (1951).
51. Tomboulian, D. H., and Bedo, D. E., *Rev. Scient. Instrum.* **26**, 747 (1955).
52. Ejiri, A., and Sasaki, T., *J. Phys. Soc. Japan* **20**, 876 (1965).
53. Hunter, W. R., *Optical Properties and Electronic Structure of Metals and Alloys*, North-Holland, Amsterdam, 1966, p. 136.
54. Sasaki, T., and Ejiri, A., *Optical Properties and Electronic Structure of Metals and Alloys*, North-Holland, Amsterdam, 1966, p. 417.
55. Kroger, H., and Tomboulian, D. H., *Phys. Rev.* **130**, 152 (1963).
56. Sorokin, O. M., *Optika Spektrosk.* **24**, 976 (1968) [*Optics Spectrosc. USSR* **24**, 524 (1968)].
57. Tittel, H. O., *Phys. Letters* **26** A, 145 (1968).
58. Daudé, A., Priol, M., and Robin, S., *Compt. Rend.* B, **264**, 1489 (1967).
59. Toots, J., Fowler, H. A., and Marton, L., *Phys. Rev.* **172**, 670 (1968).
60. Ehrenreich, H., and Philipp, H. R., *Phys. Rev.* **128**, 1622 (1962).
61. Ruthemann, G., *Naturwissenschaften* **29**, 648 (1941).
62. Marton, L., and Leder, L. B., *Phys. Rev.* **94**, 203 (1954).
63. Kleinn, W., *Optik* **11**, 226 (1954).
64. Watanabe, H., *J. Phys. Soc. Japan* **9**, 1035 (1954); **11**, 112 (1956).
65. Powell, C. J., and Swan, J. B., *Phys. Rev.* **115**, 869 (1959); **116**, 81 (1959); **118**, 640 (1960).
66. Mendlowitz, H., *Proc. Phys. Soc. London* **75**, 664 (1960).
67. Powell, C. J., *Proc. Phys. Soc. London* **76**, 593 (1960).
68. la Villa, R., and Mendlowitz, H., *Phys. Rev. Letters* **9**, 149 (1962).
69. Kunz, C., *Z. Phys.* **167**, 53 (1962); **180**, 127 (1964).
70. Kunz, C., and Raether, H., *Solid State Commun.* **1**, 214 (1963).
71. Swanson, N., *J. Opt. Soc. Am.* **54**, 1130 (1964).
72. Schmuser, P., *Z. Phys.* **180**, 105 (1964).
73. Miller, W. R., and Axelrod, N., *Solid State Commun.* **3**, 133 (1965).
74. Otto, A., and Swan, J. B., *Z. Phys.* **206**, 277 (1967).
75. Thirlwell, J., *J. Phys. Chem. (Proc. Phys. Soc.) London* **1**, 979 (1968).
76. Powell, C. J., *Phys. Rev.* **175**, 972 (1968).
77. Steinmann, W., and Skibowski, M., *Phys. Rev. Letters* **16**, 989 (1966).
78. Steinmann, W., *Phys. Status Solidi* **28**, 437 (1968).
79. Feuerbacher, B., Godwin, R. P., and Skibowski, M., *Phys. Letters* **26** A, 595 (1968).
80. Fischer, E. I., Fujita, I., and Weissler, G. L., *Tech. Rep. Vac. UV* 104 (1965).
81. Schüler, C. C., *Optical Properties and Electronic Structure of Metals and Alloys*, North-Holland, Amsterdam, 1966, p. 221.
82. Hartley, B. M., *Phys. Letters* **27**A, 499 (1968).
83. Rudberg, E., *Proc. Roy. Soc. London* A, **127**, 111 (1930).
84. Fano, U., and Cooper, J. W., *Rev. Mod. Phys.* **40**, 441 (1968).
85. Ditchburn, R. W., and Hudson, R. D., *Proc. Roy. Soc. London* A, **256**, 53 (1960); Garton, W. R. S., and Codling, K., *Proc. Phys. Soc.* **75**, 87 (1960), and *J. Phys.* B, II, **1**, 106 (1968).
86. Mosteller, L. P., and Wooten, F., *Phys. Rev.* **171**, 743 (1968).
87. Arakawa, E. T., Hamm, R. N., Hanson, W. F., and Jelinek, T. M., *Optical Properties and Electronic Structure of Metals and Alloys*, North-Holland, Amsterdam, 1966, p. 374.

88. KANDARE, S., ROBIN-KANDARE, S., and ROBIN, J., *Compt. Rend.* **262**, 1302 (1966).
89. ROBIN-KANDARE, S., ROBIN, J., and KANDARE, S., *J. Phys. Radium* **25**, 218 (1964).
90. JELINEK, T. M., HAMM, R. N., ARAKAWA, E. T., and HUEBNER, R. H., *J. Opt. Soc. Am.* **56**, 185 (1966).
91. STARK, R. W., and FALICOV, L. M., *Phys. Rev. Letters* **19**, 795 (1967).
92. PRADAL, F., thèse, Toulouse, 1959.
93. GAUTHÉ, B., *Phys. Rev.* **114**, 1265 (1959).
94. ROBINS, J. L., *Proc. Phys. Soc.* **78**, 1177 (1961).
95. LEDER, L. B., MENDLOWITZ, H., and MARTON, L., *Phys. Rev.* **101**, 1460 (1956).
96. LEDER, L. B., *Phys. Rev.* **107**, 1569 (1957).
97. HERICKHOFF, R. J., ARAKAWA, E. T., and BIRKHOFF, R. D., *Phys. Rev.* **137 A**, 1433 (1965).
98. KEETON, S. C., and LOUCKS, T. L., *Phys. Rev.* **168**, 672 (1968).
99. BLODGETT, A. J., SPICER, W. E., and YU, A. Y.-C., *Optical Properties and Electronic Structure of Metals and Alloys*, North-Holland, Amsterdam, 1966, p. 246.
100. DIMMOCK, J. O., FREEMAN, A. J., and WATSON, R. E., *Optical Properties and Electronic Structure of Metals and Alloys*, North-Holland, Amsterdam, 1966, p. 237.
101. BERGLUND, C. N., and SPICER, W. E., *Phys. Rev.* **136 A**, 1044 (1964).
102. SPICER, W. E., *Phys. Rev. Letters* **11**, 1 (1963).
103. DIMMOCK, J. O., and FREEMAN, A. J., *Phys. Rev. Letters* **13**, 750 (1964).
104. HODGSON, J. N., and CLEYET, B., *J. Phys. Chem. (Proc. Phys. Soc.) London* **2**, 97 (1969).
105. AUWÄRTER, M., *Z. tech. Phys.* **11**, 457 (1937).
106. AUWÄRTER, M., *J. Appl. Phys.* **10**, 705 (1939).
107. SABINE, G. B., *Phys. Rev.* **55**, 1064 (1939).
108. WALKER, W. C., WAINFAN, N., and WEISSLER, G. L., *J. Appl. Phys.* **24**, 1318 (1953); **26**, 1366 (1955).
109. ROBIN, S., *Compt. Rend.* **236**, 674 (1953).
110. FABRE, D., and ROMAND, J., *Compt. Rend.* **242**, 893 (1956).
111. IVANOVA, M. K., LOMONOSOVA, T. N., and YAKOVLEVA, A. V., *Optika Spektrosk.* **4**, 535 (1957).
112. AGARWAL, B. K., and GIVENS, M. P., *J. Phys. Chem. Solids* **6**, 178 (1958).
113. HASS, G., and TOUSEY, R., *J. Opt. Soc. Am.* **49**, 593 (1959).
114. AXELROD, N. N., and GIVENS, M. P., *Phys. Rev.* **120**, 1205 (1960).
115. MADDEN, R. P., and CANFIELD, L. R., *J. Opt. Soc. Am.* **51**, 838 (1961).
116. RUSTGI, O. P., WALKER, W. C., and WEISSLER, G. L., *J. Opt. Soc. Am.* **51**, 1357 (1961).
117. EHRENREICH, H., PHILIPP, H. R., and OLECHNA, D. J., *Phys. Rev.* **131**, 2469 (1963).
118. SPICER, W. E., *Phys. Rev. Letters* **11**, 243 (1963).
119. JACOBUS, G. F., MADDEN, R. P., and CANFIELD, L. R., *J. Opt. Soc. Am.* **53**, 1084 (1963).
120. MADDEN, R. P., *Physics of Thin Films*, Academic Press, New York, 1963, vol. 1, p. 169.
121. LE BLANC, L. J., FARELL, J. S., and JUENKER, D. W., *J. Opt. Soc. Am.* **54**, 956 (1964).
122. WHEATON, J. E. G., *J. Opt. Soc. Am.* **54**, 1287 (1964).
123. DRUMMETER, L. F., and HASS, G., *Physics of Thin Films*, Academic Press, New York, 1964, vol. 2, p. 337.
124. HODGES, L., EHRENREICH, H., and LANG, N. D., *Phys. Rev.* **152**, 505 (1966).
125. BLODGETT, A. J., and SPICER, W. E., *Phys. Rev. Letters* **15**, 29 (1965).
126. KIRILLOVA, M. M., CHARIKOV, V. A., *Fizika Metall.* **19**, 495 (1965).
127. SEGALL, B., *Phys. Rev.* **125**, 109 (1962).
128. STANFORD, J. L., ARAKAWA, E. T., and BIRKHOFF, R. D., *Oak Ridge Natn. Lab. TM* 1392 (1966).
129. BLODGETT, A. J., and SPICER, W. E., *Phys. Rev.* **146 A**, 390 (1966); **158**, 514 (1967).
130. YU, A. Y.-C., and SPICER, W. E., *Phys. Rev. Letters* **17**, 1171 (1966).
131. ROBIN, S., *Optical Properties and Electronic Structure of Metals and Alloys*, North-Holland, Amsterdam, 1966, p. 202.
132. SPICER, W. E., *Optical Properties and Electronic Structure of Metals and Alloys*, North-Holland, Amsterdam, 1966, p. 296.

133. Cairns, R. B., and Samson, J. A. R., *J. Opt. Soc. Am.* **56**, 1568 (1966).
134. Comes, F. J., and Wenning, U., *Phys. Letters* **23**, 537 (1966).
135. Kirillova, M. M., Bolotin, G. A., and Maevskij, V. M., *Fizika Metall.* **24**, 95 (1967).
136. Vilesov, F. I., Azgrubskii, A. A., and Kirillova, M. M., *Optika Spektrosk.* **23**, 153 (1967) [*Optics Spectrosc. USSR* **23**, 79 (1967)].
137. Hass, G., Jacobus, G. F., and Hunter, W. R., *J. Opt. Soc. Am.* **57**, 758 (1967).
138. Samson, J. A. R., Padur, J. P., and Sharma, A., *J. Opt. Soc. Am.* **57**, 966 (1967).
139. Vehse, R. C., Stanford, J. L., and Arakawa, E. T., *Phys. Rev. Letters* **19**, 1041 (1967).
140. Breen, W. M., Wooten, F., and Huen, T., *Phys. Rev.* **159**, 475 (1967).
141. Cuthill, J. R., McAlister, A. J., Williams, M. L., and Watson, R. E., *Phys. Rev.* **164**, 1006 (1967).
142. Vehse, R. C., and Arakawa, E. T., *Bull. Am. Phys. Soc.* **13**, 59 (1968).
143. Vehse, R. C., and Arakawa, E. T., *Oak Ridge Natn. Lab. TM* 2240 (1968).
144. Lapeyre, G. J., and Kress, K. A., *Phys. Rev.* **166**, 589 (1968).
145. Girault, P., Seignac, A., Priol, M., and Robin, S., *Compt. Rend.* **266** B, 688 (1968).
146. Yu, A. Y.-C., Donovan, T. M., and Spicer, W. E., *Phys. Rev.* **167**, 670 (1968).
147. Yu, A. Y.-C., and Spicer, W. E., *Phys. Rev.* **167**, 674 (1968); **169**, 497 (1968).
148. Yu, A. Y.-C., Spicer, W. E., and Hass, G., *Phys. Rev.* **171**, 834 (1968).
149. Eastman, D. E., and Krolikowski, W. F., *Phys. Rev. Letters* **21**, 623 (1968).
150. Calcott, T. A., MacRae, A. U., *Phys. Rev.* **178**, 966 (1969).
151. Sonntag, B., Haensel, R., and Kunz, C., *Solid State Commun.* **7**, 597 (1969).
152. Beattie, J. R., and Conn, G. K. T., *Phil. Mag.* **46**, 989 (1955).
153. Shkliarevskii, I. N., and Padalka, V. G., *Optika Spektrosk.* **6**, 78 (1959) [*Optics Spectrosc.* **6**, 45 (1959)].
154. Dold, B., and Mecke, R., *Optik* **22**, 453 (1965).
155. Lenham, A. P., and Treherne, D. M., *J. Opt. Soc. Am.* **56**, 1137 (1966).
156. Lang, W., *Optik* **3**, 233 (1948).
157. Watanabe, H., *J. Phys. Soc. Japan* **9**, 920 (1954).
158. Marton, L., Leder, L. B., and Mendlowitz, H., *Advances in Electronics and Electron Physics*, Academic Press, New York, 1955, vol. 7, p. 183
159. Robins, J. L., and Swan, J. B., *Proc. Phys. Soc. London* **76**, 857 (1960).
160. Bearden, J. A., and Burr, A. F., *Rev. Mod. Phys.* **39**, 125 (1967).
161. Phillips, J. C., *Phys. Rev.* **140** A, 1254 (1965).
162. Hagstrum, H. D., and Becker, G. E., *Phys. Rev.* **159**, 572 (1967).
163. Mott, N. F., *Optical Properties and Electronic Structure of Metals and Alloys*, North-Holland, Amsterdam, 1966, p. 314.
164. Cuthill, J. R., McAlister, A. J., and Williams, M. L., *Phys. Rev. Letters* **16**, 933 (1966).
165. Nesbet, R. K., and Grant, P. M., *Phys. Rev. Letters* **19**, 222 (1967).
166. Lynch, M. J., and Swan, J. B., *Australian J. Phys.* **21**, 811 (1968).
167. Priol, M., Seignac, A., and Robin, S., *Compt. Rend.* **258**, 5398 (1964).
168. Canfield, L. R., and Hass, G., *J. Opt. Soc. Am.* **55**, 61 (1965).
169. Beaglehole, D., *Proc. Phys. Soc. London* **85**, 1007 (1965).
170. Cooper, B. R., Ehrenreich, H., and Philipp, H. R., *Phys. Rev.* **138** A, 494 (1965).
171. Beaglehole, D., *Optical Properties and Electronic Structure of Metals and Alloys*, North-Holland, Amsterdam, 1966, p. 154.
172. Garfinkel, M., Tiemann, J. J., and Engeler, W. E., *Phys. Rev.* **148**, 695 (1966).
173. Gerhardt, U., Beaglehole, D., and Sandrock, R., *Phys. Rev. Letters* **19**, 309 (1967).
174. Gerhardt, U., *Phys. Rev.* **172**, 651 (1968).
175. Haensel, R., Kunz, C., Sasaki, T., and Sonntag, B., *Appl. Opt.* **7**, 301 (1968).
176. Krolikowski, W. F., Techn. Reps. 5281-1, Stanford University, 1967.
177. Snow, F. C., *Phys. Rev.* **171**, 785 (1968).
178. Creuzburg, M., *Z. Phys.* **196**, 433 (1966).
179. Thirlwell, J., *Proc. Phys. Soc. London* **91**, 552 (1967).

180. Gournay, L. S., Rense, W. A., Scientific Report no. 1, University of Colorado, 1963.
181. Huebner, R. H., Arakawa, E. T., MacRae, R. A., and Hamm, R. N., *J. Opt. Soc. Am.* **54**, 1434 (1964).
182. Huebner, R. H., Arakawa, E. T., Hamm, R. N., and MacRae, R. A., *Oak Ridge Natn. Lab. TM* 1104 (1965).
183. Ehrenreich, H., *Optical Properties and Electronic Structure of Metals and Alloys*, North-Holland, Amsterdam, 1966, p. 109.
184. Lewis, P. E., and Lee, P. M., *Phys. Rev.* **175**, 795 (1968).
185. Hopfield, J. J., *Phys. Rev.* **139** A, 419 (1965).
186. Steinmann, W., *Phys. Rev. Letters* **5**, 470 (1960).
187. Brown, R. W., Wessel, P., and Trounson, E. P., *Phys. Rev. Letters* **5**, 472 (1960).
188. Geiger, J., *Z. Phys.* **161**, 243 (1961).
189. Steinmann, W., *Z. Phys.* **163**, 92 (1961).
190. Hattori, M., and Yamada, K., *J. Phys. Soc. Japan* **18**, 200 (1963).
191. Yamaguchi, S., *J. Phys. Soc. Japan* **18**, 266 (1963).
192. Arakawa, E. T., Davis, N. O., and Birkhoff, R. D., *Phys. Rev.* **135** A, 224 (1964).
193. Brambring, J., and Raether, H., *Phys. Rev. Letters* **15**, 882 (1965).
194. Otto, A., *Z. Phys.* **185**, 232 (1965).
195. Boersh, H., and Sauerbrey, G., *Optical Properties and Electronic Structure of Metals and Alloys*, North-Holland, Amsterdam, 1966, p. 386.
196. Geiger, J., and Wittmaack, K., *Z. Phys.* **195**, 44 (1966).
197. Schreiber, P., and Raether, H., *Z. Naturforsch.* **21a**, 2116 (1966).
198. Emeric, N., and Emeric, A., *Thin Solid Films* **1**, 13 (1967).
199. Brambring, J., and Raether, H., *Z. Phys.* **199**, 118 (1967).
200. Daniels, J., *Z. Phys.* **203**, 235 (1967).
201. Schulz, D., and Zurheide, M., *Z. Phys.* **211**, 165 (1968).
202. Schreiber, P., *Z. Phys.* **211**, 257 (1968).
203. Daniels, J., *Z. Phys.* **213**, 227 (1968).
204. Dobberstein, P., Hampe, A., and Sauerbrey, G., *Phys. Letters* **27**A, 256 (1968).
205. Fujimoto, F., Ishida, K., and Komaki, K., *J. Phys. Soc. Japan* **26**, 1070 (1969).
206. Schröder, E., *Z. Phys.* **222**, 33 (1969).
207. Emerson, L. C., *Physics To-day* 95, Dec. 1968.
208. Canfield, L. R., Hass, G., and Hunter, W. R., *J. Phys.* **25**, 124 (1964).
209. Cole, T. T., and Oppenheimer, F., *Appl. Opt.* **1**, 709 (1962).
210. Arakawa, E. T., Emerson, L. C., Hammer, D. C., and Birkhoff, R. D., *Phys. Rev.* **131**, 719 (1963).
211. Lukirskii, A. P., Savinov, E. P., Erskov, O. A., and Shepelev, F., *Optika Spektrosk.* **16**, 310 (1964) [*Optics Spectrosc.* **16**, 168 (1964)].
212. Jaeglé, P., and Missoni, G., *Compt. Rend.* **262**, 71 (1966).
213. Robins, J. L., *Proc. Phys. Soc. London* **79**, 119 (1962).
214. Zeppenfeld, K., *Z. Phys.* **223**, 32 (1969).
215. Hunter, W. R., and Tousey, R., *J. Phys.* **25**, 148 (1964).
216. Lemonnier, J. C., Stéphan, G., Le Calvez, Y., and Robin, S., *J. Phys. Chem. Solids* **30**, 1147 (1969).
217. Balabanova, L. A., Bredov, M. M., and Kotov, B. A., *Fizika tverd. Tela* **4**, 86 (1962) [*Soviet Phys. Solid State* **4**, 60 (1962)].
218. Herickhoff, R. J., Arakawa, E. T., and Birkhoff, R. D., *Oak Ridge Natn. Lab. TM* 1107 (1965).
219. Codling, K., Madden, R. P., Hunter, W. R., Angel, D. W., *J. Opt. Soc. Am.* **56**, 189 (1966).
220. Lemonnier, J. C., and Robin, S., *Compt. Rend.* **265**, 661 (1967).
221. MacRae, R. A., Arakawa, E. T., and Williams, M. W., *Phys. Rev.* **162**, 615 (1967).
222. Watanabe, H., *J. Electron Microsc. Japan* **4**, 24 (1956).
223. Sueoka, O., *J. Phys. Soc. Japan* **20**, 2203 (1965).

224. Rasigni, G., Codaccioni, J. P., Michaud-Bonnet, J., Abba, F., and Petrakian, J. P., *Compt. Rend.* **262** B, 772 (1966).
225. Cardona, M., and Greenaway, D. L., *Phys. Rev.* **133** A, 1685 (1964).
226. Lemonnier, J. C., Le Calvez, Y., Stéphan, G., and Robin, S., *Compt. Rend.* **264**, 1599 (1967).
227. Toots, J., and Marton, L., *J. Opt. Soc. Am.* **59**, 1305 (1969).
228. Sandström, A. E., *Handbuch der Physik* **30**, 226 (1957).
229. Hunter, W. R., Angel, D. W., and Tousey, R., *Appl. Opt.* **4**, 891 (1965).
230. Cohen, M. H., and Heine, V., *Adv. Phys.* **7**, 395 (1958).
231. Ziman, J. M., *Adv. Phys.* **10**, 13 (1961).
232. Muldawer, L., and Goldman, H. J., *Optical Properties and Electronic Structure of Metals and Alloys*, North-Holland, Amsterdam, 1966, p. 574.
233. Fukutani, H., and Sueoka, O., *Optical Properties and Electronic Structure of Metals and Alloys*, North-Holland, Amsterdam, 1966, p. 565.
234. Vishnubhatla, S. S., and Jan, J. P., *Phil. Mag.* **16**, 45 (1967).
235. Spadling, D. R., and Metherell, A. J. F., *Phil. Mag.* **18**, 41 (1968).
236. Myers, H. P., Wallden, L., and Karlsson, Å., *Phil. Mag.* 18, 725 (1968).
237. Feinleib, J., Scouler, W. J., and Ferretti, A., *Phys. Rev.* **165**, 765 (1968).
238. Lye, R. G., and Logothetis, E. M., *Phys. Rev.* **147**, 622 (1966).
239. Haensel, R., Radler, K., Sonntag, B., and Kunz, C., *Solid State Commun.* **7**, 1495 (1969).

Additional bibliography

METALS OBEYING THE FREE-ELECTRON THEORY

Alkali metals

Anderegg, M., Feuerbacher, B., and Fitton, B., Optically excited longitudinal plasmons in potassium, *Phys. Rev. Letters* **27**, 1565 (1971).

Dresselhaus, G., Wilson, A. R., and Young, C. Y., The Fermi surface and optical properties of potassium, *Solid State Commun.* 8, 2125 (1970).

Karaskashian, A. S., and Bardasis, A., Optical absorption of sodium, *Phys. Rev.* B, **4**, 404 (1971).

Lindau, I., Löfgren, H., and Wallden, L., Observation of characteristic energy losses in u.v. photoemission spectra from Li and Al, *Phys. Letters* **36** A, 293 (1971).

Mathewson, A. G., and Myers, H. P., The optical absorption spectrum of lithium metal in the range 0.5—4 eV, *Phil. Mag.* **25**, 853 (1972).

Mayer, H., Blanaru, D. L., and Steffen, H., Photoelectric and optical properties of thin alkali metal films, *Thin Solid Films* **5**, 407 (1970).

Oswald, R. G., and Calcott, T. A., Photoemission from core states of Cs and Rb, *Phys. Rev.* B, **4**, 4122 (1971).

Palmer, R. E., and Schnatterly, S. E., Observation of surfaces plasmons, measurement of the optical constants for sodium and potassium, *Phys. Rev.* B, **4**, 2329 (1971).

Payan, R., Optical absorption of alkaline metals in thin layers in relation to their electronic structure. *Ann. Phys. Paris* **4**, 543 (1969).

Payan, J. C., and Roux, D., Transmission optique, en lumière polarisée de couches très minces de sodium sous incidence oblique, *Optics Commun.* **4**, 144 (1971).

Richard, J. C., Saget, P., and Boutry, G. A., Contribution à l'étude des propriétés photoélectriques du sodium pur, préparé et conservé dans l'ultra-vide, *Compt. Rend.* **271**, 1098 (1970).

Smith, N. V., Optical constants of rubidium and caesium from 0.5 to 4.0 eV, *Phys. Rev.* B, **2**, 2840 (1970).

SMITH, N. V., and FISHER, G. B., Photoemission studies of the alkali metals: II, Rubidium and caesium, *Phys. Rev.* B, **3**, 3662 (1971).

SMITH, N. V., and SPICER, W. E., Photoemission studies of the alkali metals: I, Sodium and potassium. *Phys. Rev.* **188**, 593 (1969).

SUTHERLAND, J. C., HAMM, R. N., and ARAKAWA, E. T., Extinction coefficient and imaginary part of the dielectric constant for sodium and potassium above the plasma energy, *J. Opt. Soc. Am.* **59**, 1581 (1969).

THOMAS, S., and HAAS, T. W., Observation of surface plasmon losses in K and Rb overlayers, *Solid State Commun.* **11**, 193 (1972).

VAN OIRSCHOT, TH. G. H., VAN DEN BRINK, M., and SACHTLER, W. M. H., Photoelectric emission and phase transitions of evaporated metastable sodium, potassium and sodium-potassium alloys films, *Surf. Sci.* **29**, 189 (1972).

WATABE, M., On plasmon contribution to the ultraviolet absorbing power of the alkali metals, *J. Phys. Soc. Japan* **28**, 655 (1970).

WHANG, U. S., ARAKAWA, E. T., and CALCOTT, T. A., Collective effects in the optical properties of Cs above the plasma frequency, *Phys. Rev. Letters* **25**, 646 (1970).

WHANG, U. S., ARAKAWA, E. T., and CALCOTT, T. A., Optical properties of Cs for photons of energy 3.6-9.6 eV., *J. Opt. Soc. Am.* **61**, 740 (1971).

WHANG, U. S., ARAKAWA, E. T., and CALCOTT, T. A., Optical properties of Rb between 3.3 and 10.5 eV, *Phys. Rev.* B, **5**, 2118 (1972).

WHITEFIELD, R. J., and BRADY, J. J., New value for work function of sodium and the observation of surface plasmon effects, *Phys. Rev. Letters* **26**, 380 (1971).

YAMAGUCHI, S., and HANYU, T., Optical properties of potassium. *J. Phys. Soc. Japan* **31**, 1431 (1971).

Aluminium, magnesium, beryllium

AVETYAN, K. T., and BOROVKII, I. B., Dependance of the frequency of plasma oscillations on the thickness of aluminium films, *Soviet Phys. Solid State* **11**, 1415 (1969).

BRAUNDMEIER, A. J., JR., and ARAKAWA, E. T., Observed radiation from surface plasmon in aluminium, *Z. Phys.* **239**, 337 (1970).

BRAUNDMEIER, A. J., WILLIAMS, M. W., ARAKAWA, E. T., and RITCHIE, R. H., Radiative decay of surface plasmons from Al, *Phys. Rev.* B, **5**, 2754 (1972).

CAZAUX, J., Oscillations de plasma: les pertes d'énergie caractéristiques des électrons aux petits angles de diffusion dans le magnésium et le bismuth monocristallin, *Compt. Rend.* **269**, 1263 (1969).

DAUDÉ, A., SAVARY, A., and ROBIN, S., Effect of different roughnesses on the excitation by photons of the surface plasmon of aluminium, *J. Opt. Soc. Am.* **62**, 1 (1972).

ENDRIZ, J. G., and SPICER, W. E., Study of aluminium films: I, Optical studies of reflectance drops and surface oscillations on controlled roughness film, *Phys. Rev.* B, **4**, 4144 (1971).

ENDRIZ, J. G., and SPICER, W. E., Study of aluminium films: II, Photoemission studies of surface plasmon oscillations on controlled roughness films, *Phys. Rev.* B, **4**, 4159 (1971).

ENDRIZ, J. G., and SPICER, W. E., Experimental evidence for the surface photoelectric effect in aluminium, *Phys. Rev. Letters* **27**, 570 (1971).

FEUERBACHER, B., and FITTON, B., Investigation of collective electron oscillations in Cd, Mg, and Zn by photoemission, *Phys. Rev. Letters* **24**, 499 (1970).

GESELL, T. F., and ARAKAWA, E. T., Attenuation length for photoelectrons excited in aluminium by 21.2 eV photons, *Phys. Rev. Letters* **26**, 377 (1971).

GESELL, T. F., and ARAKAWA, E. T., Volume plasmon of Mg from photoelectric measurements, *Phys. Letters* **36** A, 77 (1971).

HUEN, T., and WOOTEN, F., Photoemission from aluminium, *Solid State Commun.* **9**, 871 (1971).

HUNTER, W. R., OSANTROWSKI, G. F., and HASS, G., Reflectance of aluminium overcoated with MgF_2 and LiF in the wavelength region from 1600 Å to 300 Å at various angle of incidence, *Appl. Opt.* **10**, 540 (1971).

HUTCHESON, E. T., HASS, G., and COULTER, J. K., A direct comparison of the visible and ultraviolet reflectance of aluminium films evaporated in coventional and ultra-high vacuum systems, *Optics Commun.* **3**, 213 (1971).

LINDAU, I., LÖFGREN, H., and WALLDEN, L., Observation of characteristic energy losses in u.v. photoemission spectra from Li and Al, *Phys. Letters* **36** A, 293 (1971).

MACEK, CH., OTTO, A., and STEINMANN, W., Resonant photoemission from aluminium films at 5 eV photon energy due to non radiative surface plasma waves, *Phys. Status Solidi* **61**, K 59 (1972).

MATHEWSON, A. G., and MYERS, H. P., Optical absorption in aluminium and the effect of temperature, *J. Phys. F-Metal Physics* **2**, 403 (1972).

POWELL, C. J., Analysis of optical and inelastic electron-scattering data: II, Application to Al, *J. Opt. Soc. Am.* **60**, 78 (1970).

POWELL, C. J., Analysis of optical and inelastic electron-scattering data; III, Reflectance data for beryllium, germanium, antimony and bismuth, *J. Opt. Soc. Am.* **60**, 214 (1970).

ROSEI, R., and LYNCH, D. W., Thermomodulation spectra of Al, Au, and Cu, *Phys. Rev.* B, **5**, 3883 (1972).

SOROKIN, O. M., and BLANK, V. A., Optical transmission and excitation of plasma oscillation in free films of Al, In, Sn, Mg, Ge, Se and Te in the vacuum ultraviolet region of the spectrum, *Optika Spektrosk.* **28**, 1178 (1970) [*Optics Spectrosc.* **28**, 634 (1970)].

DIVALENT METALS

Alkaline earth metals

GAUDART, L., and RIVOIRA, R., Propriétés photoélectriques des couches minces de calcium, *Appl. Opt.* **10**, 2336 (1971).

GAUDART, L., and RIVOIRA, R., Propriétés photoélectriques des couches minces de calcium en liaison avec leur structure, *Compt. Rend.* **272** B, 855 (1971).

HELMS, C. R., and SPICER, W. E., Photoemission studies of the oxidation of strontium, *Phys. Rev. Letters* **28**, 565 (1972).

KRESS, K. A., and LAPEYRE, G. J., Photoemission measurement of the *d* band energies in calcium, *Solid State Commun.* **9**, 827 (1971).

Cadmium, zinc

BARTLETT, R. J., LYNCH, D. W., and ROSEI, R., Optical properties of single crystal cadmium, *Phys. Rev.* B, **3**, 4074 (1971).

FEUERBACHER, B., and FITTON, B., Investigation of collective electron oscillations in Cd, Mg and Zn by photoemission, *Phys. Rev. Letters* **24**, 499 (1970).

LEMONNIER, J. C., GIRAULT, P., and ROBIN, MME S., Propriétés optiques et électroniques du zinc dans l'ultraviolet extrême, *Compt. Rend.* **269** B, 329 (1969).

MOTULEVICH, G. P., and SHUBIN, A. A., The optical constants of zinc, *Soviet Phys. JETP* **29**, 24 (1969).

RUBLOFF, G. W., Normal incidence reflectance, optical properties and electronic structure of Zn, *Phys. Rev.* B, **3**, 285 (1971).

SCHWARZ, H., Optical properties of cadmium single crystal, *Phys. Status Solidi* **43**, 755 (1971).

WEAVER, J. H., LYNCH, D. W., and ROSEI, R., Optical properties of single crystal zinc, *Phys. Rev.* B, **5**, 2829 (1972).

RARE-EARTHS

BAKULIN, E. A., BALABANOVA, L. A., STEPIN, E. V., and SHERBININA, V. V., Characteristic energy losses of electrons in Ho, Er, Tu and Yb rare earth metals, *Fizika tverd. Tela* **13**, 241 (1971) [*Soviet Phys. Solid State* **13**, 189 (1971)].

EASTMAN, D. E., Photoelectric work functions of transitions, rare-earth, and noble metals, *Phys. Rev.* B, **2**, 1 (1970).

NEMCHENOK, R. L., STRAKOVSKAYA, S. E., and TITENSKII, A. I., Energy distribution of photoelectrons of gadolinium and terbium, *Soviet Phys. Solid State* **11**, 2181 (1970).

PETRAKIAN, J. P., Direct transitions at optical frequencies in rare-earth metals, *J. Opt. Soc. Am.* **62**, 401 (1972).

TSVERMAN, E. V., REDKIN, V. S., ZASKKVARA, V. V., and KORSUNSKII, M. I., Spectra of characteristic energy losses of electrons in Gd and Dy, *Fizika tverd. Tela* **13**, 2793 (1971) [*Soviet Phys. Solid State* **13**, 2339 (1971)].

TRANSITION METALS

BAKULIN, E. A., BALABANOVA, L. A., and BREDOV, M. M., Collective oscillations of electrons in *d*-transition metals, *Fizika tverd. Tela* **12**, 72 (1970) [*Soviet Phys. Solid State* **12**, 57 (1970)].

BAKULIN, E. A., BALABANOVA, L. A., BREDOV, M. M., OSTROUMOVA, E. G., STEPIN, E. V., and SHERBININA, V. V., Characteristic energy loss in titanium, *Fizika tverd. Tela* **12**, 63 (1970) [*Soviet Phys. Solid State* **12**, 49 (1970)].

BAKULIN, E. A., BALABANOVA, L. A., and VASILEV, V. A., Characteristic electron energy losses in Ni and NiO, *Fizika tverd. Tela* **13**, 542 (1971).

BAKULIN, E. A., BREDOV, M. M., and VASILEV, V. A., Formation of the electron energy loss spectra in iron, cobalt and nickel, *Soviet Phys. Solid State* **13**, 3114 (1972).

BOS, L. W., and LYNCH, D. W., Optical properties of antiferromagnetic chromium and dilute Cr-Mn and Cr-Re alloys, *Phys. Rev.* B, **2**, 4567 (1970).

COX, J. T., HASS, G., and HUNTER, W. R., Optical properties of evaporated rhodium films deposited at various substrate temperatures in the vacuum ultra-violet from 150 to 2000 Å, *J. Opt. Soc. Am.* **61**, 360 (1971).

COX, J. T., HASS, G., RAMSEY, J. B., and HUNTER, W. R., Reflectance of evaporated rhenium and tungsten films in the vacuum ultraviolet from 300 to 2000 Å, *J. Opt. Soc. Am.* **62**, 781 (1972).

EASTMAN, D. E., Photoelectric work functions of transition, rare-earth, and noble metals, *Phys. Rev.* B, **2**, 1 (1970).

FIERMANS, L., and VENNIK, J., Characteristic energy loss studies of V_2O_5 and vanadium, *Phys. Status Solidi* **41**, 621 (1970).

GOLOVASHKIN, A. I., LEKSINA, I. E., MOTULEVICH, G. P., and SHUBIN, A. A., The optical properties of niobium, *Soviet Phys. JETP* **29**, 27 (1969).

HASS, G., RAMSEY, J. B., and HUNTER, W. R., Reflectance of semitransparent platinum films on various substrates in the vacuum ultraviolet, *Appl. Optics* **8**, 2255 (1969).

JANAK, J. F., EASTMAN, D. E., and WILLIAMS, A. R., Direct transition analysis of new photoemission data for palladium, *Solid State Commun.* **8**, 271 (1970).

KIRILLOVA, M. M., Electron interband transitions in nickel, *Zh. ėksp. teor. Fiz.* **61**, 336 (1971) [*Soviet Phys. JETP* **34**, 178 (1972)].

KIRILLOVA, M. M., NOMEROVANNAYA, L. V., and NOSKOV, M. M., Optical properties of molybdenum single crystals, *Soviet Phys. JETP* **33**, 1210 (1971).

KIRILLOVA, M. M., NOMEROVANNAYA, L. V., and NOSKOV, M. M., Optical properties of molybdenum monocrystals, *Zh. ėksp. teor. Fiz.* **60**, 2252 (1971).

KORABLEV, Y. Y., Characteristic energy loss spectra of electrons in tungsten single crystals, *Fizika tverd. Tela* **12**, 1298 (1970).

KORSUNSKII, M. I., GENKIN, Y. E., and ZAVODINSKII, V. G., Characteristic energy loss in transition metals, *Fizika tverd. Tela* **13**, 2550 (1972).

KRESS, K. A., and LAPEYRE, G. J., Photoemission and optical investigation of the electronic structure of ruthenium, *Phys. Rev.* B, **2**, 2532 (1970).

KRESS, K. A., and LAPEYRE, G. J., Optical properties of molybdenum and ruthenium. *J. Opt. Soc. Am.* **60**, 1681 (1970).

LIN, S. F., PIERCE, D. T., and SPICER, W. E., Photoemission studies of platinum, *Phys. Rev.* B, **4**, 326 (1971).

LINDAU, I., and WALLDÉN, L., Some photoemission properties of Cu, Ag, Au and Ni, *Physica Scripta* **3**, 77 (1971).

NOMEROVANNAYA, L. V., KIRILLOVA, M. M., and NOSKOV, M. M., Optical properties of tungsten single crystals, *Zh. éksp. teor. Fiz.* **60**, 748 (1971) [*Sov. Phys. JETP* **60**, 405 (1971)].

NOMEROVANNAYA, L. V., KIRILLOVA, M. M., and NOSKOV, M. M., Optical absorption in tungsten at 77 K, *Fizika tverd. Tela* **14**, 633 (1972).

PETRAKIAN, J. P., and PALMARI, J. P., Réalisation d'une enceinte pour l'étude de couches minces métalliques sous ultra-vide statique. Application à l'étude des propriétés optiques de l'yttrium, *Thin Solid Films* **4**, 423 (1969).

PETRAKIAN, J. P., PALMARI, J. P., and RASIGNI, G., Les propriétés optiques de couches minces d'yttrium sous ultra vide statique, *Appl. Opt.* **9**, 2115 (1970).

PIERCE, D. T., and SPICER, W. E., Photoemission studies of ferromagnetic and paramagnetic nickel, *Phys. Rev. Letters* **25**, 581 (1970).

PIERCE, D. T., and SPICER, W. E., Photoemission studies of rhodium, *Phys. Rev.* B, **5**, 2125 (1972).

ŚASOVSKAYA, I. I., and NOSKOV, M. M., Optical properties of nickel, *Fizika Metall.* **32**, 723 (1971).

ŚASOVSKAYA, I. I., and NOSKOV, M. M., Optical interband transition in nickel and nickel base alloys, *Fizika Metall.* **33**, 86 (1972).

SEIGNAC, A., and ROBIN, S., Propriétés optiques de couches minces de rhodium et de vanadium dans l'ultraviolet lointain, *Compt. Rend.* **271**, 919 (1970).

SEIGNAC, A., and ROBIN, S., Propriétés optiques de couches minces de platine dans l'ultraviolet lointain, *Solid State Commun.* **11**, 217 (1972).

SIMMONS, G. W., and GREEN, A. E. S., Characteristic energy loss spectra of titanium, vanadium, and chromium. *J. Appl. Phys.* **43**, 693 (1972).

UDOEV, YU. P., KAPITZA, M. L., and TERRA, A. R., Photoemission spectra distribution of tungsten and molybdenum in photon energy range 4-11 eV, *Fizika Metall.* **31**, 438 (1971).

UDOEV, YU. P., KOS'YAKOVA, N. S., and KAPITZA, M. L., Optical properties of tungsten and molybdenum, *Fizika Metall.* **31**, 439 (1971).

VEHSE, R. C., ARAKAWA, E. T., and WILLIAMS, M. W., Optical and photoemissive properties of palladium in the vacuum ultraviolet spectral region, *Phys. Rev.* B, **1**, 517 (1970).

ZASHKVARA, V. V., KORSUNSKII, M. I., LARIN, M. P., RED'KIN, V. S., MASYAGIN, V. E., KIL'DIYAROV, M. A., and CHOKIN, K. SH., Characteristic energy loss spectra for osmium, *Soviet Phys. Solid. State* **12**, 238 (1970).

ZASHKVARA, V. V., KORSUNSKII, M. I., RED'KIN, V. S., and MASYAGIN, V. E., Characteristic electron energy losses in Y, Zr, Nb, and Mo, *Soviet Phys. Solid State* **11**, 3083 (1970).

NOBLE METALS

BALDINI, G., and NOBILE, M., Thermoreflectance of Ag at the plasma frequency, *Solid State Commun.* **8**, 7 (1970).

CHRISTENSEN, N. E., Photoemission properties of gold, *Phys. Letters* **35** A, 206 (1971).

CHRISTENSEN, N. E., Primary photoemission spectra of gold and their relation to the band structure, *Phys. Status Solidi* **52**, 241 (1972).

CHRISTENSEN, N. E., and SERAPHIN, B. O., Relativistic band calculation and the optical properties of gold, *Phys. Rev.* B, **4**, 3321 (1971).

CLERC, J., GIRAUD, G., and ROUSSENQ, J., Effet photoélectrique des monocristaux de cuivre, *Compt. Rend.* B, **272**, 1346 (1971).

EASTMAN, D. E., Photoemission spectra for liquid and cristalline Au, *Phys. Rev. Letters* **26**, 106 (1971).

EASTMAN, D. E., and CASHION, J. K., Photoemission from Cu, Ag and Au in the 10 to 27 eV energy range, *Phys. Rev. Letters* **24**, 310 (1970).

EASTMAN, D. E., and GROBMAN, W. D., Photoemission energy distributions for Au from 10 to 40 eV, using synchrotron radiation, *Phys. Rev. Letters* **28**, 1327 (1972).

FONG, C. Y., COHEN, M. L., ZUCCA, R. M. L., STOKES, J., and SHEN, Y. R., Wavelength modulation spectrum of copper, *Phys. Rev. Letters* **25**, 1486 (1970).

GARRIGOS, R., KOFMAN, R., JOLIVET, A., and DONNADIEU, A., Electroreflexion des métaux nobles, *Compt. Rend.* **272** B, 1078 (1971).

GROBMAN, W. D., and EASTMAN, D. E., Temperature modulated photoemission spectra for gold, *Phys. Rev. Letters* **28**, 1038 (1972).

IRANI, G. B., HEUN, T., and WOOTEN, F., Optical properties of Ag and α phase Ag-Al alloys, *Phys. Rev.* B, **3**, 2385 (1971).

JASPERSON, S. N., and SCHNATTERLY, S. E., Photon-surface plasmon coupling in thick Ag foils, *Phys. Rev.* **188**, 759 (1969).

KOFMAN, R., RICHARD, J., and SAISSY, A., Piezoreflectivity in Ag films, *Solid State Commun.* **10**, 1251 (1972).

KROLIKOWSKI, W. F., and SPICER, W. E., Photoemission studies of the noble metals: I, Copper, *Phys. Rev.* **185**, 882 (1969).

KROLIKOWSKI, W. F., and SPICER, W. E., Photoemission studies of the noble metals: II, Gold, *Phys. Rev.* B, **1**, 478 (1970).

LINDAU, I., and WALLDÉN, L., Optical transition between conductions bands in copper studied by photoemission, *Solid State Commun.* **9**, 1147 (1971).

LINDAU, I., and WALLDÉN, L., Some photoemission properties of Cu, Ag, Au, and Ni, *Physica Scripta* **5**, 77 (1971).

MARFAING, J., and PAYAN, F., Effet photoélectrique des couches minces d'or et d'argent dans l'ultraviolet lointain, *Compt. Rend.* B, **271**, 995 (1970).

MARFAING, J., and PAYAN, F., Sur quelques propriétés photoélectriques de l'argent dans l'ultraviolet lointain, *Compt. Rend.* B, **272**, 1343 (1971).

MARFAING, J., and PAYAN, F., Sur quelques propriétés photoélectriques du cuivre dans l'ultra violet lointain, *Compt. Rend* B, **273**, 1024 (1971).

MARFAING, J., and PAYAN, F., Propriétés optiques des couches minces de cuivre dans l'ultra-violet lointain, *Compt. Rend* B, **274**, 1155 (1972).

NILSSON, P. O., LINDAU, I., and HAGSTRÖM, S. B., Optical plasma resonance absorption in thin films of silver and some silver alloys, *Phys. Rev.* B, **1**, 498 (1970).

NILSSON, P. O., NORRIS, C., and WALLDÉN, L., The electronic structure of gold studied by photoemission, *Physik. Kondens. Mater.* **11**, 220 (1970).

NILSSON, P. O., and SANDELL, B., Piezo optical properties of Ag and Ag-Au alloys, *Solid State Commun.* **8**, 721 (1970).

PHILLIPS, J. C., Nature of optical transitions in copper and in Cu-Ni alloys, *Phys. Rev.* **187**, 1175 (1969).

ROSEI, R., and LYNCH, D. W., Thermomodulation spectra of Al, Au and Cu, *Phys. Rev.* B, **5**, 3883 (1972).

RUPKEN, K., The optical constants of Au, Ag and their alloys in the energy region from 2.4 to 4.4 eV, *Z. Phys.* **250**, 228 (1972).

SCHLÜTER, M., The optical properties of gold, silver and gold-silver alloys from 2 to 40 eV by energy loss spectroscopy of electrons, *Z. Phys.* **250**, 87 (1972).

SHKLYAREVSKY, N., KORNEEVA, T. I., and GREBENNIK, T. G., Optical constants of thin gold films, *Optika i Spektrosk.* **31**, 414 (1971) [*Optic Spectrosc.* **31**, 221 (1971)].

SKILLMAN, D. C., Optical constants of silver, *J. Opt. Soc. Am.* **61**, 1264 (1971).

SMITH, N. V., Photoemission properties of caesiated copper, *Phys. Rev. Letters* **23**, 1452 (1969).

SMITH, N. V., Photoelectron energy spectra and the band structures of the noble metals, *Phys. Rev.* B, **3**, 1862 (1971).

STOKES, J., SHEN, Y. R., TSANG, Y. W., COHEN, M. L., and FONG, C. Y., Wave length modulation spectra of single crystals of silver and gold, *Phys. Letters* **38** A, 347 (1972).

THÈYE, M. L., Investigation of optical properties of Au by means of thin semi-transparent films, *Phys. Rev.* B, **2**, 3060 (1970).

OTHER METALS

Cazaux, J., Oscillations de plasma: les pertes d'énergie caractéristiques des électrons aux petits angles de diffusion dans le magnésium et le bismuth monocristallin, *Compt. Rend.* B, **269**, 1263 (1969).

Choyke, W. J., Vosko, S. H., and O'Keeffe, T. W., A comparison of the optical properties of single crystal and liquid mercury in the energy range 0.5-8.25 eV, *Solid State Commun.* **9**, 361 (1971).

Gauthé, B., Wehenkel, C., Characteristic electron energy losses in thin bismuth films, *Phys. Letters* **39** A, 171 (1972).

Koyama, R. Y., and Spicer, W. E., Photoemission studies of indium, *Phys. Rev.* **4**, 4318 (1971).

Liljenwall, H. G., and Mathewson, A. G., The optical properties of lead in the energy range 0.6-6 eV, *Phil. Mag.* **22**, 243 (1970).

Norris, E., and Walldén, L., Photoemission from Pb, *J. Phys. F-Metal Phys.* **2**, 180 (1972).

Powell, C. J., Analysis of optical and inelastic-electron scattering: III, Reflectance data for beryllium, germanium, antimony and bismuth, *J. Opt. Soc. Am.* **60**, 214 (1970).

Schwarz, H., Optical properties and electronic structure of white tin, *Phys. Status Solidi* **44**, 603 (1971).

Sorokin, O. M., and Blank, V. A., Optical transmission and excitation of plasma oscillation in free films of Al, In, Sn, Mg, Ge, Se and Te in the vacuum ultraviolet region of the spectrum, *Optika Spektrosk.* **28**, 1178 (1970) [*Optics Spectrosc.* **28**, 634 (1970)].

Wang, P. Y., and Jain, A. L., Optical interband transitions in antimony, *Opt. Commun.* **3**, 412 (1971).

ALLOYS AND METALLIC COMPOUNDS

Bakulin, E. A., Bredov, M. M., and Vasilev, V. A., Electron energy loss in alloys of aluminium with 3*d* transitions metals, *Fizika tverd. Tela* **13**, 2536 (1972).

Beaglehole, D., Optical absorption in virtual bound states alloys, *Phys. Letters* **40** A, 209 (1972).

Beaglehole, D., and Erlbach, E., Band behaviour in noble metal-noble metal alloys: single bands and separated bands, *Solid State Commun.* **8**, 255 (1970).

Beaglehole, D., and Erlbach, E., Electronic structure of noble metal-metal alloys, *Phys. Rev.* B, **6**, 1209 (1972).

Beaglehole, D., and Wihl, W., Optical studies of the electronic structure of aluminium-manganese and aluminium-copper alloys, *J. Phys. F-Metal Phys.* **2**, 43 (1972).

Bernert, B., and Zacharias, P., The optical constants of silver-gallium alloys near the plasma frequency, quantitative comparison with electron energy loss spectra, *Z. Phys.* **241**, 205 (1971).

Fukutami, H., Optical constants of silver-gold alloys, *J. Phys. Soc. Japan* **30**, 399 (1971).

Golovashkin, A. I., Levchenko, I. S., and Motulevich, G. R., Optical properties and electron characteristics of vanadium-gallium alloys obtained by evaporation in vacuum, *Sov. Phys. JETP* **30**, 44 (1970).

Green, E. L., and Muldawer, L., Optical properties of the alpha phase alloys Ag-Zn and Ag-Cd, *Phys. Rev.* B, **2**, 330 (1970).

Irani, G. B., Heun, T., and Wooten, F., Optical properties of Ag and α phase Ag-Al alloys, *Phys. Rev.* B, **3**, 2384 (1971).

Nilsson, P. O., and Sandell, B., Piezo optical properties of Ag and Ag-Au alloys, *Solid State Commun.* **8**, 721 (1970).

Phillips, J. C., Nature of optical transitions in copper and in Cu-Ni alloys, *Phys. Rev.* **187**, 1175 (1969).

Rupken, K., The optical constants of Au, Ag and their alloys in the energy region from 2.4 to 4.4 eV, *Z. Phys.* **250**, 228 (1972).

Schlüter, M., The optical properties of gold, silver and gold-silver alloys from 2 to 40 eV by energy loss spectroscopy of electrons, *Z. Phys.* **250**, 87 (1972).

Schmidt, B. F., and Lynch, D. W., Optical properties and piezoreflectance of siver-palladium alloys, *Phys. Rev.* B, **3**, 4015 (1971).

Seib, D. H., and Spicer, W. E., Nature of optical transitions in copper and Cu:Ni alloys, *Phys. Rev.* **187**, 1176 (1969).

Seib, D. H., and Spicer, W. E., Photoemission and optical studies of Cu-Ni alloys: I, Cu-rich alloys, *Phys. Rev.* B, **2**, 1676 (1970).

Seib, D. H., and Spicer, W. E., Photoemission and optical studies of Cu-Ni alloys: II, Ni-rich and nearly equiatomic alloys, *Phys. Rev.* B, **2**, 1694 (1970).

Siskind, B., Boiani, J., and Rice, S. A., Reflectivities of Hg-In liquid alloys from 0.31 to 9 eV, *Phys. Rev. Letters* **29**, 259 (1972).

van Oirschot, Th. G. J., van den Brink, M., and Sachtler, W. M. H., Photoelectric emission and phase transitions of evaporated metastable sodium, potassium and sodium-potassium alloy films, *Surf. Sci.* **29**, 189 (1972).

CHAPTER 3

Optical properties of ionic insulators

G. STEPHAN and S. ROBIN (Mrs)
Laboratoire de Spectroscopie, Université de Rennes, 35031 Rennes-Cedex, France

Contents

In classifying solids, Seitz[1] characterized ionic crystals by "a good ionic conductivity at high temperatures, strong infrared absorption spectra and good cleavage". To these properties must be added the existence of a large electronic forbidden band gap which makes them insulators at ordinary temperature. We will limit our study of these ionic insulators to those having a forbidden energy gap greater than 4 eV; to this category belong the alkali and alkaline earth halides as well as certain compounds such as oxides. The study of their optical properties was first carried out in Göttingen[2] towards 1930, and this was a starting point for a vast literature concerning colour centres[3] and the influence of temperature on the characteristic absorption edge. The work of Eby *et al.*[4] marked an important step in the understanding of electronic resonances in alkali halides, and particularly of the excitons[5] in the spectral range 5-11 eV. In recent years, the techniques of measuring absorptance of thin films and reflectance of single crystals have improved considerably, and, in conjunction with band-structure calculations, have enabled the interpretation of the resonance spectra of several ionic insulators. In this chapter, we first discuss briefly the various types of resonance and then give the results obtained for the alkali and alkaline earth halides and for some compounds such as oxides and other halides. In some cases, the longitudinal resonance spectra deduced from optical experiments will be compared with the electron energy-loss spectra.

1. Vacuum ultraviolet absorption processes for ionic insulators: electronic resonances

The absorption spectrum of pure ionic insulators is characterized by absorption in infrared due to phonons, and a complete transparence in the visible region followed by absorption bands in the u.v. up to the X-ray region. Although the latter are due to polarons, still in the first approximation, the electron-phonon coupling can be neglected and the resonances related to absorption in the u.v. have a purely electronic nature. The many-body problem (10^{28} electrons/m^3) to be solved is first approximated by a band theory which is based on the individual behaviour of electrons, and describes the direct interband transitions.(7) The collective behaviour of electrons gives rise to the excitons, the plasmons, and the so-called "non-direct" interband transitions of Spicer.(15)

The first fundamental absorption band is due to the real resonance absorption of photons, creating electron-hole bound pairs called excitons, which can propagate inside the crystals.(5) This occurs because in ionic insulators the dielectric constant is small and, in contrast to metals, the screening effect is weak. The exciton is a collective phenomenon which may be transversal or longitudinal; the two resonances are close and separated by an interval of energy $\Delta E = kf/\varepsilon\omega$(6) where k is a constant for the solid in question, f the oscillator strength of the exciton, ω its frequency, and ε an appropriate dielectric constant.(6) As the excitons are destroyed by interaction with lattice vibrations, a decrease in temperature lengthens their lifetimes and consequently sharpens the absorption bands, shifts them towards higher energies, and augments their maxima. When the excitation energy is sufficiently large, the electron and the hole are decoupled and, in the ε_2 spectrum, one finds the van Hove discontinuities(7–9) characterizing the critical points of the joint density of states. However, there is a competition between this non-local excitation where the hole is shared among the different sites of the crystal and a local excitation where the hole remains localized and even self-trapped. In the latter case, the shape of ε_2 is given by the "optical" density of states in the various bands.(15)

Miyakawa(13) has shown that in the exciton representation, the dispersion relation for the plasmon is identical to that for metals found by Wilson(10) if the term due to free electrons occurring there is removed. The conclusion of Wilson on the displacement of plasmons by interband transitions can therefore be adapted to insulators. Moreover, the condition for the existence of a longitudinal wave being expressed by $\varepsilon_1 = 0$, Miyakawa shows that a plasmon is possible only after a discontinuity M_2 or M_3. Further, this condition can only be satisfied for discontinuities due to transitions having a large oscillator strength. These theoretical results will be illustrated with potassium halides as examples.

Second-order resonances—i.e. excitation of two electrons by a photon—have been investigated experimentally[11,12] and theoretically.[13,14] However, systematic experiments are needed for their study.

To conclude this brief summary of electronic resonances, we may mention that certain insulators, like the fluorides of sodium or potassium, do not possess, in their longitudinal spectra, a plasmon for the valence electrons. In metals, the electron-electron interaction gives rise to the plasmon[16] so that the individual and the collective behaviours are well separated. For the fluorides in question this is not the case, and the band model is probably not as good an approximation as for metals or even for semiconductors.

2. The alkali halides

The alkali halides are the best representatives of the ionic insulators. In each the forbidden band gap is large, varying from about 6 eV for iodides[4] to 13.6 eV for LiF;[17] the latter crystal, because of its consequent transparency, is very often used in v.u.v. optics.

In 1938, Pohl wrote a review of the work already done:[22] Optical absorption of thin films was known down to 1100 Å,[20] and the influence of the temperature on the displacement of excitons recognized.[21] Some attempts had been made to measure reflectances,[18] but the indices were known only to 2000 Å.[19] More recently, the absorption by thin films of iodides and bromides has been studied by Fisher in 1954[23] and by Martienssen in 1957.[24] The principal papers are those of Teegarden[25] in 1957 and of Eby *et al.*[4] in 1959. The latter contains the absorption curves of thin films at 80 K for all the alkali halides (except LiF) between 5 and 11.4 eV. Table 1 shows the energy for the first doublet absorption band due to Γ excitons. Photoemissivity measurements[26] have shown that the threshold of interband transitions occurs after the first exciton, so that the energy of the forbidden band may be approximated from this table. More recently, absorption measurements have been extended to very low temperatures[29] and down to 600 Å at room temperature.[11,28] However, the study of optical properties based upon absorption measurements contains the drawback that thin films do not have exactly the same properties as bulk materials. The recent tendency is towards measuring the reflectance of freshly cleaved crystals.[30,31]

In alkali halides the valence band is constructed from the p states of the halogen: $2p$ F^-, $3p$ Cl^- ... The energy maximum of this band is generally found at the Γ point, where the corresponding wave function belongs to the representation Γ_{15}, split by spin-orbit interaction into two components Γ_8^- and Γ_6^-. The lowest conduction band is formed by the s states of the alkali ions: $2s$ Li^+, $3s$ Na^+ ... In general, the minimum of energy of this band is also found at the Γ point, where the wave function has the symmetry Γ_1. The absorption spectrum therefore starts generally with a Γ exciton, as shown in Table 1, split into two components, theoretically[27] separated by 0.047 eV for fluorides, 0.103 eV for chlorides,

TABLE 1. EXCITON LEVELS IN THE LITHIUM, SODIUM, POTASSIUM, AND RUBIDIUM HALIDES (eV)

	$\Gamma_{15(3/2)}$	$\Gamma_{15(1/2)}$
LiF	12.7	
LiCl	8.8	8.9
LiBr	7.20	7.72
LiI		6.8
NaF	10.6	
NaCl	7.96	8.09
NaBr	6.68	7.2
NaI	5.56	6.73
KF	9.95	
KCl	7.76	7.87
KBr	6.77	7.26
KI	5.80	6.68
RbF	9.45	
RbCl	7.51	7.64
RbBr	6.60	7.08
RbI	5.70	6.47

The experimental values are taken from the 80 K absorption data of Eby *et al.*[4] and, for LiF, from ref. 17.

0.432 eV for bromides, and 0.899 eV for iodides. It continues by the interband transitions and other excitons at the Brillouin zone limit; internal excitons are found at very high energies.

2.1. LITHIUM HALIDES

LiF

This has the largest band gap of all the crystals studied. It was first studied experimentally in its region of transparency,[32–34] and Ewing and Seitz[35] have investigated its band structure. The investigation of the electronic spectrum started only with the measurements of Kato,[36, 37] giving the optical constants between 10 and 13.5 eV, and those of Walker[38] on the reflectance spectra up to 25 eV. The excitonic spectrum in the 10-13 eV region has also been studied by some Japanese,[39, 40] French,[41, 42] and American[43, 44] groups. Optical constants have been deduced from reflectance measurements up to 30 eV[45] and on freshly cleaved crystals,[47] and the influence of the surface conditions was demonstrated.[46] These investigations permitted Kunz *et al.*[48] to interpret the optical data in terms of the calculated band structure represented in Fig. 1. It is remarkable that, unlike other alkali halides, the minimum of the conduction band occurs at the L point. The authors give the following explanation for this exception: around the Li^+ ion, the electrons having s symmetry wave function experience the repulsive potential due to $1s$ electrons of Li^+. Hence the wave function of the representation L_1 (having a p symmetry) is lower in energy than that of the representation Γ_1 (having *s* symmetry). On the other hand, around F^- ions there is a diminution in energy of electrons with *p* symmetry. This fluoride is, therefore, different from other halides of lithium, and the first peak is an exciton L ($L'_3 - L_1$). Table 2 displays the energy of the principal transitions and the theoretical assignments.

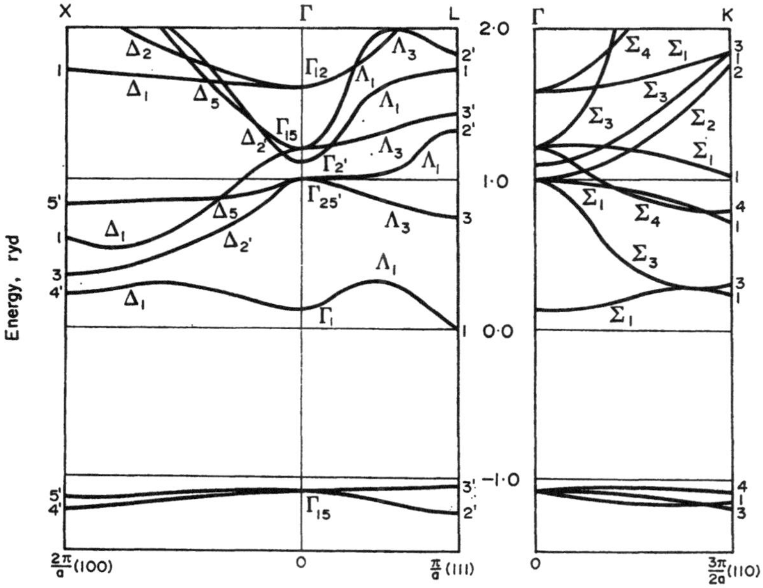

FIG. 1. The highest valence and low conduction bands of LiF are given. Energies are in Rydbergs. Spin-orbit and relativistic effects are not included.[48]

TABLE 2. ENERGY AND TYPE OF PRINCIPAL INTERBAND TRANSITIONS FROM THE VALENCE BAND OF LiF[48]

Transitions	Energy (eV)	Types
$L'_3 \to L$	14.6	M_0
$\Gamma_{15} \to \Gamma_1$	16.7	M_0
$X'_5 \to X_3$	20.1	M_0
$X'_5 \to X_1$	23.4	M_3
$L'_3 \to L_3$	25	M_0
$\Gamma_{15} \to \Gamma'_{25}$	28.6	M_3
$\Lambda_3 \to \Lambda_1$	18.6	M_3

LiCl

As LiCl is very hygroscopic, it has been studied by only a few authors. Its absorption spectrum at 80 K[4] has received a tentative interpretation[49] and has been re-examined at 10 K.[29] The reflectance spectrum up to 35 eV of crystals cleaved in vacuum[50] is reproduced in Fig. 2. A comparison of these results with the band-energy calculations of Kunz[51] shows that the Γ exciton is at 8.7 eV and the X (or L) exciton at 9.5 eV.

The peaks of reflectance at about 15 eV are attributed to transitions occurring principally at the zone boundaries. The plasma oscillation is located at 17.2 eV, in agreement with the electron energy-loss spectrum recorded by Creuzburg.[52] The excitation of the 2s electrons of Cl^- occurs immediately after this energy at about 20 eV.

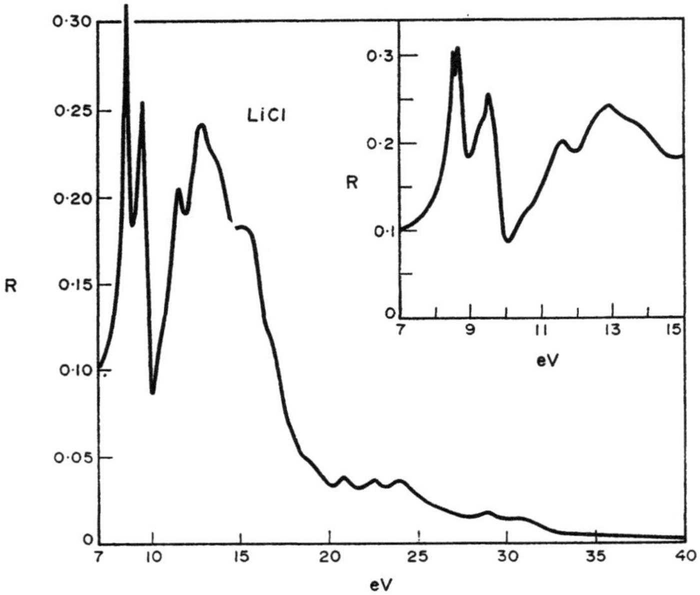

FIG. 2. Near-normal reflectance spectrum of a single crystal of LiCl cleaved in vacuum for $T = 293$ K.[50]

LiBr

The absorption spectra of thin films are reported in refs. 4, 11, 20, 28, and 29, and a partial interpretation has been given.[29, 49, 53] It is remarkable that only Γ excitons are observed experimentally. Kunz[53] has calculated the band structure by the orthogonalized plane-wave method (OPW).

LiI

This compound is difficult to prepare and, in spite of some work both on absorption[24, 54, 55] and on reflection spectra,[2, 55] certain characteristic parameters, such as the separation of the excitonic Γ doublet, so far have not been determined. As in the case of LiBr, the spectrum does not exhibit definitively the presence of a second exciton. This is in agreement with the theory of Kunz *et al.*[48] on the evolution of L excitons in lithium halides.

2.2. SODIUM HALIDES

Being less hygroscopic than the preceding compounds (except LiF), these have been widely studied.

NaF

The first exciton band is situated between 9 and 11.5 eV and has been studied by Onaka *et al.*,[39] Fabre,[42, 56–58] Eby *et al.*;[4] Sano (see ref. 59) has clearly observed the splitting of this band at 78 K. Metzger[28] extended the absorption curve from 10 to 22 eV. The reflectance spectrum[50] between 12 and 30 eV consists of three principal bands centred at 16, 20, and 23.5 eV respectively. These bands are not very strong, and the conditions necessary for obtaining a plasmon, according to Miyakawa[6] are not satisfied simultaneously. To our knowledge NaF is the only compound exhibiting non-collective resonances apart from those of excitons. Above 33 eV the existence of an exciton peak and a structure arising from the excitation of the $3p$ electrons of Na^+ may be discerned.

NaCl

As early as 1929, Forro[60] measured the position of the absorption edge of pure and doped crystals as a function of temperature. The absorption spectrum of thin films[2, 4, 20, 28, 29, 34, 61–63] and reflectances[18, 60, 64–72] have been the object of numerous researches. The most recent work is that of Roessler and Walker,[71] who have calculated the complex dielectric constant from reflectance measurements. Their results agree with those of the group at Rennes[72] who extended the data up to 35 eV. The band structure calculation has been undertaken by various authors;[51, 73, 74] the discrepancies between the various estimates have been explained by Kunz *et al.*[75] Electron energy-loss experiments compare favourably with the optical data, and the plasmon of valence electrons is located at 16 eV.

NaBr

The thin-film absorption spectrum may be found in refs. 2, 4, 20, 29, 62, 76, and 78. As in the case of other bromides (see Table 1), the first exciton is split and the doublet separation is 0.52 eV at room temperature. The band-energy calculation has been made by Kunz who assigned an X origin to the second exciton band at 8.5 eV.[53] Electron energy-losses[52] show a plasmon at 14.3 eV.

NaI

The absorptances[2, 4, 20, 24, 28, 29, 54, 55, 62, 77, 78] and reflectances[55] of thin films and of single crystals have been measured at various temperatures, and the spectra up to 12 eV have been tentatively discussed by Taft and Philipp[55] and by Phillips.[49] However, two peaks (at 8.25 and 9.15 eV)—attributed by the latter author to L excitons—could also be due to X excitons in the light of the band structure calculated by Kunz.[79]

2.3. POTASSIUM HALIDES

With these start the heavy halides which have spectra generally better studied and understood than those preceding.

KF

The absorption spectrum may be found in the general articles of refs. 4, and 28. The reflectances of freshly cleaved single crystals were measured first by Roessler[(80)] and those of thin films by Stephan and Robin.[(81)] These incomplete and imprecise data have been improved by Miyata and Tomiki[(59)] in the exciton region and by Stephan and Robin[(82)] from 7 to 45 eV. The latter authors have operated with crystals cleaved *in vacuo* since KF is extremely hygroscopic. The band structure of KF has also been calculated by Kunz *et al.*[(83)] The electronic resonance spectrum is made remarkable by the absence of plasma oscillation of the valence electrons and by the presence of a repeated structure due to the identical nature of the selection rules for transitions arising from the 3*p* K^+ and 2*p* F^- bands. This repeated structure is illustrated in Table 3 and Fig. 3, showing the dielectric constants ε_1, and ε_2 from 5 to 45 eV. The longitudinal resonance spectrum shows a peak at about 31 eV, characteristic of a collective mode, and due to a polarization wave attached to the 3*p* K^+ electrons.

TABLE 3. ENERGY POSITIONS (IN eV) OF THE PEAKS AND DISCONTINUITIES OF ε_2 FOR KF (SEE FIG. 3)

2*p* F^- (valence) electron	A 9.61	B 12.00	C 13.10	D 13.75	E 16.70	F 18.70
3*p* K^+ electrons	A′ 19.75	B′ 21.75	C′ 22.75	D′ 23.5	E′ 26.4	F′ 28.5
Energy differences between corresponding maxima	A′-A 10.14	B′-B 9.75	C′-C 9.65	D′-D 9.75	E′-E 9.7	F′-F 9.80
Interpretation	Γ excitons	X excitons	$L_3' \rightarrow L_1$	$X_5' \rightarrow X_3$	$X_4' \rightarrow X_1$	$\Gamma_{15} \rightarrow \Gamma_{25}'$

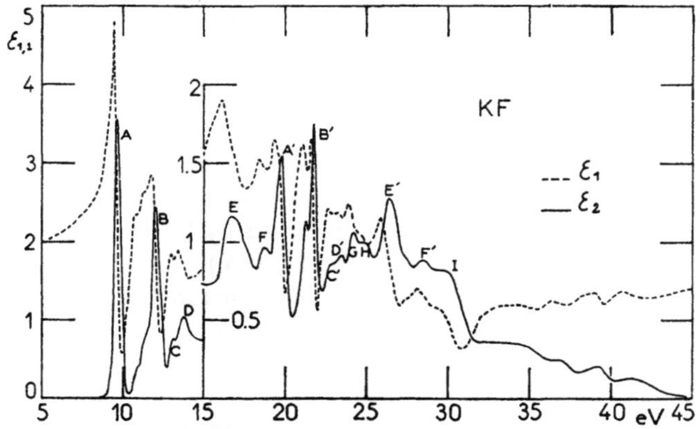

FIG. 3. Dielectric parameters for a single crystal of KF.[(82)]

KCl

This is probably the most studied of all alkali halides, and as early as 1928 Pfund[64] discovered a metallic reflection band at 1500 Å. Since then, the absorptance of thin films has been measured by several authors,[2, 4, 11, 28, 29, 34, 60, 62, 65] even down to liquid helium temperature.[66, 68] Mollwo[84] has followed the evolution of the absorption edge with temperature. The reflectance has been measured below 12 eV,[30, 66, 67, 68, 85–88] and Tomiki[89, 90] has studied its variation with

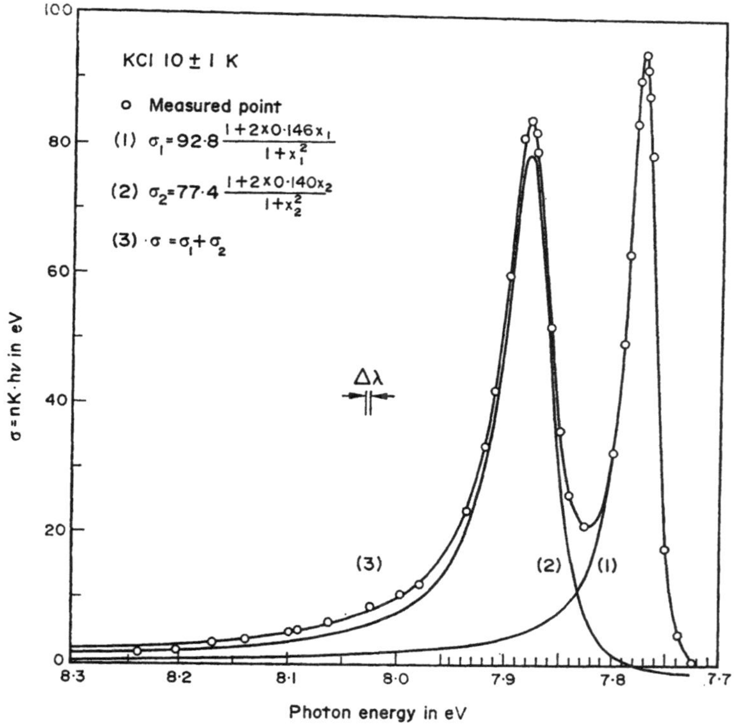

FIG. 4. The conductivity spectrum of the first exciton band, and its representation by the combination of two asymmetric Lorentzian functions. $T = 10$ K.[90]

temperature; he has verified Urbach's law for the absorption edge and has also demonstrated that the first exciton doublet results from the superposition of two assymetric Lorentzian bands as predicted by Toyozawa.[91] This is illustrated by Fig. 4 which shows the conductivity σ in the 7.7-8.3 eV region; here σ is defined by Tomiki as $\sigma = nkE = \varepsilon_2 E/2$, E being the photon energy in eV. The exciton spectrum of KCl has also been discussed by Onodera *et al.*[92] and Miyakawa *et al.*[93] Roessler and Walker[71] have deduced the complex dielectric constant from the reflectance measurements between 5 and 27 eV at room temperature and between 5 and 11.4 eV at 77 K. However, their interpretation, based upon the first band structure calculations,[94, 95] has been questioned by Kunz.[51] Stephan *et al.*

have recently measured the reflectance of crystals, freshly cleaved *in vacuo*.[88] With the aid of a K.-K. transform, they have deduced the optical parameters and the loss function; the latter is represented in Fig. 5, and, for purpose of comparison, the characteristic energy-loss spectrum from Creuzburg[52] is also given. The first peak corresponds to a Γ-point exciton and the second to a X-point exciton. Two analogous excitons are found again at about 20 eV, in the resonance spectrum, due to $3p$ K^+ electrons, which permits the determination of

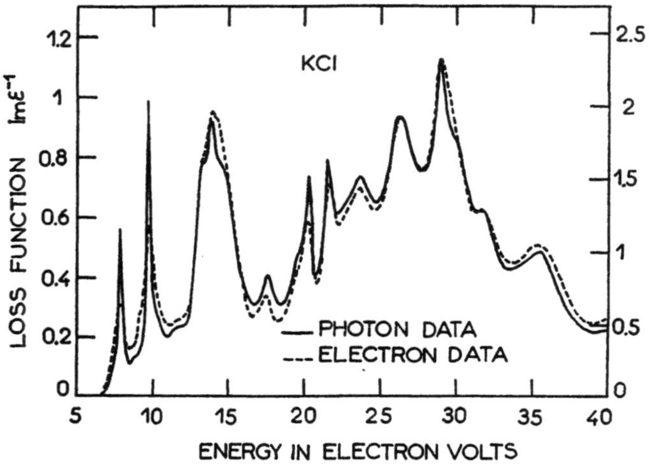

FIG. 5. Longitudinal spectra of KCl: left scale refers to photon data[88], right scale to electron data.[52]

certain band parameters.[88] The collective longitudinal excitation of valence electrons occurs at about 14 eV immediately after an M_2 discontinuity, in agreement with Miyakawa's theory.[6] At about 29 eV (Fig. 5) a second longitudinal oscillation takes place similar to that at about 31 eV in KF. The transverse and longitudinal optical absorption spectra have been studied by Ejiri[96] and confirm the other results.

KBr

Several authors have studied the absorption spectrum[2, 4, 11, 20, 28, 60, 77, 83, 90–93, 96, 98] and the reflectance[30, 31, 86, 101, 102] up to 45 eV.[102] The electronic resonance spectrum resembles that of KCl. Martienssen[97] has verified the Urbach rule and Frohlich *et al.*[99, 100] have made two-photon absorption measurements. The loss function calculated from the dielectric constant[102] is in good agreement with the results obtained from the characteristic energy losses of electrons. Kunz[53] has calculated the band structure by the OPW method without spin-orbit interaction, and the spectrum is partially interpreted.[29–31, 49, 102–3]

KI

Description of the absorption by thin films of potassium iodides can be found in references 2, 24, 25, 77, 84, 104-7. Haupt[105] has studied the displacement of the first exciton as a function of the temperature. Watanabe *et al.*,[86] Roessler and Walker,[47] Philipp and Ehrenreich,[30] and Baldini and Bosacchi[31] have studied the reflectance of cleaved single crystals, the last-named authors down to a temperature of 55 K.[31] Watanabe and Kato[87] have studied the reflectance of thin films. The optical parameters can be found in refs. 19, 30, 32, and 47. These studies are limited to 30 eV. Onodera *et al.*[108, 109] have given an interpretation based upon

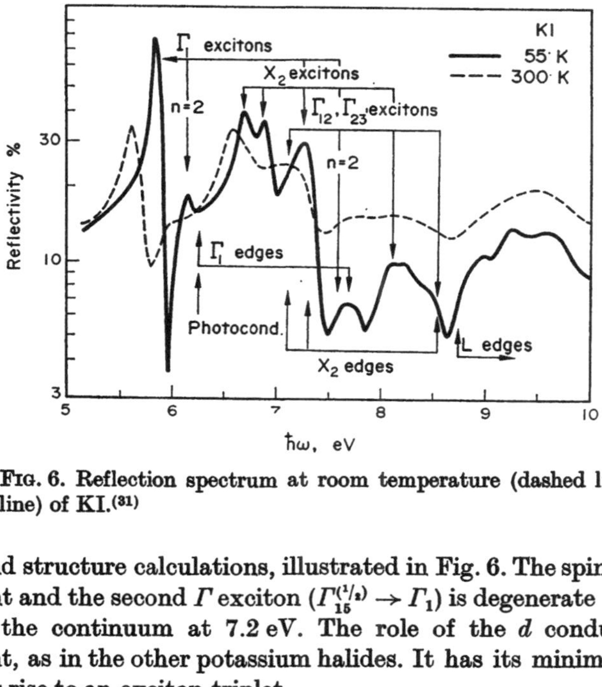

FIG. 6. Reflection spectrum at room temperature (dashed line) and at 55 K (full line) of KI.[31]

their band structure calculations, illustrated in Fig. 6. The spin-orbit interaction is important and the second Γ exciton ($\Gamma_{15}^{(1/2)} \rightarrow \Gamma_1$) is degenerate with the transition towards the continuum at 7.2 eV. The role of the *d* conduction band is also important, as in the other potassium halides. It has its minimum at the *X* point and gives rise to an exciton triplet.

Electronic excitations in potassium halides

Figure 7 shows the band structure of KCl[51] in the Λ and the Δ directions. This diagram may serve equally as a model for other potassium halides and rubidium halides. In Table 4 are collected certain band parameters for KF, KCl, and

TABLE 4. BAND PARAMETERS FOR KF, KCl, AND KBr AT $T = 300$ K (eV)

Exciton $\Gamma_{15(3/2)} \rightarrow \Gamma_1$	Exciton $X_5' \rightarrow X_3$	Exciton $\Gamma_{15} \rightarrow \Gamma_1$ 3*p* K+ hole	Exciton $X_5' \rightarrow X_3$ 3*p* K+ hole	Intervals between 3*p* K+ and *np* X		Γ_{15}-X_5' valence band		Γ_1-X_3 conduction band	
				At Γ	At X	Exp.	Theor.	Exp.	Theor.
KF /9.61	12	19.75	21.75	10.14	9.75	0.4	0.12	2	3.82
KCl /7.65	9.4	20.1	21.3	12.45	11.9	0.55	0.13	1.2	2.25
KBr/6.56	8.5	19.85	21.15	13.29	12.65	0.64	0.1	1.3	1.7

KBr, deduced from the repeated structure of the Γ and X excitons, and assuming that the binding energies of all these excitons are nearly equal. (This has also an advantage of small energy separation between Γ and X for the $3p$ K^+ band.)

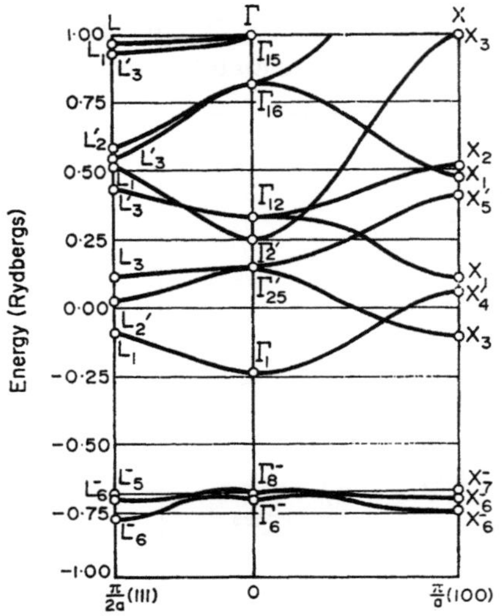

Fig. 7. The band structure of KCl. Calculated points are shown by circles. The bands have been drawn using the compatibility relations and the results of other calculations.[51]

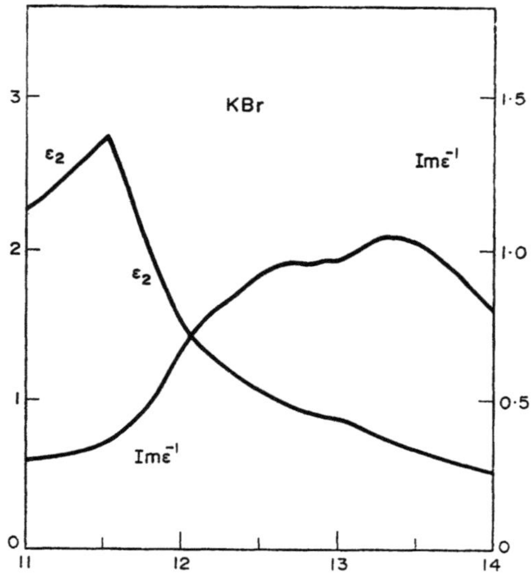

Fig. 8. ε_2 and Im ε^{-1} spectra in the plasmon region.[102]

The differences between theoretical and experimental values show the importance of the different binding energies of the various excitons.

Three theoretical results obtained by Miyakawa[6] may be illustrated with the aid of the compounds in Table 4.

(a) There is a relationship between longitudinal polarization waves and interband transitions: the theoretical plasmons (counting six valence free electrons per molecule) at 11.6 eV for KCl and 10.7 eV for KBr are located experimentally at 14.1 and 13.3 eV. These polarization waves resemble longitudinal interband transitions with a weak damping.

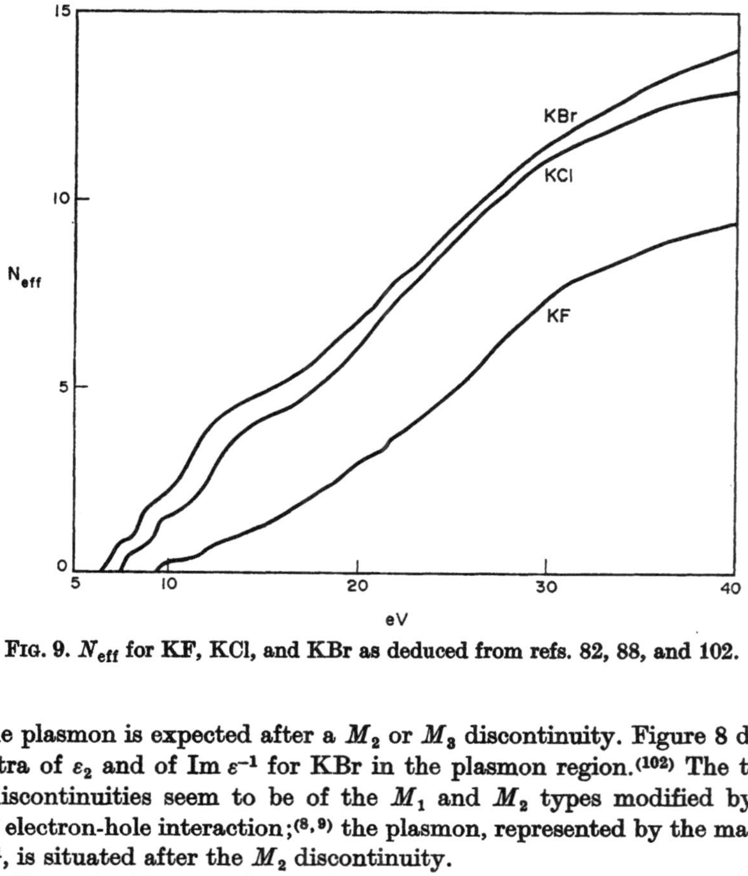

FIG. 9. N_{eff} for KF, KCl, and KBr as deduced from refs. 82, 88, and 102.

(b) The plasmon is expected after a M_2 or M_3 discontinuity. Figure 8 displays the spectra of ε_2 and of Im ε^{-1} for KBr in the plasmon region.[102] The two observed discontinuities seem to be of the M_1 and M_2 types modified by the g factor of electron-hole interaction;[8, 9] the plasmon, represented by the maximum of Im ε^{-1}, is situated after the M_2 discontinuity.

(c) The discontinuities preceding a plasmon must have a large oscillator strength. In Fig. 9, N_{eff} is represented for KF, KCl, and KBr. It can be seen that below 20 eV for the fluoride, the oscillator strengths are approximately three times weaker than those for the other compounds. As a consequence, potassium fluoride does not possess a plasmon for its valence electrons.

2.4. RUBIDIUM HALIDES

The band structures of potassium and rubidium halides resemble each other very closely, as do their electronic resonance spectra.

RbF

No measurement of reflectance has yet been made, but the absorption spectrum may be found in the papers of Schneider and O'Bryan,[20] of Eby *et al.*[4] (from 9 to 11.4 eV), Metzger[28] (from 12 to 21 eV), and Teegarden and Baldini[29] (from 9 to 12 eV at 10 K). The forbidden band gap is about 10 eV; as in the case of NaF and KF, the first exciton is well separated from the interband transition continuum. The energy loss spectrum of electrons[52] shows no evidence of a valence plasmon.

RbCl

Forro[60] was the first to determine the position of the absorption edge of RbCl containing 10^{-4} part of terbium impurity. The absorption spectra of thin films are found in references 2, 4, 19, 20, 28, 29, 54, and 62. Baldini and Bosacchi[31] measured the reflectance of single crystals at 55 K and at room temperature. The spectrum and the variation of n[32] are analogous to those for KCl; between 7 and 10 eV RbCl exhibits the usual Γ- and X-point excitons. The spin-orbit parameters have been evaluated by Kunz.[79]

RbBr

The absorption spectra of thin films have been studied at different temperatures;[2, 4, 21, 28, 29, 62, 97, 110] the reflectance spectrum at 55 K, from 6 to 10eV[31] has given rise to an interpretation which is in agreement with the calculations of Kunz and which is identical to that for KBr.

RbI

Rudibium iodide has been the object of numerous studies, both in absorption[2, 4, 20, 29, 54, 55, 107, 140] and in reflection.[31, 47, 55, 87, 111] Roessler and Walker[47] give the reflectance and the deduced dielectric constant of freshly cleaved

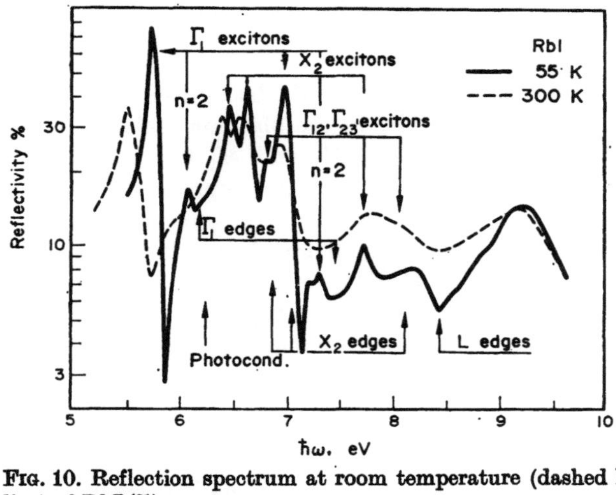

FIG. 10. Reflection spectrum at room temperature (dashed line) and at 55 K (full line) of RbI.[31]

crystals between 5 and 11.3 eV and also calculate the binding energy of the exciton using the hydrogenic model; they find a value of 0.5 eV. Figure 10 represents the reflection spectrum as well as the assignments of the observed discontinuities between 5 and 10 eV.[31] There are striking similarities with the reflectance spectrum of KI.

2.5. CAESIUM HALIDES

Under ordinary conditions, all the caesium halides have the crystalline structure of the CsCl type (b.c.c.) except for CsF which has the same structure as NaCl (f.c.c.). The optical studies are comparatively few. The absorption spectra are given in the previously cited references 2, 4, 20, 28, 54, and 62, and the absorption edge position of the pure and coloured halides in refs. 112, 113, and 115. Gavini and Cardona[114] have measured the reflectance of thin films of CsCl and CsBr between 5 and 8 eV. However, evaporated thin films on LiF or on glass, have f.c.c. and b.c.c. structures respectively, and hence the spectra are different. The spectrum of CsI, studied by Taft and Philipp[55] has been interpreted by Onodera[116] in the exciton region. The caesium halide spectra have also been discussed in refs. 29, 52, 117-21.

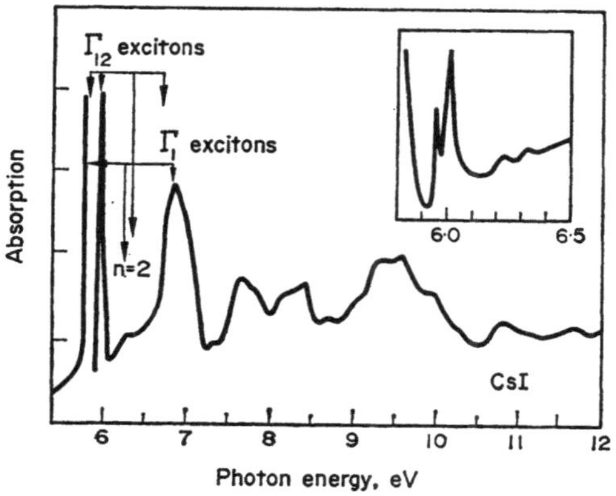

FIG. 11. Absorption spectrum of thin CsI film at 10 K from Teegarden and Baldini.[29] The inset shows the double-step absorption observed at 14 K by Fisher and Hilsch[54] (from ref. 116).

It is interesting to note that in b.c.c. structure the d levels are lowered considerably, and the Γ_{12} excitons, which are hardly seen in the alkali iodine spectra (Fig. 6), are as important here as the Γ_1 excitons (Fig. 11). Onodera[116] has also shown that optical excitation to the exciton triplet state of $\Gamma_{15'}$ resulting from the $\Gamma_8^- - \Gamma_8^-$ transition, becomes allowed, due to strong coupling with the singlet state; this occurs at 5.92 eV (Fig. 11).

3. Solid solutions of alkali halides

Numerous experimental studies have been made on solid solutions or on mixed crystals of alkali halides; to a first approximation, the substitution of the halide, in such systems, produces a modification of the valence band while the substitution of the alkali affects the lowest conduction band. For example, in the KBr-NaBr system[134] one observes simply a displacement represented in Fig. 12 and a slight broadening of the first exciton bands with increased concentration of the alkali. In contrast, in the KCl-KBr mixed crystals[123] the exciton bands of Cl^-

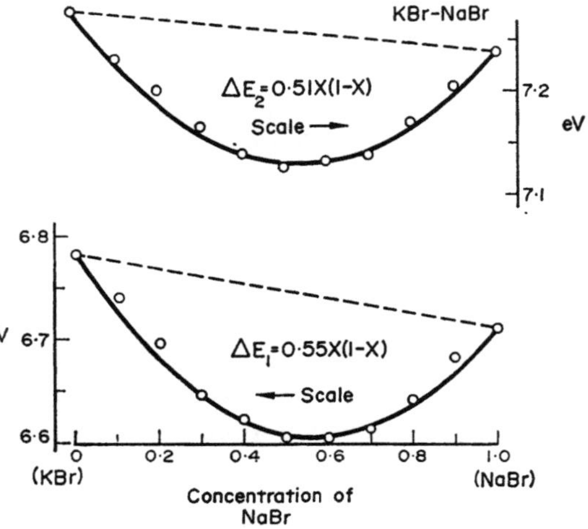

FIG. 12. Positions of the doublet peak (excitons Γ) at 80 K in the lowest energy region of the fundamental absorption of KBr-NaBr solid solutions are plotted as functions of concentration. The circles indicate the observed values of the peak positions, and the solid curves are drawn using the equations shown in the figure; ΔE means the deviation of the peak position from the straight line connecting the peak positions of pure materials, and X is the concentration of NaBr in the samples (from ref. 134).

and Br^- may be identified separately; their intensities vary with the concentration and their positions with the lattice constant. These systems are called "persistence type" by Onodera and Toyozawa[92] while the previous ones are "amalgamation type". A theory of solid solutions has been put forward by these authors.

The compounds KI-KCl,[123-6] KBr-KCl,[68, 86, 123, 127] KI-NaI or RbI,[126, 128] KI-KF,[23, 129] RbCl-1% RbI,[125] and NaCl-NaBr[86] have been generally studied at room temperature or at 80 K in the form of evaporated films. The reflectances of single crystals KBr-KCl have been measured between 5-25 eV.[86] The alkali-substituted crystals studied[86, 122, 130-3] are, in general, the chlorides or bromides of Na, K, and Rb. Murata[134] discusses his results on KBr-RbBr, KBr-NaBr, and KBr-KCl and finds a good agreement with theory.[92]

4. Alkaline earth halides

The work of Tubbs,[135-7] Fesefeldt,[21] and Schneider and O'Bryan[20] concerns the thin-film absorption of different metallic halides. Of these, the alkaline earth fluorides are ionic insulators with the largest band gap.

MgF_2

Magnesium fluoride is an uniaxial crystal having the rutile structure. The position of its absorption edge has been known for long time.[20] In the form of thin films, MgF_2 has been studied by Fabre,[42, 56–58, 138] who calculated its optical constants by an original method. Reflectances of thin films and single crystals and, thereby, the optical constants, have also been determined.[139–41] However, there remains the problem of evaluating independently the ordinary and extraordinary complex optical constants with the help of the method developed recently[218] (see Chapter 1). As in the case of NaF, the spectrum of MgF_2 contains only a single exciton band, but there is a plasma oscillation at 25 eV conforming well with Horie's formula.[139, 142]

CaF_2

For a long time, fluorite (CaF_2) has been the dispersive material most frequently used in the v.u.v., since pure natural crystals are easily available. In 1936 Tousey[143] calculated the optical constants from 1605 to 927 Å, by applying the Fresnel formulae to the reflectance of polished natural single crystals measured by photographic photometry. Various authors[20, 42, 59, 144–53] have studied the spectrum generally up to 13 eV. The recent work of Stephan *et al.*[141] extends from 10 to 45 eV and shows an interesting reflection band at about 30 eV arising from the excitation of $3p$ Ca^{++} electrons.

SrF_2

The absorption edge position of slightly coloured SrF_2 crystals, measured by Chernevskaya,[150] is situated at 1350 Å, while S. Robin *et al.*[151] find it at 1298 Å for pure crystals at room temperature. The latter authors have also studied the reflectance of single crystals and the absorptance of thin films up to 14 eV. Tomiki and Miyata[154] determined the reflectance from 9 to 13.4 eV, giving a partial interpretation based on the transitions at Γ and X points of the energy schema proposed by Onodera.[154]

BaF_2

Various authors[42, 46, 115, 138, 150–2, 155–7] have studied the optical properties of BaF_2 up to about 14 eV, but the most recent work has been on the reflectance (9-11.8 eV) of single crystals at 78 K.[59] As in the case of SrF_2, one observes splitting of the first exciton band into five components, attributed by Miyata and Tomiki[59] to two Γ excitons (Γ_8^-- and $\Gamma_6^- \rightarrow \Gamma_1$) and three X excitons (X_6^- and $X_7^- \rightarrow X_3$ and $X_2' \rightarrow X_3$).

5. Oxides and other compounds

5.1. ALKALINE EARTH OXIDES

BeO

Beryllium oxide has the Wurtzite structure and is uniaxial. O'Sullivan[158, 159] has calculated its band structure while the position of its reflection edge has been determined by Doyle,[160] and its reflectance up to 15 eV by Loh.[161] However, the most complete work is that of Roessler *et al.*[162, 163] who give the dielectric constant—calculated from the reflectance measurements—up to 20 eV at room temperature and to 11 eV at low temperature. They notice a doubling of the exciton peak, attributed to an exciton-phonon interaction quite different from that in the alkali halides; this difference arises partly from the very dissimilar dimensions of the excitons in the two cases (30 Å for the oxide but 3 Å in alkali halides).

MgO

The optical constants and the electronic spectrum may be found in numerous papers.[47, 161, 163–76] The most complete experimental spectra are those in refs. 173 and 174, where a very detailed interpretation and a comparison between the experimental and theoretical ε_2 is given.

CaO, SrO, BaO

The reflectance spectrum of single crystals of CaO has been studied by Whited and Walker,[177] but the results do not agree with the Toyozawa and Hermanson[178] hypothesis of an exciton-phonon particle. The absorptance of CaO thin films has been measured between 5.5 and 11.3 eV.[179]

The absorptance of SrO thin films has been measured by Zollweg,[180] Saum and Hensley,[181] and Sproull;[182] BaO is treated in refs. 180, 181, 183-6.

5.2. OTHER COMPOUNDS

In general, the oxides are semi-ionic, that is to say their chemical bands are partially covalent. Consequently, it is difficult to embrace them in the same category since some are excellent insulators (SiO, SiO_2, GeO_2, Se_2O_3) and others good conductors (ReO_3,[187]). Absorptance measurements from the visible to 7 eV on the alkali oxides Li_2O, Na_2O, K_2O, Rb_2O, and Cs_2O, demonstrate that the forbidden band-gap increases as the alkali atom becomes lighter. The other main ionic compounds studied are several iodides, chlorides, bromides, and fluorides; in each the absorption spectrum begins with an exciton band. In all cases where the measurements have been extended to sufficiently high energies, a plasmon is found. Table 5 gives the references for the different compounds.

TABLE 5

SiO	188, 189
SiO_2	146, 147, 153, 161, 190-8
GeO_2	199-202
Al_2O_3	146, 147, 153, 161, 203-7
Sc_2O_3	208, 209
Li_2O, Na_2O, K_2O, Rb_2O, Cs_2O	211
CdI_2, PbI_2	135-7
CdF_2	152-212
Pb, Tl, Cd, Ni and Ca iodides, bromides, chlorides; AgCl, InCl, ZnI_2, $BaBr_2$	21
MnF_2, ZnF_2	213
$MgCl_2$, $BaCl_2$, $MgBr_2$, $ZnCl_2$, CuCl, AgCl	20
Ca, Cd, Fe, Mn, Pb, Zn, Bi, Sb iodides	137
SnO, SnO_2, PbO, BeO, ZnO_2, Ta_2O_5, As_2O_3, Sb_2O_3, Bi_2O_3	160, 210
Ba and Sr oxides, sulphides, selenides, and tellurides	180
BaO, SrO, BaS, SrS, CaS, BaSe, SrSe, CaSe, MgSe, BaTe, SrTe, CaTe, MgTe	181

6. Conclusion

In enumerating the different references on ionic insulators it has not been possible for us to do justice to all the theoretical work on the subject and, in particular, to papers discussing the existence of hyperbolic excitons (refs. 248-52).

Moreover, the resonance electronic spectra observed raise unsolved problems. In the case of the alkali halides, for example, the exciton region has been interpreted but the complexity of the bands still makes it difficult to assign the discontinuities to various transitions. The valence plasmons have now been observed but the longitudinal resonances at about 30 eV in potassium halides still need clarification. The ionic insulators involve the problem of a bound Fermion gas in which the electron and the hole have a strong correlation. Thus the study of these materials in the extreme ultraviolet continues to have a great interest to experimenter and theorist alike.

References

1. SEITZ, F., *Modern Theory of Solids*, McGraw-Hill, New York, 1940, ch. 1.
2. HILSCH, R., and POHL, R. W., *Z. Phys.* **57**, 145 (1929); **59**, 812 (1930); **48**, 384 (1928).
3. SEITZ, F., *Rev. Mod. Phys.* **26**, 7 (1954).
4. EBY, J. F., TEEGARDEN, K. J., and DUTTON, D. B., *Phys. Rev.* **116**, 1099 (1959).
5. KNOX, R. S., Theory of Exciton, *Solid St. Phys.*, Suppl. **5** (1963).
6. MIYAKAWA, T., *J. Phys. Soc. Japan* **24**, 768 (1968).
7. PHILLIPS, J. C., *Solid St. Phys.*, **18**, 55 (1966).
8. VELICKY, B., and SAK, J., *Phys. Status Solidi* **16**, 147 (1966).
9. KUNDRAT, V., *Czech. J. Phys.* B, 18, 1003 (1968).

10. Wilson, C. B., *Proc. Phys. Soc. London* **80**, 481 (1960).
11. Duckett, S. W., and Metzger, D. H., *Phys. Rev.* **137** A, 953 (1965).
12. Sagawa, T., *et al.*, *J. Phys. Soc. Japan* **21**, 2587 (1966).
13. Miyakawa, T., *J. Phys. Soc. Japan* **17**, 1898 (1962).
14. Hermanson, J., *Phys. Rev.* **177**, 1234 (1969); **166**, 893 (1968).
15. Spicer, W. E., *Phys. Rev.* **154**, 385 (1967).
16. Ziman, J. M., *Electrons and Phonons*, Clarendon Press, Oxford, 1962, ch. 4.
17. Roessler, D. M., and Walker, W. C., *J. Phys. Chem. Solids* **28**, 1507 (1967).
18. Smith, A., *Phys. Rev.* **44**, 520 (1933).
19. Kublitzky, A., *Ann. Phys.* **20**, 793 (1934).
20. Schneider, E. G., and O'Bryan, H. M., *Phys. Rev.* **51**, 293 (1937).
21. Fesefeldt, H., *Z. Phys.* **64**, 623 (1930); **64**, 741 (1930).
22. Pohl, R. W., *Z. Phys.* **39**, 36 (1938).
23. Fisher, F., *Z. Phys.* **139**, 328 (1954).
24. Martienssen, W., *J. Phys. Chem. Solids* **2**, 257 (1957).
25. Teegarden, K. J., *Phys. Rev.* **105**, 1222 (1957).
26. Philipp, H. R., and Taft, E. A., *Phys. Rev.* **106**, 671 (1957); *J. Phys. Chem. Solids* **1**, 159 (1956).
27. Knox, R. S., and Inchauspé, N., *Phys. Rev.* **116**, 1093 (1959).
28. Metzger, P. H., *J. Phys. Chem. Solids* **26**, 1879 (1965).
29. Teegarden, K. J., and Baldini, G., *Phys. Rev.* **155**, 896 (1967).
30. Philipp, H. R., and Ehrenreich, H., *Phys. Rev.* **131**, 2016 (1963).
31. Baldini, G., and Bosacchi, B., *Phys. Rev.* **116**, 863 (1968).
32. Gyulai, Z., *Z. Phys.* **46**, 80 (1928).
33. Powell, W. M., *Phys. Rev.* **45**, 154 (1934).
34. Schneider, E. G., *Phys. Rev.* **45**, 152 (1934); **49**, 341 (1936).
35. Ewing, D. H., and Seitz, F., *Phys. Rev.* **50**, 760 (1936).
36. Kato, R., Nakamura, K., Matsui, A., and Uchida, Y., *J. Phys. Soc. Japan* **15**, 2105 (1960).
37. Kato, R., *J. Phys. Soc. Japan* **16**, 1476 (1961); **16**, 2525 (1961).
38. Walker, W. C., *J. Opt. Soc. Am.* **52**, 223 (1962).
39. Onaka, R., Fukuda, A., and Ejiri, A., *J. Phys. Soc. Japan* **16**, 340 (1961).
40. Uchida, Y., Kato, R., and Matsui, E., *J. Quant. Spectrosc. Radiative Transfer* **2**, 589 (1962).
41. Robin, S., *Rev. Opt.* **33**, 193 (1954).
42. Fabre, D., *Rev. Opt.* **43**, 394 (1964); and **43**, 504 (1964).
43. Milgram, A., and Givens, M. P., *Phys. Rev.* **125**, 1506 (1962).
44. Lynch, D. W., *Solid State Commun.* **4**, 159 (1966).
45. Lemonnier, J. C., Stephan, G., and Robin, S., *Compt. Rend.* **261**, 2463 (1965).
46. Davis, R. J., *J. Opt. Soc. Am.* **56**, 837 (1966).
47. Roessler, D. M., and Walker, W. C., *J. Opt. Soc. Am.* **57**, 835 (1967); **57**, 677 (1967).
48. Kunz, A. B., Miyakawa, T., and Oyama, S., *Phys. Status Solidi* **34**, 581 (1969).
49. Phillips, J. C., *Phys. Rev.* **136** A, 1705 (1964).
50. Stephan, G., and Robin, S., *J. Phys.* **32**, Supp. C4, 246 (1971).
51. Kunz, A. B., *Phys. Rev.* **175**, 1147 (1968).
52. Creuzburg, M., *Z. Phys.* **196**, 433 (1966).
53. Kunz, A. B., *Phys. Status Solidi* **29**, 115 (1968).
54. Fisher, F., and Hilsch, R., *Nachr. Akad. Wiss. Göttingen, Math.-Phys. Kl.* 8, 241 (1959).
55. Taft, E. A., and Philipp, H. R., *J. Phys. Chem. Solids* **3**, 1 (1957).
56. Fabre, D., Romand, J., and Vodar, B., *Optica Acta* **9**, 73 (1962).
57. Fabre, D., and Kieffer, J., *Compt. Rend.* **257**, 2444 (1963).
58. Fabre, D., Romand, J., and Vodar, B., *J. Phys.* **25**, 55 (1964).
59. Miyata, T., and Tomiki, T., *J. Phys. Soc. Japan* **24**, 954 (1968); **24**, 1408 (1968).

60. Forro, M., *Z. Phys.* **56**, 534 (1929).
61. O'Bryan, H. M., *Phys. Rev.* **57**, 995 (1940).
62. Saito, H., Saito, S., Onaka, R., and Ikeo, B., *J. Phys. Soc. Japan* **24**, 1095 (1968).
63. Uzan, E., Roncin, J.-Y., Chandrasekharan, V., and Damany-Astoin, N., *Compt. Rend.* **267** B, 708 (1968).
64. Pfund, A. H., *Phys. Rev.* **32**, 39 (1928).
65. Smith, A., *Phys. Rev.* **39**, 1013 (1932); **44**, 520 (1933).
66. Hartmann, P. L., Nelson, J. R., and Siegfried, J. G., *Phys. Rev.* **105**, 123 (1957).
67. Taylor, J. N., and Hartmann, P. L., *Phys. Rev.* **113**, 1421 (1959).
68. Hartmann, P. L., *J. Quant. Spectrosc. Radiative Transfer* **2**, 579 (1962).
69. Miyata, T., Sano, R., and Tomiki, T., *J. Phys. Soc. Japan* **19**, 2233 (1964).
70. Miyata, T., and Tomiki, T., *J. Phys. Soc. Japan* **22**, 209 (1967); **24**, 1286 (1968).
71. Roessler, D. M., and Walker, W. C., *Phys. Rev.* **166**, 599 (1968).
72. Stephan, G., Thesis, Rennes, 1970.
73. Fong, C. Y., and Cohen, M. L., *Phys. Rev. Letters* **21**, 22 (1968).
74. Clark, T. D., and Kliever, K. L., *Phys. Letters* **27** A, 167 (1968).
75. Kunz, A. B., Fowler, W. B., and Schneider, P. M., *Phys. Letters* **28**, 553 (1969).
76. Bauer, G., *Annln Physik* **19**, 434 (1934).
77. Fujita, G., Gähviller, C., Brown, F. C., *Phys. Rev. Letters* **22**, 1369 (1969).
78. Haensel, R., Kunz, C., and Sonntag, B., *Phys. Rev. Letters* **20**, 501 (1968).
79. Kunz, A. B., *Phys. Rev.* **151**, 620 (1966); **159**, 738 (1967).
80. Roessler, D. M., and Lempka, H. J., *Br. J. Appl. Phys.* **17**, 1553 (1966).
81. Stephan, G., and Robin, S., *Solid State Commun.* **5**, 883 (1967).
82. Stephan, G., and Robin, S., *Compt. Rend.* **267**, 1286 (1968).
83. Kunz, A. B., Miyakawa, T., and Fowler, W. B., *Mem. Soc. r. Sci. Liège* **20**, 263 (1970).
84. Mollwo, E., *Z. Phys.* **124**, 118 (1947).
85. Smith, A., *J. Opt. Soc. Am.* **50**, 862 (1960).
86. Watanabe, M., Nakamura, Y., Nakai, Y., and Murata, T., *J. Phys. Soc. Japan* **24**, 428 (1968); **26**, 1014 (1969).
87. Watanabe, M., and Kato, R., *Japan J. Appl. Phys.* **7**, 21 (1968).
88. Stephan, G., and Robin, S., *Optics Commun.* **1**, 40 (1969).
89. Kobayashi, K., and Tomiki, T., *J. Phys. Soc. Japan* **15**, 1982 (1960); **16**, 1417 (1961).
90. Tomiki, T., *J. Phys. Soc. Japan* **20**, 1518, (1965); **21**, 403 (1966); **22**, 463 (1967); **23**, 1280 (1967); **26**, 738 (1969).
91. Toyozawa, Y., *Prog. Theoret. Phys. Japan* **20**, 53 (1958); **27**, 89 (1962); see also *J. Phys. Chem. Solids* **25**, 59 (1964).
92. Onodera, Y., and Toyozawa, Y., *J. Phys. Soc. Japan* **22**, 833 (1967); **24**, 341 (1968).
93. Miyakawa, T., and Oyama, S., *J. Phys. Soc. Japan* **24**, 996 (1968).
94. de Cicco, P. D., *Phys. Rev.* **153**, 931 (1967).
95. Oyama, S., and Miyakawa, T., *J. Phys. Soc. Japan* **20**, 264 (1965); **21**, 868 (1966).
96. Ejiri, A., *J. Phys. Soc. Japan* **23**, 901 (1967); **24**, 1181 (1968).
97. Martienssen, W., *Nachr. Akad. Wiss. Göttingen, Math.-Phys. Kl.* **11** A, 257 (1955).
98. Huggett, G. R., and Teegarden, K. J., *Phys. Rev.* **141**, 797 (1966).
99. Fröhlich, D., and Staginus, B., *Phys. Rev. Letters* **19**, 496 (1967).
100. Fröhlich, D., Staginus, B., and Schonherr, E., *Phys. Rev. Letters* **19**, 1032 (1967).
101. Greenaway, D. L., and Nitsche, R., *J. Phys. Chem. Solids* **26**, 1445 (1965).
102. Stephan, G., Garignon, E., and Robin, S., *Compt. Rend.* B, **268**, 408 (1969).
103. Keil, P., *Z. Phys.* **214**, 266 (1968).
104. Apker, L., and Taft, E., *Phys. Rev.* **81**, 698 (1951).
105. Haupt, U., *Z. Phys.* **157**, 232 (1959).
106. Givens, M. P., and Teegarden, K. J. *J. Quant. Spectrosc. Radiative Transfer* **2**, 629 (1962).
107. Ramamurti, J. A., and Teegarden, K. J., *Phys. Rev.* **145**, 698 (1966).
108. Onodera, Y., Okasaki, M., and Inui, T., *J. Phys. Soc. Japan* **21**, 816 (1966); **11**, 2229 (1966).

109. ONODERA, Y., OKASAKI, M., *J. Phys. Soc. Japan* **21**, 1273 (1966).
110. SMAKULA, A., *Z. Phys.* **63**, 762 (1930).
111. WATANABE, M., and KATO, R., *J. Phys. Soc. Japan* **21**, 191 (1966).
112. AVAKIAN, P., and SMAKULA, A., *Phys. Rev.* **120**, 2007 (1960).
113. AVAKIAN, P., and SMAKULA, A., *J. Appl. Phys.* **31**, 1720 (1960).
114. GAVINI, A. A., and CARDONA, M., *Phys. Letters* **27** A, 112 (1968).
115. BALLARD, S. S., COMBES, L. S., and MACKARTHY, K. A., *J. Opt. Soc. Am.* **41**, 772 (1952); **42**, 684 (1952).
116. ONODERA, Y., *J. Phys. Soc. Japan* **25**, 469 (1968).
117. OVERHAUSER, A. W., *Phys. Rev.* **101**, 1702 (1956).
118. DEXTER, D. L., *Phys. Rev.* **83**, 435 (1951); **108**, 707 (1957).
119. FISHER, F., *Z. Phys.* **160**, 194 (1960).
120. KNOX, R. S., *Phys. Rev.* **133** A, 498 (1964).
121. PHILLIPS, J. C., *Phys. Rev.* **136** A, 1721 (1964).
122. NAKAI, Y., MURATA, T., and NAKAMURA, K., *Japan J. Appl. Phys.*, suppl. 1, **4**, 616 (1965).
123. MAHR, H., *Phys. Rev.* **122**, 1464 (1961); **125**, 1510 (1962).
124. NAKAMURA, K., FUKUDA, K., KATO, R., MATSUI, A., and UCHIDA, Y., *J. Phys. Soc. Japan* **16**, 1262 (1961).
125. BALDINI, G., and TEEGARDEN, K., *J. Phys. Chem. Solids* **27**, 943 (1966).
126. NAGASAWA, N., NAKAGAWA, H., and NAKAI, Y., *J. Phys. Soc. Japan*, **24**, 356 (1968); **22**, 667 (1967).
127. MURATA, T., and NAKAI, Y., *J. Phys. Soc. Japan* **23**, 904 (1967).
128. NAKAMURA, K., and NAKAI, Y., *Proceedings of the International Conference on Localized Excitations in Solids*, Plenum Press, New York, 1968, p. 651.
129. HILSCH, R., and MARTIENSSEN, W., *Nuovo Cim.* **7** (1), 480 (1958).
130. NAKAI, Y., MURATA, T., and NAKAMURA, K., *J. Phys. Soc. Japan* **18**, 1481 (1963).
131. NAKAMURA, K., *J. Phys. Soc. Japan* **22**, 511 (1967).
132. NAKAMURA, K., and NAKAI, Y., *J. Phys. Soc. Japan* **23**, 455 (1967).
133. MURATA, T., NAKAI, Y., OZAWA, K., and KUBO, K., *Phys. Letters* **26** A, 213 (1968).
134. MURATA, T., *J. Phys. Soc. Japan* **25**, 1632 (1968).
135. TUBBS, M. R., *Proc. Roy. Soc. London* A, **280**, 566 (1964).
136. TUBBS, M. R., and FORTY, A. J., *J. Phys. Chem. Solids* **26**, 711 (1965).
137. TUBBS, M. R., *J. Phys. Chem. Solids* **27**, 1667 (1966); **29**, 1191 (1968).
138. FABRE, D., and ROMAND, J., *Compt. Rend.* **250**, 1226 (1960); **256**, 5523 (1963).
139. STEPHAN, G., LE CALVEZ, Y., LEMONNIER, J. C., and ROBIN, S. *Compt. Rend.* **264**, 1667 (1967).
140. WILLIAM, M. W., MACRAE, R. A., and ARAKAWA, E. T., *J. Appl. Phys.* **38**, 1701 (1967).
141. STEPHAN, G., LE CALVEZ, Y., LEMONNIER, J. C., and ROBIN, S., *J. Phys. Chem. Solids* **30**, 601 (1969).
142. HORIE, C., *Progr. Theoret. Phys. Japan* **21**, 113 (1959).
143. TOUSEY, R., *Phys. Rev.* **50**, 1057 (1936); **57**, 1060 (1940).
144. O'BRYAN, H. M., *Phys. Rev.* **38**, 32 (1931).
145. POWELL, N. N., *Phys. Rev.* **45**, 154 (1934).
146. KLEVENS, H. B., and PLATT, J. R., *J. Am. Chem. Soc.* **69**, 3055 (1947).
147. BEAUPLÉ, R., GILLES, A., ROMAND, J., and VODAR, B., *J. Opt. Soc. Am.* **40**, 788 (1950).
148. KUNDSON, A. R., and KUPPERIAN, J. E., *J. Opt. Soc. Am.* **47**, 440 (1957).
149. GERASIMOVA, G., IVANOVA, M. K., and KULIKOV, S. A., *Fizika Zborn. Lvov* 146 (1958).
150. CHERNEVSKAYA, S. E. G., *Optika i Spektrosk.* **11**, 513 (1961) [*Optics Spectrosc.* **11**, 277 (1961)].
151. ROBIN-KANDARE, S., ROBIN, J., and QUEMA, Y., *Compt. Rend.* **259**, 549 (1964).
152. ROBIN-KANDARE, S., ROBIN, J., *Compt. Rend.* **262**, 1020 (1966); **262**, 1211 (1966).
153. STEPHAN, G., LEMONNIER, J. C., and ROBIN, S., *J. Opt. Soc. Am.* **57**, 486 (1967).
154. TOMIKI, T., and MIYATA, T., *J. Phys. Soc. Japan* **25**, 635 (1968).

155. Johannin-Gilles, A., and Vodar, B., *Rev. Opt.* **32**, 1 (1953).
156. Chubb, T. A., *J. Opt. Soc. Am.* **46**, 362 (1956).
157. Stepanov, J. V., Sinioukova, J. A., and Chernevskaya, *Optika i Spektrosk.* **4**, 272 (1958) (English translation *OTS* and *SLA*).
158. O'Sullivan, W. J., *Univ. Pennsylvania, Tech. Rep. TN* 58–474.
159. Glasser, M. L., *J. Phys. Chem. Solids* **10**, 229 (1959).
160. Doyle, W. P., and Clarke, F. J. P., *Br. J. Appl. Phys.* **12**, 574 (1961).
161. Loh, E., *Solid State Commun.* **2**, 269 (1964).
162. Walker, W. C., Roessler, D. M., and Loh, E., *Phys. Rev. Letters* **20**, 847 (1966).
163. Roessler, D. M., Walker, W. C., and Loh, E., *J. Phys. Chem. Solids* **30**, 157 (1969).
164. Molnar, J. P., and Hartmann, C. C., *Phys. Rev.* **79**, 1015 (1950).
165. Weber, H., *Z. Phys.* **130**, 392 (1951).
166. Townsend, J. R., *Phys. Rev.* **92**, 556 (1953).
167. Haendler, H. M., Wheeler, G. M., and Bernard, W. J., *J. Opt. Soc. Am.* **43**, 215 (1953).
168. Johnson, P. D., *Phys. Rev.* **94**, 845 (1954).
169. Nelson, J. R., *Phys. Rev.* **99**, 1902 (1955).
170. Reiling, G. H., and Hensley, E. B., *Phys. Rev.* **112**, 1106 (1958).
171. Roessler, D. M., *Br. J. Appl. Phys.* **16**, 1359 (1965).
172. Roessler, D. M., and Walker, W. C., *Phys. Rev. Letters* **17**, 319 (1966).
173. Roessler, D. M., and Walker, W. C., *Phys. Rev.* **159**, 733 (1967).
174. Cohen, M. L., Lin, P. J., Roessler, D. M., and Walker, W. C., *Phys. Rev.* **155**, 992 (1967).
175. Williams, M. W., and Arakawa, E. T., *J. Appl. Phys.* **38**, 5272 (1967).
176. Fong, C. Y., Saslow, W., and Cohen, M. L., *Phys. Rev.* **168**, 992 (1968).
177. Whited, R. C., and Walker, W. C., *Phys. Rev. Letters* **22**, 1428 (1969).
178. Toyozawa, Y., and Hermanson, J., *Phys. Rev. Letters* **21**, 1637 (1968).
179. Glasscok, H. R., and Hensley, E. B., *Phys. Rev.* **131**, 649 (1963).
180. Zollweg, R. J., *J. Phys. Rev.* **111**, 113 (1958); **97**, 288 (1955).
181. Saum, G. A., and Hensley, E. B., *Phys. Rev.* **113**, 1019 (1959).
182. Sproull, R. L., *Phys. Rev.* **78**, 630 (1950).
183. Tyler, W. W., *Phys. Rev.* **76**, 1887 (1949).
184. Okumura, K., *Phys. Rev.* **96**, 1704 (1954).
185. Jahoda, F. C., *Phys. Rev.* **107**, 1261 (1957).
186. Krumhansl, J. A., *Proceedings of the Conference on Photoconductivity, Atlantic City*, Wiley, New York, 1964.
187. Feinleb, J., Scouler, W. J., and Feretti, A., *Phys. Rev.* **165**, 765 (1968).
188. Astoin, N., and Vodar, B., *J. Phys. Radium* **14**, 424 (1953).
189. Iguchi, Y., *Sci. Light Tokyo* **13**, 37 (1964).
190. Gilles, A., and Vodar, B., *J. Opt. Soc. Am.* **42**, 783 (1952).
191. Foex, M., and Johannin-Gilles, A., *Compt. Rend.* **237**, 1668 (1953).
192. Groth, W., and Weyssenhoff, H., *Z. Naturforsch.* **11** A, 165 (1956).
193. Nelson, C. M., and Weeks, R. A., *J. Appl. Phys.* **12**, 883 (1961).
194. Laikin, M., *J. Opt. Soc. Am.* **51**, 238 (1961).
195. Hunter, W. R., *Japan J. Appl. Phys.* **4**, 520 (1965).
196. Philipp, H. R., *Solid State Commun.* **4**, 73 (1966).
197. Platzoder, K., *Phys. Status Solidi* **29** K, 63 (1968).
198. Platzoder, K., and Steinmann, W., *J. Opt. Soc. Am.* **58**, 588 (1968).
199. Papazian, H. A., *J. Appl. Phys.* **27**, 1253 (1956).
200. Cohen, A. J., and Smith, H. L., *J. Phys. Chem. Solids* **7**, 301 (1958).
201. Madden, R. P., *Physics of Thin Films*, Academic Press, New York, 1963.
202. Pajasova, L., *Czech. J. Phys.* **19**, 1265 (1969).
203. Freed, S., MacMurray, M. L., and Rosenbaum, E. J., *J. Chem. Phys.* **7**, 853 (1939).
204. Vodar, B., Freymann, R., and Yeou, Ta, *J. Chim. Phys.* **8**, 349 (1940).

205. FRIEDMAN, H., *Physics of the Upper Atmosphere*, Academic Press, New York, 1960, p. 160.
206. MALITSON, I. H., *J. Opt. Soc. Am.* **52**, 1377 (1962).
207. FREEMAN, G. H. C., *Br. J. Appl. Phys.* **16**, 927 (1965).
208. COMPANION, A. L., *J. Phys. Chem. Solids* **25**, 357 (1964).
209. TIPPINS, H. H., *J. Phys. Chem. Solids* **27**, 1069 (1966).
210. DOYLE, W. P., *J. Phys. Chem. Solids* **4**, 144 (1958).
211. RAUCH, W., *Z. Phys.* **116**, 652 (1940).
212. PRENER, J. S., and KINGSLEY, J. D., *J. Chem. Phys.* **35**, 2256 (1961).
213. PARKINSON, W. W., and WILLIAMS, F. E., *J. Chem. Phys.* **18**, 534 (1950).
214. MARPLE, D. T. F., and EHRENREICH, H., *Phys. Rev. Letters* **8**, 87 (1962).
215. PHILLIPS, J. C., *Phys. Rev. Letters* **10**, 329 (1963).
216. JAIN, K. P., *Phys. Rev.* **139** A, 544 (1965).
217. HERMANSON, J., *Phys. Rev. Letters* **18**, 170 (1967).
218. CHANDRASEKHARAN, V., and DAMANY, H., *Appl. Opt.* **8**, 671 (1969).

Additional bibliography

ELECTRONIC RESONANCES

BALDINI, G., BOSACCHI, A., and BOSACCHI, B., Exciton-phonon interactions in alkali halides, *Phys. Rev. Letters* **23**, 846 (1969).

BROST, D. M., and FRY, J. L., Energy bands of LiF by a self-consistent tight-binding procedure, *Bull. Am. Phys. Soc.* **15**, 1378 (1970).

CAZAUX, J., Dielectric constant and electron energy losses in solids having two separated groups of filled bands (energy range 3—30 eV), *Optics Commun.* **3**, 221 (1971).

CHANEY, R. C., LAFON, E. E., and LIN, C. C., Energy band structure of Li fluoride crystals by the method of tight binding, *Phys. Rev.* B, **4**, 2734 (1971).

DOW, J. D., and REDFIELD, D., Theory of exponential absorption edges in ionic and covalent solids, *Phys. Rev. Letters* **26**, 762 (1971).

FONG, C. Y., and COHEN, M. L., Pseudopotential calculation of NaCl and KCl bands, *Phys. Rev.* **185**, 1168 (1969).

FOWLER, W. B., and KUNZ, A. B., Conduction bands in caesium fluoride, *Phys. Rev.* **186**, 956 (1969).

KUNZ, A. B., Local orbital equations for excited states, *Phys. Status Solidi* b, **46**, 697 (1971).

LANG, M., Excitons, plasmons and polaritons in insulating crystals, *Phys. Rev.* B, **2**, 4022 (1970).

LIPARI, N. O., and FOWLER, W. B., Correlation effects in the energy bands of insulating crystals, *Solid State Commun.* **8**, 1395 (1970).

PAGE, L. J., and HYGH, E. H., Calculation of energy bands in alkali halides, *Phys. Rev.* B, **1**, 3472 (1970).

STAROSTIN, N. V., Zone structure of fluoride type crystals, *Soviet Phys. Solid State* **11**, 1317 (1969).

THE ALKALI HALIDES

AHRENKIEL, R. K., and TEEGARDEN, K. J., Magnetoreflectance of the first exciton in KI and RbI, *Phys. Status Solidi* b, **51**, 603 (1972).

AKAGI, M., and YOSHIDA, S., Energy distribution of transmitted secondary electrons from porous KCl films, *J. Appl. Phys.* **43**, 880 (1972).

BALDINI, G., and BOSACCHI, B., Optical properties of Na and Li halides crystals at 55 K, *Phys. Status Solidi* **38**, 325 (1970).

BALDINI, G., and CANOSSI, G., Faraday rotation of excitons in alkali-halides, *Solid State Commun.* **10**, 373 (1972).

BALDINI, G., and RIGALDI, L., Ultraviolet optical constants of thin films determined by reflectance measurements, *J. Opt. Soc. Am.* **60**, 495 (1970).

BASSANI, F., and GIULIANO, P., Pseudopotential approach to the band structure of alkali halides, *Nuovo Cim.* **8B**, 193 (1972).

BLECHSCHMIDT, D., KLUCKER, R., and SKIBOWSKI, N., Dielectric properties of KCl, KBr and KI single crystals in the extreme u.v. up to 35 eV, *Phys. Status Solidi* **36**, 625 (1969).

BLECHSCHMIDT, D., SKIBOWSKI, M., and STEINMANN, W., Photoelectric yield of the potassium halides in the photon energy range 10 to 35 eV, *Optics Commun.* **1**, 275 (1970).

BLECHSCHMIDT, D., SKIBOWSKI, M., and STEINMANN, W., Photoemission from the potassium halides in the photon energy range 7 to 30 eV, *Phys. Status Solidi* **42**, 61 (1970).

BLECHSCHMIDT, D., HAENSEL, R., KOCH, E. L., NIELSEN, V., and SKIBOWSKI, M., Fine structure of the extreme u.v. spectra of the potassium halide crystals at 10 K, *Phys. Status Solidi* b, **44**, 787 (1971).

BLECHSCHMIDT, D., SAILE, V., SKIBOWSKI, M., and STEINMANN, W., Excitation of the Cs-5p core level in Cs halides, *Phys. Letters* A, **4**, 221 (1971).

BOURSEY, E., and RONCIN, J. Y., Absorption spectra of caesium halides at 20 K from 12 to 20 eV, *Solid State Commun.* **9**, 1049 (1971).

BROWN, F. C., GÄHWILLER, C., KUNZ, A. B., and LIPARI, N. O., Soft X-ray spectra of the lithium halides and their interpretation, *Phys. Rev. Letters* **25**, 927 (1970).

BROWN, F. C., GÄHWILLER, C., FUJITA, H., KUNZ, A. B., SCHEIFLEY, W., and CARRERA, N., Extreme ultraviolet spectra of ionic crystals, *Phys. Rev.* B, **2**, 2126 (1970).

CARDONA, M., HAENSEL, R., LYNCH, D. W., and SONNTAG, B., Optical properties of the rubidium and caesium halides in the extreme u.v., *Phys. Rev.* B, **2**, 1117 (1970).

DROST, D. M., and BRENER, N., Tight-binding calculations for alkali halides *Bull. Am. Phys. Soc.* **16**, 405 (1971).

ERSHOW, O. A., and BURTSEVA, V. M., Analysis of the long wavelength X-ray reflection spectrum of KCl by the Kramers-Kronig method, *Optika Spektrosk.* **28**, 167 (1970) [*Optics Spectrosc.* **28**, 84 (1970)].

FASSLER, D., and ZIMMERMANN, G., Absolute reflexion vermögen von alkali halogeniden im kurzweiligen ultraviolette, *Z. Phys. Chem.* **246**, 33 (1971).

FOWLER, W. B., and KUNZ, A. B., ENDOR, the F-center ground state, and conduction bands in alkali halide crystals, *Phys. Status Solidi* **40**, 249 (1970).

FRÖHLICH, D., STAGINNUS, B., and ONODERA, Y., Two photon spectroscopy in CsI and CsBr, *Phys. Status Solidi* **40**, 547 (1970).

FUJITA, H., GÄHWILLER, C., and BROWN, F. C., Far ultraviolet spectra due to 4*d* electrons in the alkali iodides, *Phys. Rev. Letters* **22**, 1369 (1969).

GOUT, C., FRONDON, R., and SADACA, J., Calculation of the valence band of NaF by the tight-binding method, *Phys. Letters* A, **29**, 656 (1969).

HATTORI, A., TOMURA, M., and NISHIMURA, H., Diffusion of excitons in RbI, *J. Phys. Soc. Japan* **31**, 611 (1971).

IGUCHI, Y., SASAKI, T., SUGARAWA, H., SATO, S., NASU, T., EJIRI, A., ONARI, S., KOJIMA, K., and OYA, T., Photoelectric emission of RbCl in the extreme u.v., *Phys. Rev. Letters* **26**, 82 (1971).

KRISTIANPOLLER, N., and DAVIDSON, Z., On the u.v. excitation of thermally stimulated conductivity in KBr and KI crystals, *J. Phys.* C, **5**, 279 (1972).

KUDYKINA, T. A., and TOLPYGO, K. B., Influence of exciton-phonon interaction in alkali-halide crystals, *Soviet Phys. Solid State* **13**, 2402 (1972).

KUDYKINA, T. A., and TOLPYGO, K. B., An exciton of high symmetry in alkali halide crystals, *Soviet Phys. Solid State* **14**, 626 (1972).

KUDYKINA, T. A., and TOLPYGO, K. B., Singlet exciton bands in NaCl crystals, *Soviet Phys. Solid State* **13**, 921 (1971).

KUNZ, A. B., Self consistent local orbitals for NaCl, NaBr, KCl and KI, *Phys. Rev.* B, **4**, 609 (1971).

KUNZ, A. B., Electron energy bands and the optical properties of potassium iodide and rubidium iodide, *J. Phys. Chem. Solids* **31**, 265 (1970).

KUNZ, A. B., Energy bands and the optical properties of LiCl, *Phys. Rev.* B, **2**, 5015 (1970).

KUNZ, A. B., Hartree-Fock energy bands in LiCl, *J. Phys.* C, **3**, 1542 (1970).

KUNZ, A. B., and LIPARI, N. O., Electronic structure of NaBr, *Phys. Rev.* B, **4**, 1374 (1971).

KUNZ, A. B., and LIPARI, N. O., Hartree-Fock determination of the energy bands and the optical properties of lithium bromide, *J. Phys. Chem. Solids* **32**, 1141 (1971).

LAMASH, H., ROSSEL, J., and SAURER, E., The optical constants of CsI in the excitonic region, *Phys. Status Solidi* b, **49**, 311 (1972).

LAMATSCH, H., ROSSEL, J., and SAURER, E., Localized and self-trapped excitons in CsI, *Phys. Status Solidi* **41**, 605 (1970).

LE COMTE, A., SAVARY, A., MORLAIS, M., and ROBIN, S., Absorption de couches minces de NaF et de NaCl entre 25 et 72 eV, *Optics Commun.* **4**, 296 (1971).

LIPARI, N. O., and KUNZ, A. B., Energy bands and optical properties of NaCl, *Phys. Rev.* B, **3**, 491 (1971).

LIPARI, N. O., and KUNZ, A. B., Energy bands for KCl, *Phys. Rev.* B, **4**, 4639 (1971).

MELVIN, J. S., SMITH, T., Resolution of apparent APW—OPW energy band differences in NaCl, *Phys. Status Solidi* b, **49**, 173 (1972).

MIYATA, T., Line shape analysis of the Γ-exciton spectra of NaCl, NaBr and NaI single crystals, *J. Phys. Soc. Japan* **31**, 529 (1971).

NAGASAWA, N., Localized excitons and band excitons in alkali chloride-alkali iodide solid solution, *J. Phys. Soc. Japan* **27**, 1535 (1969).

NAKAI, S., ISHII, T., and SAGAWA, T., Core exciton absorption in sodium halides, *J. Phys. Soc. Japan* **30**, 428 (1971).

NOSENZO, L., REGUZZONI, E., and SAMOGGIA, G., Thermoreflectance in alkali halides, *Phys. Rev. Letters* **28**, 1388 (1972).

OVERHOF, H., Energy bands of potassium halides (Green's function method), *Phys. Status Solidi* b, **43**, 575 (1971).

OVERHOF, H., Calculation of the optical constants for the potassium halides from Korringa—Kohn—Rostoker energy bands, *Phys. Rev.* B, **5**, 4079 (1972).

PEIMANN, C. J., and SKIBOWSKI, M., Dielectric properties of the rubidium halides crystals in the extreme u.v. up to 30 eV, *Phys. Status Solidi* b, **46**, 655 (1971).

PEROT, F., Hartree-Fock band structure of alkali fluorides and chlorides, *Phys. Status Solidi* b, **52**, 163 (1972).

PETROFF, Y., PINCHAUX, R., CHEKROUN, C., BALKANSKI, M., and KAMIMURA, H., Excitonic structure of alkali halides, *Phys. Rev. Letters* **27**, 1377 (1971).

PONG, W., Exciton induced photoemission from KCl films, *J. Appl. Phys.* **42**, 564 (1971).

RUBLOFF, G. W., Far ultraviolet reflectance spectra and the electronic structure of ionic crystals, *Phys. Rev.* B, **5**, 662 (1972).

RUBLOFF, G. W., FREEDUF, J., FRITZSCHE, H., and MURASE, K., Far ultraviolet optical spectra of ionic crystals, *Phys. Rev. Letters* **26**, 1317 (1971).

SAITO, H., SAITO, S., and ONOKA, R., Absorption spectra of alkali fluorides in the X.u.v. region, *J. Phys. Soc. Japan* **28**, 699 (1970).

SAITO, H., WATANABE, M., EJIRI, A., SATO, S., YAMASHITA, H., SHIBAGUCHI, T., NISHIDA, H., and YAMAGUCHI, S., Absorption spectra of KCl and RbCl in the X.u.v. region, *Solid State Commun.* **8**, 1861 (1970).

SAITO, H., and ONAKA, R., Multiplet absorption band of CsI in the X.u.v. region, *J. Phys. Soc. Japan* **28**, 1380 (1970).

SASAKI, T., IGUCHI, Y., SUGAWARA, H., SATO, S., NASU, T., EJIRI, A., ONARI, S., KOJIMA, K., and OYA, T., Photoelectric emission of potassium halides in the extreme u.v., *J. Phys. Soc. Japan* **30**, 580 (1971).

SASAKI, T., IGUCHI, Y., SUGAWARA, H., SATO, S., NASU, T., EJIRI, A., ONARI, S., KOJIMA, K., and OYA, T., Photoelectric emission of rubidium halides in the extreme u.v., *J. Phys. Soc. Japan* **30**, 581 (1971).
SHULISHENKO, B. V., ZAVM, G. S., and KRISTOFEL, N. N., Density of states for KCl in the two bands model, *Soviet Phys. Solid State* **12**, 2107 (1970).
SUGARAWA, H., and SASAKI, T., Photoemission of potassium chloride in the photon energy region 10 to 40 eV, *Optics Commun.* **2**, 333 (1970).
SYDOR, M., Effects of surfaces in absorption spectrum of RbI crystals, *Phys. Rev. Letters* **27**, 1286 (1971).
STÉPHAN, G., and ROBIN, S., Spectres électroniques longitudinaux et transversaux d'halogénures alcalins dans l'ultraviolet extrême, *J. Phys. Paris* **32**, C4-246 (1971).
THOMAS, J. M., ADAMS, I., and BARBER, M., Experimental determination of valence band width and configuration interaction in NaCl, *Solid State Commun.* **9**, 1571 (1971).
WATANABE, M., EJIRI, A., YAMASHITA, H., SAITO, H., SATO, S., SHIBAGUCHI, T., and NISHIDA, H., Absorption spectra of rubidium halides in the extreme ultraviolet, *J. Phys. Soc. Japan* **31**, 1085 (1971).
WATANABE, M., and NAKAMURA, K., Reflection spectrum of RbCl in the extreme u.v., *J. Phys. Soc. Japan* **30**, 1764 (1971).
WILLIAMS, R. T., and KABLER, M. N., Excitons perturbed by self-trapped excitons in alkali-iodide crystals, *Solid State Commun.* **10**, 49 (1972).

OXIDES AND OTHER COMPOUNDS

BACHRACH, R. Z., and BROWN, F. C., Exciton optical properties of TlBr and TlCl, *Phys. Rev.* **1**, 818 (1970).
BENNET, A. J., and ROTH, L. M., Calculation of the optical properties of amorphous SiO_x materials, *Phys. Rev.* B, **4**, 2686 (1971).
BIRTCHER, R. C., DEUTSCH, P. W., WANDELKEN, J. F., and KUNZ, A. B., Valence band structure in silver fluoride, *J. Phys.* C, **5**, 562 (1972).
BLANK, V. A., and SOROKIN, O. M., External photoeffect of LiF and MgF_2 layers in 15-220 nm region *Fizika tverd. Tela* **12**, 1931 (1970) [*Soviet Phys. Solid State*, **12**, 1536 (1970)].
BROWN, F. C., GÄHWILLER, C., and KUNZ, A. B., Extreme u.v. absorption and band density of NiO, *Solid State Commun.* **9**, 487 (1971).
CARRERA, N. J., and BROWN, F. C., Optical response of AgCl and AgBr in the near and extreme u.v., *Phys. Rev.* B, **4**, 3651 (1971).
CHRYSSOU, E., and HOLZAPFEL, G., Exoelectron emission from Al_2O_3, *Phys. Status Solidi* A, **8**, K47 (1971).
DEB, S. K., Exciton absorption and photoconductivity in mercurous halides, *Phys. Rev.* B, **2**, 5003 (1970).
DI STEFANO, T. H., and EASTMAN, D. E., Photoemission measurements of the valence levels of amorphous SiO_2, *Phys. Rev. Letters* **27**, 1560 (1971).
DI STEFANO, T. H., and EASTMAN, D. E., The band edge of amorphous SiO_2 by photo injection and photo-conductivity measurements, *Solid State Commun.* **9**, 2259 (1971).
FREEAUF, J., MURASE, K., and RUBLOFF, G. W., Far ultra-violet reflectance spectra of oxides, *Bull. Am. Phys. Soc.* **16**, 354 (1971).
FORMAN, R. A., HOSLER, W. R., and BLUNT, R. F., The bandgap of cadmium fluoride, *Solid State Commun.* **10**, 19 (1972).
GÖRLICH, P., and ULLMANN, P., Absorption investigation on alkaline-earth fluoride mixed crystals, *Phys. Status Solidi* **50**, 577 (1972).
HAENSEL, R., RABE, P., and SONNTAG, B., Optical absorption of Ce, CeO, Pr, PrO, Nd, NdO and Sm in the extreme u.v., *Solid State Commun.* **8**, 1845 (1970).
HANSON, W. F., ARAKAWA, E. T., and WILLIAMS, M. W., Optical properties of MgO and MgF_2 in the extreme ultra-violet region, *J. Appl. Phys.* **43**, 1661 (1972).

HAYES, W., KUNZ, A. B., and KOCH, E. E., Reflectance spectra of alkaline earth fluorides in the vacuum ultra-violet, *J. Phys.* C, **4**, 1200 (1971).

HOENERLAGE, B., and WIESNER, H., Exciton states in CaF_2, SrF_2, BaF_2, *Z. Phys.* **242**, 406 (1971).

HUNTER, W. R., and MALO, S. A., The temperature dependance of the short wavelength transmittance limit of vacuum ultraviolet window materials, *J. Phys. Chem. Solids* **30**, 2739 (1969).

HUTCHESON, E. H., COX, J. T., HASS, G., and HUNTER, W. R., Monitoring the thickness of thin MgF_2 and LiF films on Al by reflectance measurement using the 1216 Å line of hydrogen, *Appl. Opt.* **11**, 1590 (1972).

JOUANIN, C., and GOUT, C., Calcul des énergies aux points Γ, X, Z et M des bandes de valence du MgF_2, *Compt. Rend.* **274**, B, 1284 (1972).

JOUANIN, C., and GOUT, C., Valence band structure of MgF_2 by the tightbinding method, *J. Phys.* C, **5**, 1945 (1972).

KLEIN, G., and CHUN, H. V., Determination of optical interband transition in crystalline quartz from X-ray spectroscopical data, *Phys. Status Solidi* b, **49**, 167 (1972).

LE COMTE, A., and ROBIN, S., Absorption des halogénures de calcium dans l'ultra-violet extrême, *Optica Acta* **19**, 203 (1972).

MARCHETTI, A. P., and BOTTGER, G. L., Optical absorption spectrum of AgF, *Phys. Rev.* B, **3**, 2604 (1971).

MATSUI, A., and WALKER, W. C., Exciton and interband spectra of crystalline MnF_2, *J. Opt. Soc. Am.* **60**, 358 (1970).

NAGASAWA, N., Localized excitons and band excitons in alkali chloride-alkali iodide solutions, *J. Phys. Soc. Japan* **29**, 532 (1970).

OVERHOF, H., and TREUSCH, J., The energy bands of thallous chloride and bromide, *Solid State Commun.* **9**, 53 (1971).

PETRESCU, P., and GIURGEA, M., Electron emission spectra of alkaline earth fluorides, *Phys. Status Solidi* **38**, 113 (1970).

PHILIPP, H. R., Optical properties of non-crystalline Si, SiO, SiO_x and SiO_2, *J. Phys. Chem. Solids* **32**, 1935 (1971).

RABE, P., SONNTAG, B., SAGAWA, T., and HAENSEL, R., The optical absorption of MgF_2, $MgCl_2$, $MgBr_2$ in the vicinity of the Mg *L*-Shell transitions, *Phys. Status Solidi* b, **50**, 559 (1972).

RUBLOFF, G. W., FRITZSCHE, H., GERHARDT, U., and FREEDÜF, J., Far ultraviolet spectroscopy of solids in the range 6-36 eV using synchrotron radiation from an electron storage ring, *Rev. Scient. Instrum.* **10**, 1513 (1971).

SANCHEZ, C., and CARDONA, M., Piezobirefringence in the vacuum U.V.: alkali halides and alkaline—earth fluorides, *Phys. Status Solidi* A **50**, 293 (1972).

STAROSTIN, N. V., Energy structure and optical properties of MgF_2 crystals, *Optika i Spektrosk.* **29**, 298 (1970) [*Optics Spectrosc.* **29**, 557 (1970)].

STEPHAN, G., NISAR, M., and ROTH, A., Spectre électronique du fluorure de lanthane dans l'ultra-violet extrême, *Compt. Rend.* B **274**, 807 (1972).

TUBBS, M. R., Excitons in the heavy metal halides, *Phys. Status Solidi* b, **47**, K65 (1971).

WEMPLE, S. H., and DIDOMENICO, M., JR., Behavior of the electronic dielectric constant in covalent and ionic materials, *Phys. Rev.* B, **3**, 1338 (1971).

WHITED, R. C., and WALKER, W. C., Exciton and interband spectra of crystalline CaO, *Phys. Rev.* **188**, 1380 (1969).

WHITED, R. C., and WALKER, R. C., Exciton spectra of CaO and MgO, *Phys. Rev. Letters* **22**, 1428 (1969).

PART II

MOLECULAR SPECTROSCOPY

CHAPTER 4

Spectra of diatomic and inorganic polyatomic molecules

N. Damany

Laboratoire des Interactions Moléculaires et des Hautes Pressions du CNRS, Bellevue, 92190 Meudon, France

Contents

Introduction

While infrared and Raman spectroscopy provide very accurate methods for obtaining the rotational and vibrational constants of a molecule in its ground state, spectroscopy in the visible and u.v. regions is essential for provision of information on the electronically excited states. Except for the lowest valence states, excitation energies are generally higher than 7 eV, corresponding to the vacuum ultraviolet (v.u.v.) region; *a fortiori*, the Rydberg series and photoionization continua and "resonances" can be observed only in this spectral region.

Experimental difficulties long delayed the development of work on molecular spectra in the v.u.v., and even now the precision of observations is inferior to that achievable in the visible and near infrared. Nevertheless, much has been achieved in the analysis of vibrational and rotational structures of bands of assignable configurations, and many molecular constants have been calculated to good precision.[1, 2]

In association with other techniques such as mass-spectroscopic analysis, ionization chamber work, photoelectron analysers, and fluorescence measurements, v.u.v. spectroscopy constitutes a promising method for the study of ionization processes, including determination of appearance thresholds, identification of parent species, and ionization yields.[3]

Other methods, such as flash photolysis or the matrix isolation technique, make possible the study of transient species. A special method for measuring radiative lifetimes in the range 1-200 ns from "phase shifts", has been developed for molecular systems and has yielded data necessary for calculating oscillator strengths and transition moments.[4]

The purpose of Part II is to review the specific contribution of v.u.v. spectroscopy to molecular physics. The relationship between electronic spectra and electronic structures has been treated in detail in some fundamentally important books or papers by Pearse and Gaydon,[5] Herzberg,[1] Slater,[6] and Rosen[7] for diatomic molecules, and by Sponer and Teller,[8] Ramsay,[9] Slater,[6] Rosen,[7] and Herzberg[2] for polyatomic molecules; other important post-1930 works are Mulliken,[10, 11] Platt and Klevens,[12] Walsh,[13] Wallace,[14] and Wilkinson.[15] The reader will find in some of these articles—especially in refs. 1, 2, 8, 12, and 15—extensive bibliographies of both theoretical and experimental work, and in ref. 14 the band-head wavelengths of C_2, CH, CN, CO, NH, NO, O_2, OH, and their ions. To this list we should add specialized monographs on photoionization problems by Weissler,[16] Watanabe,[17] Vilesov,[18] and Marr.[3] Finally, we have to mention two bibliographies, circulated as technical reports; the first by Hirt and Schmitt[19] up to 1962, and the other, restricted to polyatomic molecules, by Johannin-Gilles,[20] to 1964.

Before reviewing the more recent progress, we may recall the historically significant observations of the pioneers of v.u.v. spectroscopy—Schumann[21] and Lyman;[22] they were followed by Leifson[23] and Hopfield,[24] who were the first to make quantitative absorption measurements with the aid of the H_2- and He_2-continua.

On account of the space available in this chapter, we shall limit ourselves to the inorganic molecules. Some particular aspects of organic molecular spectroscopy will be presented in the chapters 5 and 6.

1. Diatomic molecules

First, we shall survey the spectra of the diatomic molecules. A very extensive bibliography of the papers before 1950 is contained in the book of Herzberg,[1] and, generally, we shall avoid discussing them; on the other hand, we have tried to give a rather complete bibliography of more recent investigations (up to 1972).

1.1. H_2, D_2

The molecular spectrum of hydrogen is probably the oldest observed emission spectrum in the v.u.v. region: it was first photographed by Schumann[21] in 1892. Years after, Lyman extended this spectrum beyond the absorption limit of fluorite (see 22) and made the first wavelength measurement of its lines. In 1926, Dieke and Hopfield[25] obtained the first far u.v. absorption spectrum of H_2 beyond the dissociation limit at 845 Å. Since then, because of the simplicity of the hydrogen molecule and its basic importance to theoretical studies, precise experimental data have been sought; they are collected in ref. 26, which gives, with the wavelengths of lines, a table of all levels for H_2, DH, D_2, TH, TD, and T_2. Two atlases of emission H_2 lines are also available.[27, 28]

A repulsive state $b\,^3\Sigma_u^+$ which dissociates, as does the ground state $^1\Sigma_g^+$, into two normal (1 ^2S_g) hydrogen atoms, is the lower state of the well-known continuous spectrum of H_2, so useful as a source of u.v. light; the upper state of this transition is the stable $a\,^3\Sigma_g$[1, 29]

Six stable excited states are involved in the transitions observed to date in the v.u.v.: $B\,^1\Sigma_u^+$ (Lyman bands),[30, 31] $C\,^1\Pi_u$ (Werner bands),[30] $D\,^1\Pi_u$,[32–36] $B'\,^1\Sigma_u^+$,[36, 37] $B''\,^1\Sigma_u^+$, $D'\,^1\Pi_u$, and $D''\,^1\Pi_u$.[36, 38]

Vibrational and rotational analyses of the emission[34, 39] and absorption[35–38] spectra has permitted accurate determination of the molecular constants of the ground state and of the first excited states; the potential energy curves have been calculated by the Rydberg-Klein-Rees (R.-K.-R.) method for B, C, B', and D states.[40] The ionization potential (15.43 eV) and the dissociation energy (4.48 eV), were first measured by Beutler[32, 41] and afterwards by Herzberg and Monfils.[37]

Very few experimental determinations of absorption coefficients or oscillator strengths of H_2 transitions have been performed in the v.u.v.: Haddad *et al.*[42] have measured the absolute line strengths for transitions in four bands of the Lyman series between 1130 and 1060 Å; Weissler *et al.*[43, 44] have given the photoionization and absorption cross-sections of H_2 for about 50 wavelengths below the ionization limit. Cook and Metzger,[45] with improved techniques, have measured them for H_2 and D_2, with a resolution of 0.5 Å, and have determined the onset of the predissociation and ionization continua; Samson and Cairns have extended the measurements to 450 Å.[46] The most prominent features are the strong absorptions, due to predissociation, in the broad Hopfield bands $D\,^1\Pi_u \leftarrow X\,^1\Sigma_g^+$. Associated with each band-head is a weaker red-shaded absorption due to the various rotational levels. Some experiments have allowed the derivation of the photoionization efficiency curves (Fig. 1) and the separation of the autoionization and the non-ionizing processes above the first ionization potential.(47–49) Electron-retarding potential techniques have shown the influence of autoionization on the energy distribution of the electrons emitted;(50) theoretical work is now in progress.(51–53)

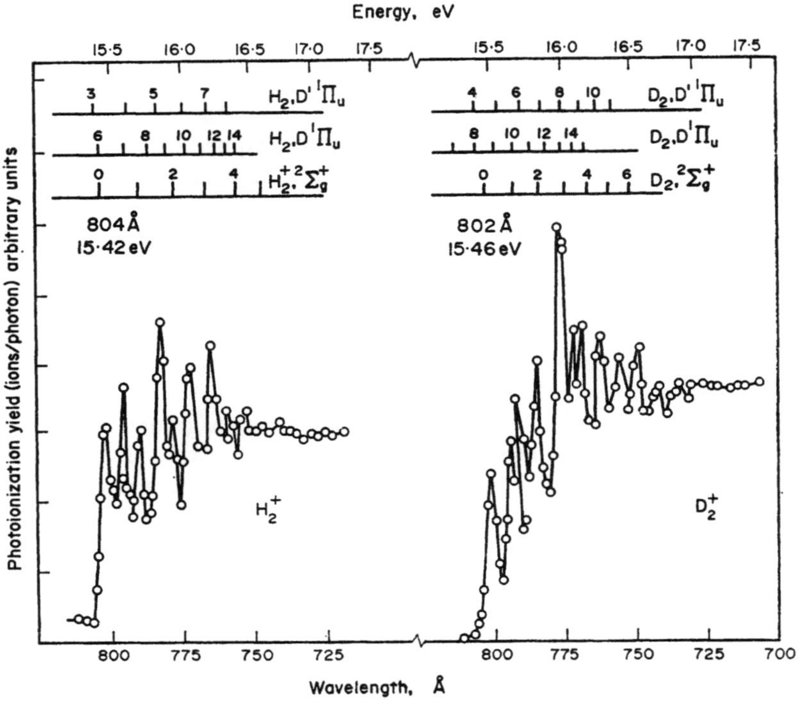

FIG. 1. Photoionization efficiency curves for H_2 and D_2.(47)

1.2. RARE GASES

The formation of diatomic rare-gas molecules was first invoked by Hopfield(54) to explain the continuous emission given by a discharge in helium. Since then, because of their practical interest as v.u.v. sources, different types of discharges in rare gases have been extensively investigated.(55) The main continuum emission is localized between 2250-1470 Å for Xe, 1850-1250 Å for Kr, 1650-1050 Å for Ar, 1000-740 Å for Ne, and 1100-600 Å for He.(56, 57) However, with appropriate excitation conditions, a weaker continuum extending approximately to 3000 Å for Ar and Xe, and to 4000 Å for He is also observed;(58) no suitable explanation of its mechanism is available to date. On the other hand, the spectral characteristics of the main continua—very similar for the five gases—strongly suggest that at least two stable excited molecular states are formed in the discharge. These states have been identified as $A\,{}^1\Sigma_u^+$ and $D\,{}^1\Sigma_u^+$ for He;(59) since the ground state is repulsive, a continuum is emitted by the transition to $X^1\Sigma_g^+$. Comparison of the calculated intensity distribution[60] with the experimental curves[61, 62] shows that the transition $A(v = 0) \rightarrow X$ can account for most of the observed continuum (Fig. 2). For the heavier rare gases, the character of the excited molecular states is not so clear, but they most probably dissociate into $^3P_0 + {}^1S_0$ atoms.[63]

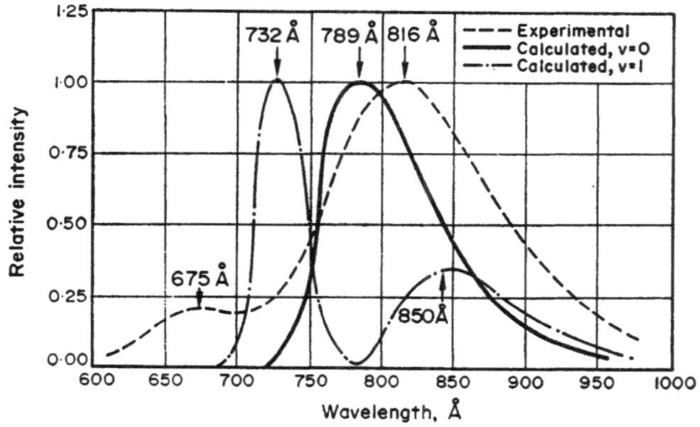

FIG. 2. Continuous emission spectra of He_2 (the maximum intensity of each spectrum has been chosen as unity).[60]

Some diffuse bands, well developed in the emission of an uncondensed discharge in helium at low pressure,[64] are tentatively explained as originating from the $A \rightarrow X$ transition, the molecule being formed in a high vibrational level.

The study of the variation of the absorption of helium and xenon with pressure and temperature, has clearly shown the existence of molecules, He_2 and Xe_2, resulting from molecular association by van der Waals forces, and has permitted measurement of the corresponding absorption cross-sections.[65, 66] More recently, discrete molecular structures have been observed in the absorption spectra of several rare gases.[64, 67]

Diatomic molecular ions of the rare gases have also been observed in photoionization experiments,[68, 69] and the ionization potentials of Xe_2, Kr_2, and Ar_2 have been determined.

1.3. ALKALI-METAL VAPOURS

The question of v.u.v. absorption spectra of the alkali-metal vapours first arose from need for corrections in the investigation of atomic absorption,[70–73] diatomic molecules being present in the alkali-metal vapours to the extent of several per cent. More recently, systematic measurement of molecular absorption cross-sections of Na_2 has been made by Hudson[74] down to 1600 Å and of K_2, Rb_2, and Cs_2 by Creek and Marr[75] down to 1100 Å. Only Na_2 exhibits a continuous absorption in the v.u.v. (above 1600 Å); the other molecules have no observed absorption between 2000 and 1100 Å.

1.4. N_2

Nitrogen is certainly the molecule which has been the most extensively studied by v.u.v. spectroscopy. For the sake of simplicity—especially for the absorption spectrum—the whole spectral range can be separated in three parts: (A) $\lambda > 1000$ Å, (B) $1000 > \lambda > 600$ Å, (C) $\lambda < 600$ Å. These regions, which require different experimental techniques, correspond also to somewhat different transitions. All the electronic states having excitation energies below 12.5 eV (1000 Å) are of valence type, and their combinations with the ground state are forbidden in electric dipole radiation; consequently the spectra are weak and usually difficult to observe.[76] On the other hand several Rydberg states[77] coexist with valence states at energies higher than 12.5 eV, with very strong interactions between neighbouring states of the same symmetry. Transitions with the ground state are generally allowed, and the bands are very strong. However, perturbations make analysis difficult.[78]

Below 600 Å, only continuous absorption has been observed, with a weak underlying structure near 500 Å[79] which involves autoionization.

For reducing the bibliographical list, only references after 1950 are indicated, and the reader should consult ref. 1 for all earlier references.

Emission spectra

To date, few emission spectra have been attributed to transitions in N_2 molecules, in the v.u.v.: Ogawa and Tanaka,[76] with medium dispersion, have analysed the vibrational structure of a system of six bands originating from a $a'\,^1\Sigma_u^-$ state. Wilkinson and his collaborators[77, 78] have investigated the emission spectrum between 1700 and 900 Å with high resolution. The rotational structure has been analysed for eight transitions. From these new data, added to those previously obtained, the molecular constants have been deduced for ten electronic states with excitation energies around 12.5 eV.

Absorption spectra $\lambda > 1000$ Å

Experimental data on absorption spectra are quite extensive: for region A, the first attempt at improved resolution was made by Tanaka *et al.*[80, 81] but only the vibrational structure was resolved. Wilkinson and his collaborators have photographed the absorption spectra at high resolution from 1520 to 1050 Å.[82–84] Accurate wavelength measurements of the rotational structure in many bands of the $a\,^1\Pi_g$,[85a] $a'\,^1\Sigma_u^-$,[85b] (see Plate 1), $B'\,^3\Sigma_u^-$,[85c] $C^3\,\Pi_u$,[85d] $\leftarrow X\,^1\Sigma_g^+$ systems and of isotopic shift in $a\,^1\Pi_g$[85e] have made it possible to identify definitely the upper states and to obtain the corresponding molecular constants and the potential energy curves.[86] The intensity distribution in five branches of the $a^1\,\Pi_g \leftarrow X\,^1\Sigma_g^+$ system[87] confirms the theory of Chiu[88] for the ratio between the transition probabilities of electric quadrupole ($\Delta J = 0, \pm 1, \pm 2$) and magnetic dipole ($\Delta J = 0, \pm 1$) transitions.

Mulliken[89] had predicted at 12.20 eV^{-1}, a $^1\Sigma_g^+$ state resulting from the association of a $3s\sigma_g$ electron with the ion core N_2^+ ($X\,^2\Sigma_g^+$); such a state can combine

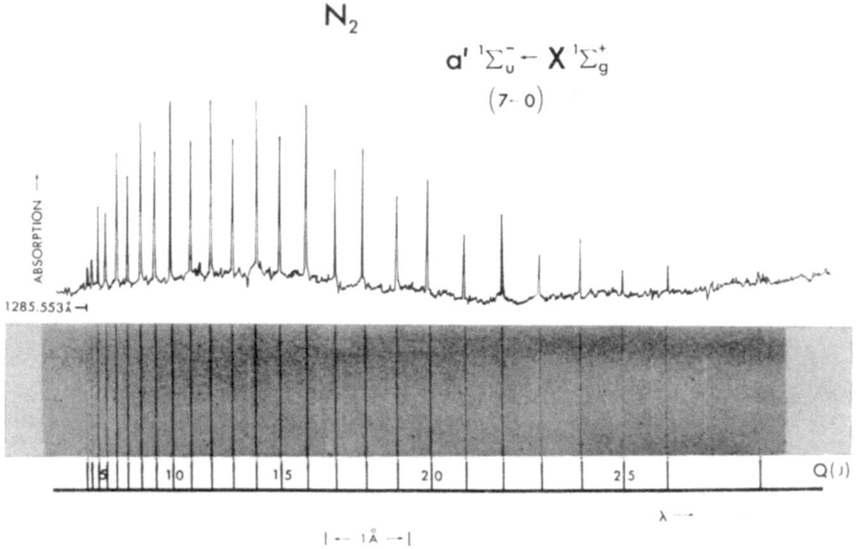

Plate 1. Microphotometer trace (*above*) and spectrogram of the $(7 \leftarrow 0)$ band of the $a'\,^1\Sigma_u^- \rightarrow X\,^1\Sigma_g^+$ transition.(85b)

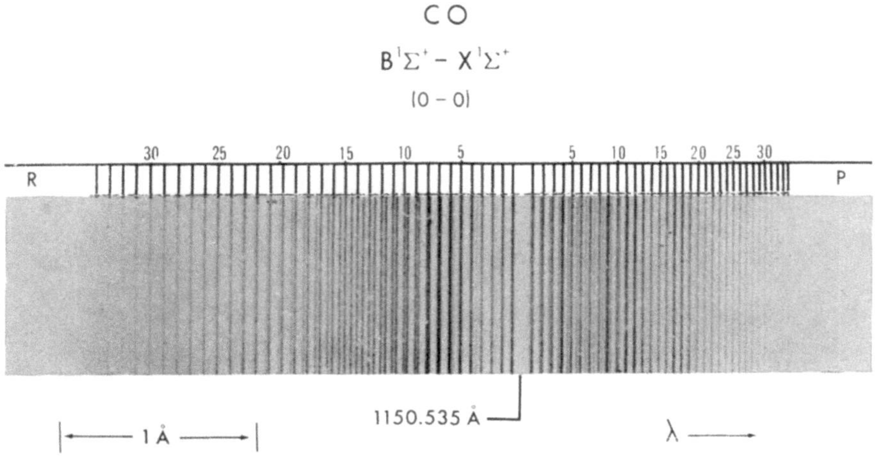

Plate 2. The (0-0) band of the $B\,^1\Sigma^+ \leftarrow X\,^1\Sigma^+$ transition in absorption. Leading lines indicate transitions in $^{12}C^{16}O$. Others are transitions in $^{13}C^{16}O$.(210)

Plate 3. Absorption spectrum of $^{15}N^{16}O$, 1568-1550 Å. Crossing bands $D^2\Sigma^+ - X^2\Pi$ (5.0) and $B^2\Pi - X^2\Pi$ (21.0). (Reproduced by permission of the National Research Council of Canada from the *Canadian Journal of Physics* **46**, 992 (1968).[237])

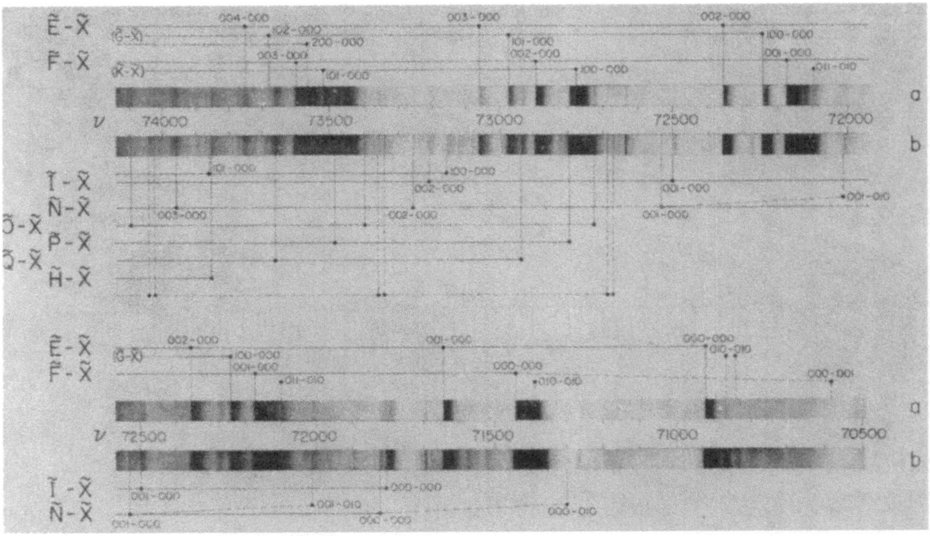

Plate 4. The absorption spectrum of OCS between 1 350 and 1 420 Å ($l = 50$ cm): (a) $p = 0.004$ torr, (b) $p = 0.012$ torr. Reproduced by permission of the National Research Council of Canada from the *Canadian Journal of Physics* **45**, 4014 (1967).[289]

with the ground state only through quadrupole or pressure-induced dipole radiation, which is difficult to observe. It was identified by Dressler and Lutz[90, 91] in the absorption of N_2 at a pressure of 40 torr and 600 cm path-length as a wide absorption band centred at 1009.77 Å, which agrees well with theoretical predictions[89, 92–95] and electron energy-loss determinations.[96, 97] Excitation energies and other spectroscopic constants have also been computed for the excited valence states of the neutral molecule and singly charged positive ion.

Absorption λ < 1 000 Å

The first attempt to study the absorption spectrum of nitrogen below 800 Å with a dispersion better than 2 Å mm^{-1}, was made by Worley in 1943;[98] the convergence of Rydberg series gives an ionization potential of 15.577 eV for the $^2\Sigma_g^+$ state of N_2^+.[99, 100] At shorter wavelengths, other Rydberg series converge to the excited states of N_2^+: $A'\,^2\Sigma_u^+$ (18.67 eV) and $A\,^2\Pi_u$.[101–4]

Until recently, the very complex spectrum between 1 000 Å and 830 Å had been interpreted in terms of twenty different states.[105–7] The work of Dressler[108] and Carroll and Collins[109] shows that the number of *ungerade* singlet excited states can be reduced to two valence states $b\,^1\Pi_u$ and $b'\,^1\Sigma_u^+$, and three Rydberg states $c\,^1\Pi_u$, $c'\,^1\Sigma_u^+$, $o^1\,\Pi_u$: the lack of regularity in the observed transitions and the anomalies in intensity distributions are evidence for very strong interactions between Rydberg and valence states of the same species. A more quantitative treatment of the $b\,^1\Pi_u - c\,^1\Pi_u$ and $b'\,^1\Sigma_u^+ - c'\,^1\Sigma_u^+$[110] perturbations requires consideration of the mutual interaction between several groups of levels. In fact, as the vibrational spacings of the valence states are small, the perturbation affects a large number of consecutive vibrational levels.

At shorter wavelengths, discrete structure is observed only in the 570-470 Å[79] region; it corresponds to two autoionizing states, members of a Rydberg series, converging to the C state of N_2^+ at 23.6 eV.

Excited nitrogen

The absorption spectrum obtained under normal conditions corresponds to transitions from the ground state of the molecule with $v'' = 0$. In the "pink" and "Lewis-Rayleigh" afterglows of activated N_2, it has been possible to photograph other strong absorption spectra between 1900 and 1 100 Å[111–14] and 1 100-600 Å.[115] The lower states of the band systems observed are the vibrational levels of the ground state of N_2, up to $v'' = 29$, the upper states being the first vibrational levels of $j\,^1\Sigma_u^+$, $b\,^1\Pi_u$, $l^1\Pi_u$, $m\,^1\Pi_u$, and $b'\,^1\Sigma_u^+$ [$b\,^1\Pi_u$, $c\,^1\Pi_u$, and $b^1\,\Sigma_u^+$ from refs. 108-9]. Such observations show that the occurrence of N_2 in high vibrational levels of the ground state is characteristic of excitation in a microwave discharge or in an uncondensed transformer discharge.

Absorption coefficients

Only fragmentary measurements of absorption coefficients had been reported in the 1450-1050 Å bands[116–19] before the work of Ching *et al.*:[120] they used the technique of pressure broadening to make integrated intensity measurements of the absorption bands of N_2 for the forbidden transitions $a\,^1\Pi_g \leftarrow C\,^3\Pi_u$, $w^1\Delta_u \leftarrow X\,^1\Sigma_g^+$, and to evaluate f values for $a\,^1\Pi$ and $C\,^3\Pi \leftarrow X$ bands. The results agree rather well with the computed Franck-Condon (F.-C.) factors(121, 122)

Because of their geophysical interest, the absorption and photoionization cross-sections of N_2 below 1000 Å have been measured fairly frequently,(116–18, 123–30) but, due to experimental difficulties—e.g. the sharpness of some lines and the pressure effect on the others—large discrepancies exist between the data in the range 1000-600 Å. More recent measurements, with improved resolution, for the systems $b\,^1\Pi_u \leftarrow X$, $l^1\Pi_u \leftarrow X$, and $p'\,^1\Sigma_u^+ \leftarrow X$(131) are in good agreement with the determination of absolute transition probabilities by the phase shift method.(4, 132)

Photoelectron spectra show that vibrational excitation in the ground state of N_2^+ is enhanced if 744-736 Å rather than 584 Å photons are used.(133)

Below 500 Å, the cross-sections are not pressure dependent and decrease monotonically from 500 to 100 Å.(46, 134)

N_2^+

To date, only one system of the N_2^+ molecule has been observed in the v.u.v.: this is the emission system $C\,^2\Sigma_u^+ \rightarrow X\,^2\Sigma_g^+$, easy to obtain by a discharge through a mixture of rare gas and nitrogen.(135–7) Inverse predissociation or charge exchange are invoked to account for the excitation process and for the influences of the discharge conditions, and the nature of foreign gas(137) on the intensity distribution. If H_2 or D_2 is added to the mixture, some emission bands of NH or ND are also observed(138) in the v.u.v.; they are assigned to $d\,^1\Sigma^+ \rightarrow b\,^1\Sigma^+$, and the vibrational constants of $b\,^1\Sigma^+$ have been revised.

Potential energy curves for N_2, N_2^+, and N_2^- are given by Gilmore.(139)

1.5. O_2

All attempts to obtain an emission spectrum of molecular oxygen in v.u.v. have been unsuccessful, probably because of the long radiative lifetime of some excited states. As a consequence, and also because of its importance in solar photochemistry, the absorption spectrum has been extensively studied. This spectrum begins with the well-known Schumann-Runge (S. R.) bands:(140–3) the 0-0 band appears at 2026 Å, but is fairly weak, then the absorption increases regularly, fixing the boundary of v.u.v. in the neighbourhood of 1950 Å. The bands have a very well-resolved structure(130, 142, 143) extending to the convergence limit at 1750.46 Å; the upper state of the transition is $B\,^3\Sigma_u^-$, but the direct dissociation is complicated by predissociation by the repulsive $^3\Pi_u$ state which intersects the left limb of the $B\,^3\Sigma_u^-$ potential curve. Values of the oscillator strengths of the transitions have been obtained by the method of pressure broadening (Fig. 3),(144,45) which is expected to be more accurate than the determination from low-pressure absorption coefficients.(147, 146–8) The F.C. factors have been calculated by the R.-K.-R. method.(149)

Beyond the discrete S.-R. bands lies an intense continuous absorption extending to about 1300 Å due to dissociation into $O^3P + O^1D$. Quantitative measurements of absorption coefficients were first made by Ladenburg and van Vorrhis[(150)] and have been often repeated both by photographic[(151–4)] and photoelectric[(117, 126, 147, 155–7)] photometry; nevertheless, some important discrepancies are apparent, even between the more recent results: for example, k_{max} varies from

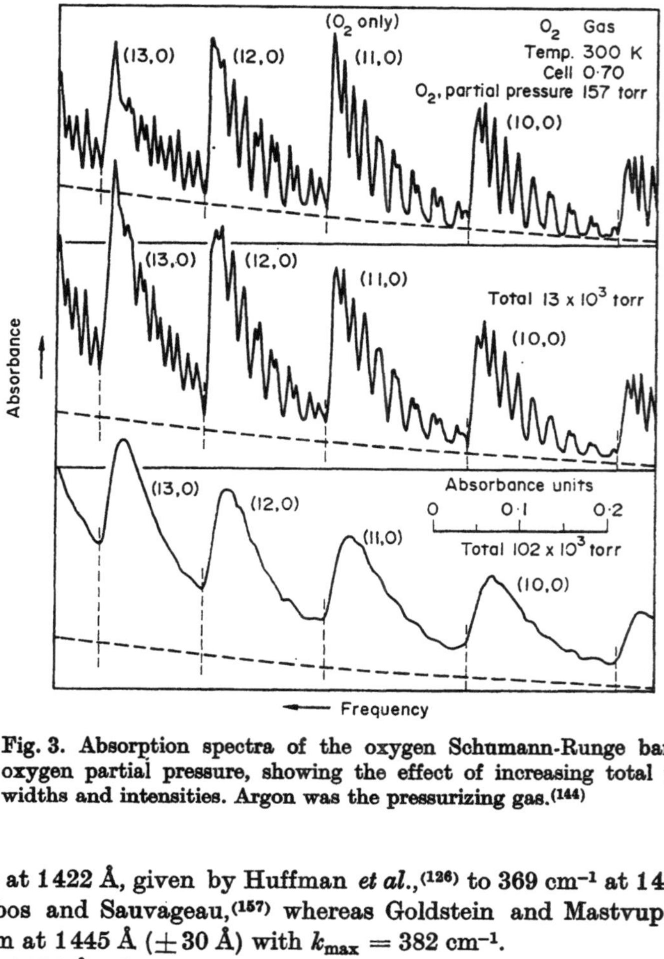

Fig. 3. Absorption spectra of the oxygen Schumann-Runge bands at constant oxygen partial pressure, showing the effect of increasing total pressure on line widths and intensities. Argon was the pressurizing gas.[(144)]

403 cm^{-1} at 1422 Å, given by Huffman *et al.*,[(126)] to 369 cm^{-1} at 1426 Å, reported by Lombos and Sauvageau,[(157)] whereas Goldstein and Mastvup[(154)] place the maximum at 1445 Å ($\pm$30 Å) with $k_{max} = 382\ cm^{-1}$.

Below 1300 Å, the spectrum is complicated by a number of strong bands superimposed on absorption continua.[(158–60)] The bands probably correspond to Rydberg states, but no identification or rotational analyses, except the very recent work of Ogawa and Yamawaki,[(163b)] has been made because of their diffuse character, and hence no Rydberg series have been identified converging on the first ionization potential. However, this has been determined by several other methods reviewed by Samson and Cairns;[160] these authors have obtained, by photoionization measurements, a very precise value: 12.063 ± 0.001 eV.

The situation is clearer below 1000 Å, where the first analysis by Price and Collins[158] has been largely confirmed and extended by Tanaka *et al.*,[161,162] Ogawa,[163a] and Yoshino and Tanaka,[164] who give very precise values for the series limits: 163,702 cm^{-1} for $v' = 0$, 164,811 cm^{-1} for $v' = 1$, and 165,885 cm^{-1} for $v' = 3$ (20.299, 20.437, and 20.570 eV); the electronic configuration of the observed Rydberg states is probably: $KK(\sigma_g 2s)^2(\sigma_u 2s)^2(\sigma_g 2p)(\pi_u 2p)^4(\pi_g 2p)^2 np\pi\, {}^3\Pi_u$. The vibrational constants of the corresponding O_2^+ state are $\omega_e = 1154$ cm^{-1} and $\omega_e x_e = 21.5$ cm^{-1}. Weak series converging to the ${}^2\Sigma_g^-$ state of O_2^+ have also been observed.[164] Near 500 Å, Codling and Madden have found four Rydberg series, two converging to a level in O_2^+ at 24.564 eV and the others to a level at 24.754 eV; the transitions involve excitation of an inner $\sigma_u 2s$ electron to outer $ns\sigma$ or $nd\pi$ orbitals; the system, broadened by autoionization, is quite weak with respect to the continuous absorption upon which it is superimposed. No other discrete absorption has been observed at shorter wavelength, but only a regularly decreasing continuum.[46]

Absorption and photoionization cross-sections have been extensively investigated[46,117,123,126,155,156,166–9] with somewhat large discrepancies even amongst the more recent data.[46,127–8,134,170,171] Huffman *et al.*[170] have compared their results in detail with those of earlier investigators; those of Matsunaga and Watanabe[171] are in better agreement with those of Cook and Metzger[127] or Samson and Cairns;[46] these also agree with measurements of de Reilhac and Damany.[134] Some high values could be due to absorption by atomic oxygen.[172]

Some metastable states have been directly observed in absorption ($a^1\Delta_g$ and $b\,{}^1\Sigma_g^+$ states[173]) and others in photoionization ($a\,{}^1\Delta_g$ state[174]).

From selected data, the potential energy curves for numerous states of O_2, O_2^-, and O_2^+ have been drawn.[139]

The photoionization efficiency curves for the formation of O_2^+ (thresholds 12.07, 16.3, and 18.4 eV) and $O^+ + O^-$ (17.25 eV) have been obtained by mass spectrometric study.[175] The vibrational structures of various electronic states of O_2^+ have been observed both with 584 Å and 744-736 Å photons, with rather large variations in the probability of vibrational excitation.[176,177] Calculations for the photoionization of π electrons from O_2[178] account for the formation of the $A\,{}^2\Pi_u$ state of O_2^+ with 736 Å photons, and for the new state at an energy of 23.7 eV as observed by Price with 304 Å photons (see ref. 178).

1.6. HALOGEN MOLECULES

The halogen molecules have hardly been investigated in the v.u.v. The absorption spectrum of fluorine has been studied by Iczkowski and Margrave;[179] they observed a Rydberg series with its first member at 952 Å; its convergence to 15.7 eV gives the ionization potential of F_2. A red-shaded band at 1035 Å corresponds to a charge transfer transition. The analysis of a progression of three bands at 834 Å leads to a dissociation energy of 3800 cm^{-1}, the possible products being F^2P_g and F^2P_u.

Only continuous absorption has been observed in the spectrum of Cl_2 in the v.u.v.,(180) but the absorption spectra of I_2(181, 182) and Br_2(182) present numerous discrete bands below 2000 Å; the vibrational structure has been studied with good resolution;(182) the analysis of a large number of Rydberg states gave a precise determination of the ionization potentials (9.40 eV for I_2 and 10.56 eV for Br_2).

Haranath and Rao(183) have studied the emission band spectra of I_2, Cl_2, and Br_2 in the spectral range 2400-1400 Å and have attempted vibrational analyses; the lower state is always identified as the ground state of the molecule. For I_2 it has been possible to assign all the observed electronic levels to terms expected from different MO electronic configurations on the basis of the scheme proposed by Mulliken.(184) Only two stable excited electronic levels have been well identified for Cl_2 and five for Br_2.

From the absorption spectra obtained by Cordes and Sponer(185) for interhalide molecules (ICl, IBr, and BrCl) between 1950 and 1600 Å, Mulliken(184) has proposed a *J-J* coupling between the $X + Y$ core and the σ^* excited electron

1.7. S_2, Se_2, Te_2

The v.u.v. spectrum of S_2 has been observed in absorption by Maeder and Miescher(186) and in emission by Tanaka and Ogawa(187) and by Lakshminarayana and Narasimham.(188) Only the vibrational structure has been studied for the seven band systems found in the region 2130-1600 Å. Two states (*D* and *C*) have been identified as $^3\Pi_u$ and four others as Σ states.

The photoionization study of S_2, Se_2, and Te_2 has been performed by Berkowitz and Chupka(189) who determined the thresholds for formation of ions and the dissociation energies.

1.8. P_2, PO, PH

Using a discharge through a helium-phosphorus mixture, Dressler(190) has observed v.u.v. emission bands identified as the *B-X* (1760-1680 Å) and *C-X* (1530-1500 Å) transitions of P_2 and the E-$X\,^2\Pi$ (1930-1820 Å) transition of PO, from which he has evaluated the heat of dissociation of PO ($D_0'' = 5.4$ eV). In flash-photolysis experiments(191) it was possible to observe in the v.u.v. three systems attributed to PH ($^1\Phi \leftarrow a^1\Delta$ near 1625 Å, $^1\Pi \leftarrow a^1\Delta$ at 1595 Å, and $^3\Pi \leftarrow X\,^3\Sigma^-$, extremely weak at 1435 Å), and several systems of P_2. The absorption spectrum of diatomic phosphorous, (in phosphorous vapour heated to 900 °C), has also been photographed at high resolution down to 1220 Å by Creutzberg(192) who has identified four $^1\Sigma_u^+$ states and four $^1\Pi_u$ states.

1.9. BH, BD

Strong absorption of BH and BD in the spectral region 1900-1340 Å has been obtained by Bauer *et al.*(193) from the flash photolysis of borine carbonyl; their analysis has led to the identification of eight new Rydberg states: values for the rotational and vibrational constants of most of these new electronic states have been derived as well as a precise determination of the ionization potential ($I_p = 9.77 \pm 0.05$ eV). The electronic configurations of the excited states are given.

1.10. HALOGEN ACIDS

Because of the experimental difficulties due to its great reactivity, no absorption spectrum of HF had been observed in the v.u.v. until the work of Waggoner *et al.*[65] who have confined the gas between two helium discharges used as spectrometric sources. Perhaps because of molecular associations at room temperature,[194] the spectrum lacks sharp vibrational structure near the first ionization potential (16.06 eV), in contrast to the absorption spectra of the three other halogen acids. These acids and their deuterated forms present very similar features: a continuous absorption, with its maximum near 1535 Å for HCl, 1785 Å for HBr, and 2150 Å for HI,[195, 196] is followed by very intense systems of relatively discrete bands. First observed by Leifson[23] for HCl and by Price[197] for HCl, HBr, and HI, the bands have been investigated in more detail by Barrow and Stamper for HBr[198] and by Stamper for DBr and DCl.[199] Rotational analyses have been performed for most of the observed bands, but little regularity is to be found amongst them, and the relationship between the spectra of isotopic species is frequently not clear. It is therefore difficult to reach detailed conclusions with regard to the nature of the upper states of the bands. Krauss *et al.* have analysed the photoionization efficiency curves of HCl and have discussed the electronic structure.[200] Frost *et al.*[201] and also Lempka *et al.* have studied the photoelectron spectrum.[202]

1.11. HALIDES

BeF

Among the monohalides of group II metals, only BeF has been studied in the v.u.v.;[203] its emission spectrum, obtained in a Schüler lamp containing BeF_2, consists of two band systems (1590-1560 and 1520-1500 Å); the states involved in these transitions are not identified.

BF (and BO)

The vacuum u.v. (2100-1200 Å) emission spectra of BF and BO have been simultaneously observed by Chretien[204] from an electrical discharge in a mixture of He and BF_3. He has identified the $A \rightarrow X$ (2100-1870 Å), $B \rightarrow X$ (1630-1480 Å), and $C \rightarrow X$ (1520-1400 Å) systems, and two bands of the $D \rightarrow X$ (1380 and 1340 Å) system of BF, and one system of BO between 1900 and 1680 Å. The vibrational structure of the $A\,^1\Pi \rightarrow X\,^1\Sigma$ band has been resolved and rotational constants determined.

All known molecular constants for the X, A, B, C, D, states are tabulated and the electronic configurations given. Comparison can be made with the theoretical values of the constants obtained from approximate Hartree-Fock calculation[95] for valence excited states or (LCAO-MO-SCF) approximation[205] for Rydberg states.

Other halides

For other halides, such as Cs halides[205] and Tl halides[206] only the photoionization has been studied in the v.u.v.—chiefly for deducing bond energies and ionization potentials; simultaneously, autoionization peaks have been observed in the ionization efficiency curves, close to the position of the corresponding peaks in the atomic absorption; the shifts are characterized by the chemical environment of the metallic ion.

1.12. CN

The emission spectrum of CN has been observed by Carroll[207] between 2100 and 1650 Å in the fourth order of a 3 m grating spectrograph. The rotational analyses of $E\,^2\Sigma \rightarrow X\,^2\Sigma$, $E\,^2\Sigma \rightarrow A\,^2\Pi$, and $J^2\Delta \rightarrow A\,^2\Pi$ allow the determination of all molecular constants of these states, and support the higher value (8.2 eV) for the dissociation energy of CN.

1.13. CO

Amongst the spectra of diatomics the case of the CO molecule has been subject to particularly extensive and frequent investigation. It is thus curious that, until recently, little work had been done in the v.u.v either in emission or in absorption. After the pioneering observations of Leifson[23] and Hopfield and Birge,[208] Read[209] analysed the rotational structure of some bands of the fourth positive system $A\,^1\Pi \rightarrow X\,^1\Sigma$ in emission. Recently, Tilford and Vanderslice[210] have studied, under high resolution, the systems $B\,^1\Sigma^+$ (Plate 2), $C\,^1\Sigma^+$, and $j\,^3\Sigma^+ \leftarrow X\,^1\Sigma^+$ down to 1060 Å.[210] Henning,[211] Tanaka,[212] and Takamine *et al.*[213] explored the absorption spectrum between 1000 and 600 Å and observed the convergence of two Rydberg series at an ionization potential of 19.58 eV. Ogawa[214] has remeasured, with improved accuracy, the wavelengths of all the lines between 690 and 630 Å and has proposed slight modifications to the assignment given in ref. 212 and some possible electronic configurations for the upper states of Rydberg transitions.

Using the continua of rare gases as a background, Tanaka *et al.*[215] have observed the $d\,^3\Pi_i \leftarrow X\,^1\Sigma^+$ bands and extended the $A\,^1\Pi$, $a\,^3\Sigma$, $a'\,^3\Sigma$, and $e\,^3\Sigma^- \leftarrow X\,^1\Sigma^+$ bands to higher v' values.

The rotational structures of some absorption bands of CO ($a'\,^3\Sigma^+$ and $e\,^3\Sigma^- \leftarrow X\,^1\Sigma^+$ transitions) have been investigated for the first time with high resolution by Herzberg and Hugo[216] as typical forbidden $^3\Sigma \leftrightarrow {}^1\Sigma$ transitions in heteronuclear diatomic molecules. The rotational and vibrational constants of the upper state, thus determined directly, are more precise than those obtained from its perturbations by other systems of CO. Other forbidden or very weak transitions have been observed and analysed;[217, 218] in particular, for the $A\,^1\Pi \leftarrow X\,^1\Sigma^+$ system,[217] electric quadrupole transitions ($\Delta J = 2$) have been observed for the first time in bands of an electric dipole allowed transition and in a heteronuclear diatomic molecule.

The analyses of v.u.v. spectra of CO have made it possible to calculate the potential energy curves for the $X\,^1\Sigma^+$, $A\,^1\Pi$, $I\,^1\Sigma^-$, $a\,^3\Pi$, $a'\,^3\Sigma^+$, $d\,^3\Delta_i$, and $e\,^3\Sigma^-$ states.(219) The spectra have also provided suitable material for testing theoretical calculations, such as the determination of Rydberg levels by the LCAO-MO-SCF method(220) and of valence excited states.(95)

Because of the sharpness of the bands, very few precise data exist for the absorption cross-sections of CO, especially at wavelengths longer than 1000 Å.(221, 222) The absolute transition probabilities have been determined only for the strong $A\,^1\Pi \leftarrow X\,^1\Sigma^+$ transition, by a somewhat different method, of combining the measurement of radiative lifetimes of the vibrational levels of the upper states with relative vibrational branching ratios.(4, 223)

The use of the Hopfield continuum as a background has allowed measurement of absorption coefficients in the 1000-600 Å region with a resolution of approximately 0.5 Å.(224, 225) Photodissociation and photoionization continua are difficult to separate from the band absorption above 600 Å. Below this wavelength only continuum absorption has been observed and measured.(34, 222, 226) This continuum is ascribed to photoionization processes; relatively strong near 500 Å, it decreases regularly towards the short wavelengths.

The photoelectron spectra have been obtained with 584 Å and 744-736 Å photons (helium and neon discharges), and the results are discussed in ref. 133.

1.14. NO

In spite of its extremely complex features, the v.u.v. absorption and emission spectra of the NO molecule are certainly the best analysed, and the variety of excited states, especially below the first ionization limit, are understood in great detail.

Briefly, the u.v. absorption spectrum begins at 2300 Å and, down to 1300 Å, it is rich in sharp bands which can be resolved into fine structure due to rotation, electron spin, and Λ coupling; however, the ordering of the observed bands into electronic and vibrational systems is by no means obvious. Even the doublet states (same multiplicity as the ground state) comprise no less than twenty-five Rydberg states (R) and five to six valence states (N-R) coexisting in the energy range 5-10 eV, and strong interactions between R and N-R states of the same symmetry occur (homogeneous interactions). Weaker interactions are also observable between states of different symmetry (heterogeneous interaction, $^2\Sigma$ — $^2\Pi$ for example).

Since the first attempts at analysis,(227–33) extensive work has been done at high resolution, and a very complete study of the interactions has been successfully accomplished by Miescher and his coworkers. We give references to only the four more important papers which review the others, and give a complete bibliography.(234–7) A complete analysis of the vibrational and rotational structures for up to four NO isotopes ($N^{14}O^{16}$, $N^{15}O^{16}$, $N^{14}O^{18}$, $N^{15}O^{18}$) has lead to satisfactory identification of the bands observed (see Plate 3).

A striking illustration of the significance of their "deperturbation" calculations is the result obtained in experiments at low temperature by Roncin *et al.*:[238] in the absorption spectrum of NO isolated in a rare gas matrix, the vibrational spacings in the $B\,^2\Pi$ and $B'\,^2\Delta$ states become perfectly regular and identical to the "deperturbed" ones on account of the disappearance of the Rydberg transitions in the condensed state (see p. 184).

In the region below 1000 Å the only published analyses are by Tanaka[239] and Huber,[240] who have observed three Rydberg-type series converging to 872 Å, 749 Å, and 677 Å, and some fainter unclassified bands. The comparison with the well-known Rydberg states of N_2, CO, and O_2 makes it possible to assign, on the basis of computed quantum defects, the electronic configurations corresponding to these α, β, and γ series (Tanaka series).

There have been only a few measurements of the absorption coefficients and photoionization yields of NO. Absolute integrated absorption intensities have been measured, above 1700 Å, on pressure broadened bands[241] as direct measurements on sharp lines with moderate resolution[242] give only approximate values. Quantitative data on the absorption of NO being of particular interest to the study of the upper atmosphere, these measurements, in association with photoionization studies, have been extended to shorter wavelength[243–5] and even below 1000 Å.[246–9] Beyond 500 Å[134] no structure is observed in the more recent experiments; the irregularities previously found have been attributed to perturbations in the line source used as a background; the absorption coefficient decreases quite smoothly down to 100 Å.

Using the electron-excitation phase-shift method, Hesser[4] has determined the absolute transition probability for the *D-X* transition, in good agreement with ref. 241.

Numerous molecular data have been determined from the v.u.v. spectra of NO and NO^+, and thus the vibrational and rotational constants of almost all the excited states of NO are known; their potential energy curves have been drawn[139] and the ionization potentials, determined with great precision, are 9.267 for IP_I ($X\,^1\Sigma^+$ of NO^+), 14.22 for IP_{II} ($^3\Sigma^+$ of NO^+), 16.56 eV for IP_{III} (probably $^3\Delta$), and 18.33 eV for IP_{IV} ($A\,^1\Pi$).[240, 250, 251]

The NO molecule, for which there is no ambiguity in the assignment of the observed Rydberg states, has been chosen as a critical test for non-empirical calculations;[204] each Rydberg state is represented by an electron configuration in which an outer electron moves in the field of a fixed molecular core constrained to be the LCAO-MO-SCF orbitals of either the NO molecule or the NO^+ ion, and, the calculations are, indeed, in good agreement with experiments.

1.15. SiO, GeO

The complete vibrational analyses of two red-degraded systems in the absorption spectra of SiO and GeO, have enabled Rowlinson and Barrow[252] to calculate the constants of the *E* and *F* states of these molecules and to fix their dissociation limits (8.0 eV for SiO and 6.8 eV for GeO).

2. Inorganic polyatomic molecules

As pointed out in the introduction to this chapter, there is very little to add to the extensive treatment by Herzberg(2) of the v.u.v. spectra of polyatomic molecules. We shall mention here only molecules for which new results have appeared since his review of 1966.

2.1. H_2O, D_2O

The v.u.v. absorption spectra of H_2O and D_2O are mainly continuous, and some structure appears only in the 1450-900 Å region. A first progression, between 1411-1256 Å, is diffuse in character and does not allow any rotational analysis. At shorter wavelengths the bands are generally more intense, and some of them well resolved; actually they form two systems with origins at 1240 and 1219 Å and with a common convergence limit at 982 Å. Rotational analyses(253) of two bands of D_2O (1240 and 1223 Å) and of one of H_2O (1223 Å) and the vibrational analyses(254) of the systems associated with the bands at 1240 and 1219 Å, have helped to ascertain the geometry of the corresponding excited state ($r \simeq 1.02$ Å, $2\alpha \simeq 110°$). Measurement of the intensity of the fluorescence produced by monochromatic wavelengths between 900 and 700 Å(255) provides evidence of the photodissociation of H_2O into electronically excited fragments:

$$H_2O \to H \ (n = 1) + OH \ (A\,^2\Sigma^+) \quad 1360 \text{ Å},$$
$$H_2O \to H_2 \ (B\,^1\Sigma_u^+) + O \ (^3P) \quad 767 \text{ Å},$$
$$H_2O \to H \ (n = 2) + OH \ (X\,^2\Pi) \quad 820 \text{ Å}.$$

The dissociation into ionic fragments (H^+, O^+, OH^+, $H_2O^+\ ^2A_1$, and $^2B_1$) has been clearly established by photoelectron spectroscopy(256) and mass spectrometric analysis of photoionization products.(257)

The absorption coefficients and photoionization yields have been measured by Wainfan *et al.*(169) and by Metzger and Cook(258) down to 500 and 600 Å respectively. At shorter wavelengths (500-100 Å) the absorption coefficients have been measured by de Reilhac and Damany.(259) For these relatively energetic photons, ionization processes largely prevail over dissociative.

The use of water as a solvent in v.u.v. spectroscopy or photochemistry has stimulated new experiments for obtaining precise values of the extinction coefficient(260-2) and its temperature dependence.(262, 263) The absorption of liquid water increases rapidly below 1950 Å, making it difficult to use as a solvent below 1600 Å.(264)

Several calculations of Rydberg states of the H_2O molecule have been attempted by somewhat different approaches,(265-7) starting with a one-centre core model of a neon-like configuration. Such calculations give results in satisfactory agreement with experiment.

2.2. H_2S, D_2S

In conformity with their structures, the molecules H_2S and H_2O exhibit very similar spectra. The spectrum of H_2S is somewhat simpler(268) because 0-0 transition is missing. The bands comprise a very sharp Q branch accompanied by weak P and R branches with rotational structure evenly spaced because of the slight change in the moment of inertia in passing from the ground to the excited states. Actually, from this point of view H_2S approximates fairly well to a symmetrical top, the axis being perpendicular to the HSH plane.

Comparative studies at low resolution, of absorption spectra of several sulphides and of mercaptan(269) have established a criterion for distinguishing Rydberg and non-Rydberg transitions.

By combining a v.u.v. monochromator and a mass spectrometer, Dibeler and Liston(270) have obtained photoionization yield curves for the molecule and the ionic fragments, and have determined both the heats of formation of the ions and radicals and the ionization thresholds. The threshold value for the formation of the H_2S^+ ion is in good agreement with that obtained by photoelectron spectroscopy(271) but is lower than that given by Rydberg extrapolation(272) from previous photoionization experiments.(273)

2.3. CO_2

CO_2 is completely transparent from the visible down to 1750 Å; from this wavelength commences a continuous absorption overlapped by many diffuse bands which do not permit analyses. Between 1400 and 1250 Å, a second group of bands, too diffuse for rotational analysis, resolves into two progressions with a frequency interval of 1225 cm^{-1}; it is expected that the corresponding upper states are bent.(274, 275)

Below 1200 Å a very strong band spectrum extends to 650 Å; most of the bands have been classified in eleven Rydberg series converging to ionization potentials 13.78, 17.31, 18.07, and 19.38 eV;(276) the remaining bands are provisionally ordered in seven progressions but need further analysis. All these ionization potentials, plus another at 21 eV, have also been observed by electron impact methods(277) and photoelectron spectroscopy.(277–81) Mass spectrometric study of photoionization(175, 282) confirms other observations, including strong autoionization in the 830-600 Å region. In addition, O^+ ions are detected at 19.1 eV from the process

$$CO_2 + h\nu \rightarrow O^+ + CO + e.$$

Absorption and photoionization coefficients have been re-measured in the whole spectral range(283, 284) and have been extended down to 230 Å by Cairns and Samson(226) and to 100 Å by de Reilhac *et al.*(134, 285) Only continuous absorption is observed below 600 Å, with a shoulder near 350 Å (36 eV), an energy a little lower than that calculated by Peters(286) for the ionization of a $2\sigma_u$ electron (39.2 eV) or for the appearance potential of CO_2^{++} (38 eV).(287)

2.4. OCS, CS_2

The most complete descriptions of the v.u.v. absorption spectra of OCS[288a] and of CS_2[288b] are those by Price and Simpson covering the entire range from the near u.v. down to 1200 Å. In spite of the relatively low dispersion employed it is clear that almost all the bands are diffuse; even some sharp ones near 1400 Å in the OCS spectrum proved, when studied at high resolution,[289] to be too diffuse to show rotational fine structure. Nevertheless, precise vibrational analyses of the spectrum in the region 1420-1350 Å (see Plate 4) have allowed Kopp to establish the characteristics of the upper states involved, and to obtain their approximate B values. In OCS two long progressions have been observed recently between 1800 and 1600 Å,[290] with vibrational spacings very close to the bending frequency ($\nu_2 = 520$ cm^{-1}).

Tanaka *et al.*[291] have made detailed photographic studies of the absorption spectra of these molecules in the wavelength region 1650-600 Å in order to determine the ionization potentials. In OCS, seven Rydberg series have been tabulated, converging to the three ionization potentials 11.23, 16.03, and 17.93 eV. Many other progressions—generally of sharp bands—have been tentatively classified, giving a prominent difference of about 750 cm^{-1}, which corresponds to the ν_2 frequency.

The CS_2 spectrum shows similar features: eight Rydberg series converge on four ionization potentials (10.07, 14.47, 16.17, and 19.5 eV); most of the remaining bands have been classified as progressions having average intervals of 490 and 590 cm^{-1}. Matsunaga and Watanabe have measured the absorption coefficients of OCS[292] in the 1700-1070 Å region. Cook and Ogawa have investigated the absorption, photoionization, and fluorescence of OCS[293a] and CS_2[293b] in the 960-600 Å region, observing in OCS a series of "apparent emission type", i.e. a series of Beutler-Fano resonances characteristic of preionization.

Mass-spectrometry study of photoionization[175] has rendered precise the nature and abundance of the ionic products, while photoelectron spectroscopy[278, 280, 281, 294] shows evidence of new ionization potentials at 12.67 eV in OCS and 17.3 eV in CS_2.

2.5. N_2O

There is little recent work on the v.u.v. spectra of N_2O; Tanaka *et al.*[291] listed up to nine Rydberg series, accompanied by some vibrational bands, between 1200 and 750 Å, and determined the corresponding ionization potentials. However, the electronic states and the configurations at the series limits are known only for the first five series. Absolute absorption and photoionization coefficients have been remeasured by Cook *et al.*[295] between 1000 and 600 Å, and by de Reilhac and Damany,[134] down to 100 Å, with a continuum background. The photodissociation and photoionization processes, of special interest to the photochemistry of the upper atmosphere, have been studied by various methods including determination of relative yields from mass analysis,[296, 297] fluorescence measurements[298, 299] and photoelectron spectroscopy.[278, 281, 300] The formation of the NO^+ ion is the predominant process.

2.6. O_3, SO_2

In spite of the similarity of their electronic and geometrical structures, O_3 and SO_2 have quite different u.v. absorption spectra. That of O_3 is almost entirely continuous from the near u.v. with only some broad maxima and very diffuse overlapping bands;[301, 302] the shorter wavelength continuum starts at 950 Å, i.e. at the ionization potential of 13 eV derived from electron impact experiments.[303]

In contrast, SO_2[289] presents many sharp absorption bands in the near u.v. followed by quite a transparent region between 1800-1350 Å, apart from three weak diffuse bands near 1550 Å. The spectrum below 1350 Å[304, 325] is interpreted in terms of a Rydberg series converging to the 12.34 eV ionization potential, obtained by photoionization,[305] but a more thorough analysis is difficult.

From photoionization measurements, Dibeler and Liston[270] have obtained the photoionization yield curves for the molecule and have analysed the ionic fragments by mass spectrometry. They have also derived the heats of formation of ions and radicals.

2.7. $HgCl_2$, $HgBr_2$

$HgCl_2$ and $HgBr_2$ show two systems of diffuse bands in the v.u.v. (1700 and 1500 Å for $HgCl_2$, 1840 and 1550 Å for $HgBr_2$) but only the systems of longer wavelengths have been studied as an example of well-developed sequences (ref. 2, p. 156). Notwithstanding the rather simple structure of the $HgCl_2$ spectrum, the previous analyses[8, 306] have been questioned in the light of a calculation of Franck-Condon factors[307] (the two-dimensional isotropic harmonic oscillators with common centre are considered as two independent one-dimensional oscillators). Comparison of the calculated and experimental relative intensities rules out the hypothesis of a $\nu_2'' = 70\ cm^{-1}$ frequency. Analogous analysis of the $HgBr_2$ spectrum leads to an identical conclusion.[308]

2.8. NH_3

Extensive studies reviewed in detail by Herzberg[2] were made long ago on the v.u.v. absorption spectrum of NH_3; we need only cite the later work of Watanabe and Sood[309] who have measured absorption coefficients and photoionization yields from 1650 to 580 Å, and that of de Reilhac and Damany,[259] who give the absorption coefficients between 400 and 100 Å.

Because of their chemical significance, the photodissociation and photoionization processes in NH_3 have been studied by various methods such as measurement of fluorescence intensities,[255, 258, 310] photoelectron spectroscopy,[271, 311] and determination of photoionization yields[257, 312] for various ions (NH_3^+ and NH_4^+) (see Fig. 4).

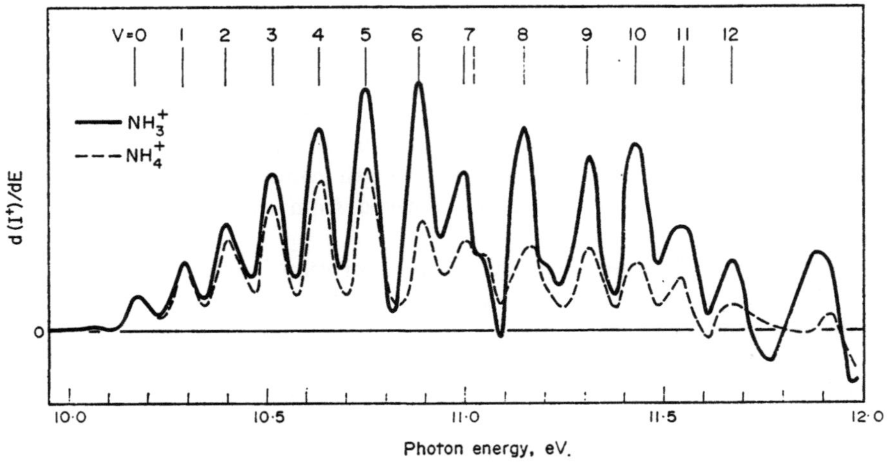

FIG. 4. The first derivative of the photoionization efficiency curves for NH_3^+ and NH_4^+ formed in NH_3 gas.(312)

2.9. C_3O_2

Infrared and electron diffraction studies show that C_3O_2 is a linear molecule with a very low fundamental frequency ($\nu_7'' = 63$ cm^{-1}); moreover it has a typical structure of overlapping orbitals particularly adapted to theoretical study. Nevertheless, the v.u.v. absorption spectrum of C_3O_2 has been investigated only recently. In fact, there is no direct evidence in this spectrum of a frequency as low as 63 cm^{-1}. In the broad 1870 Å bands, a frequency difference of 385 cm^{-1} appears.(313) The most probable explanation for this frequency is that the observed

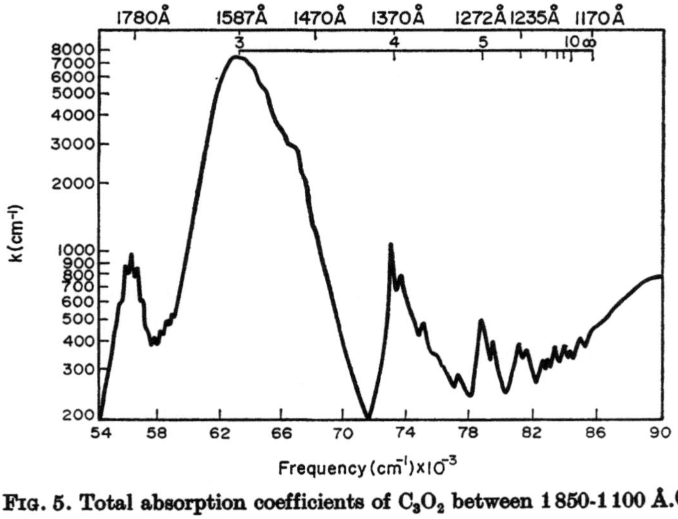

FIG. 5. Total absorption coefficients of C_3O_2 between 1850-1100 Å.(315)

structure corresponds to the v_7 sequence: then the true frequency ν_7' would be $385 + 63 = 448$ cm^{-1}, a plausible enhancement of ν_7'' for a transition of an electron from a $2\pi_u$ orbital (which favours bending) to a $2\pi_g$ orbital.

The strong, broad continuum, with a maximum ($k_\nu = 7000$ cm^{-1}) at 1590 Å, has been identified as the $n = 3$ member of a Rydberg series,(314, 315) the following members (1370 Å, 1272 Å, ...) becoming rapidly weaker and converging to 1172 Å (10.60 eV). These bands are diffuse, but show some vibrational structure corresponding to ν_1' (C = O symmetric stretch: 2100 cm^{-1}) and ν_2' (C = C symmetric stretch: 700 cm^{-1}) (Fig. 5).

2.10. SF_6

SF_6 and other hexafluorides are believed to have octahedral symmetry in their ground states; unfortunately none of these molecules exhibit in the v.u.v.,(316) the sharp band systems required for checking the special selection rules for this type of molecule. Investigations at wavelengths shorter than 1000 Å(317) have been necessary to reveal discrete structures. In this region appears a regular vibrational progression of about twenty members, between 600 and 550 Å, with a spacing of approximately 600 cm^{-1}: this may correspond to a modified symmetric stretching mode [$\nu_1(a_{1g})$] of an excited electronic state of neutral SF_6 ($\nu_1 =$ 775 cm^{-1} for the ground state). Secondly, a faint structure, forming an autoionizing Rydberg series is observed below 505 Å: this probably pertains to an excited state of SF_6 identical with that involved in the formation of SF_5^+ by photoionization(318) and by electron impact.(319) The photoelectron spectrum has been obtained by Frost *et al.*(320)

2.11. OTHER MOLECULES

Many other polyatomic molecules have been recently investigated, especially by mass spectrometric study of their photoionization products, a method which is under rapid development. For example, the nature and the abundance of the ionic fragments, as well as their appearance thresholds, have been determined for HCN,(321, 322) BF_3 and B_2F_4,(323) CF_4, C_2F_6, C_3F_6, C_4F_{10}, CF_3H, CF_3Cl, CF_3Br, and CF_3I,(324) and for many organic molecules not reviewed here. These data are of particular interest to photochemistry.

References

1. Herzberg, G., *Molecular Spectra and Molecular Structure*, vol. 1, *Spectra of Diatomic Molecules*, D. van Nostrand, New York, 1950.
2. Herzberg, B., *Molecular Spectra and Molecular Structure*, vol. III, *Electronic Spectra and Electronic Structure of Polyatomic Molecules*, D. van Nostrand, New York, 1966.
3. Marr, G. V., *Photoionization Processes in Gases*, Academic Press, New York, 1967.
4. Hesser, J. E., *J. Chem. Phys.* 48, 2518 (1968).
5. Pearse, R. W. B., and Gaydon, A. G., *Identification of Molecular Spectra*, Chapman Hall, London, 1963.

6. SLATER, J. C., *Quantum Theory of Molecules and Solids*, vol. 1, *Electronic Structure of Molecules*, McGraw Hill, New York, 1963.
7. ROSEN, B., *Encyclopedia of Physics*, Springer-Verlag, Berlin, 1964, vol. 27, pp. 221-474.
8. SPONER, H., and TELLER, E., *Rev. Mod. Phys.* **13**, 75 (1941).
9. RAMSAY, D. A., *Determination of Organic Structures by Physical Methods*, Academic Press, New York, 1962, vol. 2, pp. 245-338.
10. MULLIKEN, R. S., *Rev. Mod. Phys.* **3**, 89 (1931); **4**, 1 (1932); *Phys. Rev.* **40**, 55 (1932), **41**, 49 and 751 (1932); **43**, 279 (1933); *J. Chem. Phys.* **1**, 492 (1933); **3**, 375, 506, 564, 635, and 720 (1935); **7**, 14, 20, 201, 353, and 570 (1939); **8**, 234 and 382 (1940); **23**, 1833, 1841, and 2338 (1955); *J. Chim. Phys.* **46**, **497** and 675 (1949); *J. Am. Chem. Soc.* **72**, 600 (1950); **74**, 811 (1952); **86**, 3183 (1964); **88**, 1849 (1966); **91**, 4615 (1969).
11. MULLIKEN, R. S., and RIEKE, C. A., *Rep. Prog. Phys.* **8**, 231 (1941).
12. PLATT, J. R., and KLEVENS, H. B., *Rev. Mod. Phys.* **16**, 182 (1944).
13. WALSH, A. D., *A. Rev. Phys. Chem.* **5**, 163 (1954).
14. WALLACE, L., *Astrophys. J.*, suppl. **7**, 165 (1962).
15. WILKINSON, P. G., *J. Molec. Spectrosc.* **6**, 1 (1961); *Astrophys. J.*, **138**, 778 (1963).
16. WEISSLER, G. L., *Encyclopedia of Physics*, Springer-Verlag, Berlin, 1956, vol. 21, pp. 304-42.
17. WATANABE, K., NAKAYAMA, T., and MOTTL, J., *Final Report on Ionization Potentials of Molecules*, Univ. of Hawaii, 1959, and *J. Quant. Spectrosc. Radiative Transfer* **2**, 369 (1962).
18. VILESOV, F. I., *Usp. fiz. Nauk* **81**, 669 (1963) [*Sov. Phys. Usp.* **6**, 888 (1964)].
19. HIRT, R. C., and SCHMITT, R. G., *Bibliography of Vacuum Ultraviolet*, Document ASTIA no. AD 401,498, US Department of Commerce, Washington, 1963.
20. JOHANNIN-GILLES, A., *Bibliographie des Spectres des Molécules polyatomiques dans l'ultraviolet à vide*, Collège scientifique universitaire de Brest, France, 1966.
21. SCHUMANN, V., *Ber. wien. Akad.* **102** IIa, 415 (1893); *Wien. Akad. Anzeiger* 24 and 121 (1895); 71 (1900); Smithsonian Contributions, no. 1413, and papers listed in ref. 22, p. 151.
22. LYMAN, T., *The Spectroscopy of the Extreme Ultraviolet*, Longmans, London, 1928.
23. LEIFSON, S. W., *Astrophys. J.* **63**, 73 (1926).
24. HOPFIELD, J. J., *Astrophys. J.* **72**, 139 (1930).
25. DIEKE, G. H., and HOPFIELD, J. J., *Phys. Rev.* **28**, 849 (1926); **30**, 400 (1927).
26. DIEKE, G. H., *J. Molec. Spectrosc.* **2**, 494 (1958).
27. SCHUBERT, K. E., and HUDSON, R. D., Aerospace Corporation Report, ATN 64/9233, 2, 1963.
28. JUNKES, J., SALPETER, E. W., and MILAZZO, G., *Atomic Spectra in the Vacuum Ultraviolet*, Specola Vaticana, 1965, vol. 1.
29. WINANS, J. G., and STUECKELBERG, E. C. G., *Proc. Natn. Acad. Sci. U.S.A.* **14**, 867 (1928).
30. NAMIOKA, T., *J. Chem. Phys.* **40**, 3154 (1964).
31. WILKINSON, P. G., *Can. J. Phys.* **46**, 1225 (1968).
32. BEUTLER, H., DEUBNER, A., and JÜNGER, H. O., *Z. Phys.* **98**, 181 (1935).
33. BEUTLER, H., and JÜNGER, H. O., *Z. Phys.* **100**, 80 (1936).
34. JEPPESEN, C. R., *Phys. Rev.* **54**, 68 (1938).
35. TANAKA, Y., *Scient. Pap. Inst. Phys. Chem. Res. Tokyo* **42**, 49 (1944).
36. NAMIOKA, T., *Sci. Light, Tokyo* **2**, 73 (1953); *J. Chem. Phys.* **40**, 3154 (1964); **41**, 2141 (1964).
37. HERZBERG, G., and MONFILS, A., *J. Molec. Spectrosc.* **5**, 482 (1960).
38. MONFILS, A., *J. Molec. Spectrosc.* **15**, 265 (1965); **25**, 513 (1968).
39. HERZBERG, G., and HOWE, L. L., *Can. J. Phys.* **37**, 636 (1959).
40. NAMIOKA, T., *J. Chem. Phys.* **43**, 1636 (1965).
41. BEUTLER, H., *Z. Phys. Chem. Frankfurt* **29** B, 315 (1935).
42. HADDAD, G. N., LOKAN, K. H., FARMER, A. J. D., and CARVER, J. H., *J. Quant. Spectrosc. Radiative Transfer* **8**, 1193 (1968).

43. Po Lee, and Weissler, G. L., *Astrophys. J.* **115**, 570 (1952).
44. Wainfan, N., Walker, W. C., and Weissler, G. L., *Phys. Rev.* **99**, 542 (1955).
45. Cook, G. R., and Metzger, P. H., *J. Opt. Soc. Am.* **54**, 968 (1964).
46. Samson, J. A. R., and Cairns, R. B., *J. Opt. Soc. Am.* **55**, 1035 (1965).
47. Dibeler, V. H., Reese, R. M., and Krauss, M., *J. Chem. Phys.* **42**, 2045 (1965).
48. Comes, F. J., *Phys. Letters* **24** A, 465 (1967).
49. Chupka, W. A., and Berkowitz, J., *J. Chem. Phys.* **48**, 5726 (1968).
50. Doolittle, P. H., and Schoen, R. I., *Phys. Rev. Letters* **14**, 348 (1965).
51. Berry, R. S., *J. Chem. Phys.* **45**, 1228 (1966).
52. Bardsley, J. N., *Chem. Phys. Letters* **1**, 229 (1967).
53. Khare, S. P., *Phys. Rev.* **173**, 43 (1968).
54. Hopfield, J. J., *Phys. Rev.* **35**, 1133 (1930); **36**, 784 (1930); *Astrophys. J.* **72**, 133 (1930).
55. Samson, J. A. R., *Techniques of Vacuum Ultraviolet Spectroscopy*, Wiley, New York, 1967, p. 98.
56. Wilkinson, P. G., and Byram, E. T., *Appl. Opt.* **4**, 581 (1965).
57. Huffman, R. E., Larrabee, J. C., and Tanaka, Y., *Appl. Opt.* **4**, 1581 (1965).
58. Huffman, R. E., Tanaka, Y., and Larrabee, J. C., *J. Opt. Soc. Am.* **52**, 851 (1962).
59. Tanaka, Y., Jursa, A. S., le Blanc, F. J., *J. Opt. Soc. Am.* **48**, 304 (1958).
60. Smith, A. L., and Meriwether, J. W., *J. Chem. Phys.* **42**, 2984 (1965).
61. Huffman, R. E., Tanaka, Y., and Larrabee, J. C., *Appl. Opt.* **2**, 617 (1963).
62. Metzger, P. H., and Cook, G. R., *J. Opt. Soc. Am.* **55**, 516 (1965).
63. Wilkinson, P. G., *Can. J. Phys.* **45**, 1715 (1967).
64. Tanaka, Y., and Yoshino, K., *J. Chem. Phys.* **39**, 3081 (1963); **50**, 3087 (1969); **53**, 2012 (1970).
65. Waggoner, M. G., Chow, K. W., and Smith, A. L., *Chem. Phys. Letters* **3**, 151 (1968).
66. Shardanand, *J. Quant. Spectrosc. Radiative Transfer* 8, 1373 and 1533 (1968).
67. Castex, M.-C., and Damany, N., *Chem. Phys. Letters* **13**, 158 (1972).
68. Huffman, R. E., and Katayama, D. H., *J. Chem. Phys.* **45**, 138 (1966).
69. Samson, J. A. R., and Cairns, R. B., *J. Opt. Soc. Am.* **56**, 1140 (1966).
70. Ditchburn, R. W., Justum, P. J., and Marr, G. V., *Proc. Roy. Soc. London* A, **219**, 89 (1953).
71. Hudson, R. D., *Phys. Rev.* **135** A, 1212 (1964).
72. Hudson, R. D., and Carter, V. L., *Phys. Rev.* **139** A, 1426 (1965).
73. Marr, G. V., and Creek, D. M., *Proc. Roy. Soc. London* A, **304**, 233 (1968).
74. Hudson, R. D., *J. Chem. Phys.* **43**, 1790 (1965).
75. Creek, D. M., and Marr, G. V., *J. Quant. Spectrosc. Radiative Transfer* **8**, 1431 (1968).
76. Ogawa, M., and Tanaka, Y., *J. Chem. Phys.* **30**, 1354 (1959).
77. Wilkinson, P. G., and Houk, N. B., *J. Chem. Phys.* **24**, 528 (1956).
78. Tilford, S. G., and Wilkinson, P. G., *J. Molec. Spectrosc.* **12**, 231 (1964).
79. Codling, K., *Astrophys. J.* **143**, 552 (1966).
80. Tanaka, Y., *J. Opt. Soc. Am.* **45**, 663 (1955).
81. Tanaka, Y., Ogawa, M., and Jursa, A. S., *J. Chem. Phys.* **40**, 3690 (1964).
82. Tilford, S. G., Wilkinson, P. G., Franklin, V. B., Naber, R. H., Benesch, W., and Vanderslice, J. T., *Astrophys. J.* (suppl. ser.) **13**, 31 (1966).
83. Wilkinson, P. G., *J. Chem. Phys.* **30**, 773 (1959); *J. Quant. Spectrosc. Radiative Transfer* **2**, 343 (1962).
84. Wilkinson, P. G., and Mulliken, R. S., *J. Chem. Phys.* **31**, 674 (1959).
85. Vanderslice, J. T., Tilford, S. G., and Wilkinson, P. G., *Astrophys. J.* (a) **141**, 395 (1965); (b) **141**, 427 (1965); (c) **141**, 1226 (1965); (d) **142**, 84 (1965); (e) **142**, 1203 (1965).
86. Benesch, W., Vanderslice, J. T., Tilford, S. G., and Wilkinson, P. G., *Astrophys. J.* **142**, 1227 (1965).
87. Vanderslice, J. T., Wilkinson, P. G., and Tilford, S. G., *J. Chem. Phys.* **42**, 2681 (1965).
88. Chiu, Y. N., *J. Chem. Phys.* **42**, 2671 (1965).

89. MULLIKEN, R. S., *The Threshold of Space*, Pergamon Press, New York, 1957, p. 169.
90. DRESSLER, K., and LUTZ, B. L., *Phys. Rev. Letters* **19**, 1219 (1967).
91. LUTZ, B. L., *J. Chem. Phys.* **51**, 706 (1969).
92. LEFEBVRE-BRION, H., and MOSER, C. M., *J. Chem. Phys.* **43**, 1394 (1965).
93. DUNCAN, A. B. F., *J. Chem. Phys.* **42**, 2453 (1965).
94. DUNCAN, A. B. F., and DAMIANI, A., *J. Chem. Phys.* **45**, 1245 (1966).
95. NESBET, R. K., *J. Chem. Phys.* **43**, 4403 (1965).
96. HEIDEMAN, H. G., KUYATT, C. E., and CHAMBERLAIN, G. E., *J. Chem. Phys.* **44**, 355 (1966).
97. LASSETTRE, E. N., SKERBELE, A., and MEYER, V. D., *J. Chem. Phys.* **45**, 3214 (1966).
98. WORLEY, R. E., *Phys. Rev.* **64**, 207 (1943).
99. TANAKA, Y., and TAKAMINE, T., *Scient. Pap. Inst. Phys. Chem. Res. Tokyo* **39**, 427 (1942).
100. CARROLL, P. K., and YOSHINO, K., *J. Chem. Phys.* **47**, 3073 (1967).
101. TAKAMINE, T., SUGA, T., and TANAKA, Y., *Scient. Pap. Inst. Phys. Chem. Res. Tokyo* **34**, 854 (1938).
102. WORLEY, R. E., *Phys. Rev.* **89**, 863 (1953).
103. OGAWA, M., and TANAKA, Y., *Can. J. Phys.* **40**, 1593 (1962).
104. OGAWA, M., *Can. J. Phys.* **42**, 1087 (1964).
105. LOFTHUS, A., *The Molecular Spectrum of Nitrogen*, Spectroscopic Report no. 211, Department of Physics, University of Oslo, 1960.
106. CARROLL, P. K., *J. Quant. Spectrosc. Radiative Transfer* **2**, 417 (1962).
107. OGAWA, M., TANAKA, Y., and JURSA, A. S., *Can. J. Phys.* **42**, 1716 (1964).
108. DRESSLER, K., *Proc. Int. Conf. Spectr. Bombay* **1**, 104 (1967); *Can. J. Phys.* **47**, 547 (1969).
109. CARROLL, P. K., and COLLINS, C. P., *Can. J. Phys.* **47**, 563 (1969).
110. LEFEBVRE-BRION, H., *Can. J. Phys.* **47**, 541 (1969).
111. DRESSLER, K., *J. Chem. Phys.* **30**, 1621 (1959).
112. BASS, A. M., *J. Chem. Phys.* **40**, 695 (1964).
113. OGAWA, M., TANAKA, Y., and JURSA, A. S., *J. Chem. Phys.* **41**, 3351 (1964).
114. TANAKA, Y., INNES, F. R., and JURSA, A. S., *J. Chem. Phys.* **42**, 1183 (1965).
115. HUFFMAN, R. E., LARRABEE, J. C., and TANAKA, Y., *J. Chem. Phys.* **45**, 3205 (1966).
116. WEISSLER, G. L., LEE, P., and MOHR, E. I., *J. Opt. Soc. Am.* **42**, 84 (1952).
117. WATANABE, K., and MARMO, F. F., *J. Chem. Phys.* **25**, 965 (1956).
118. WATANABE, K., ZELIKOFF, M., and INN, E. C. Y., *Advances in Geophysics*, Academic Press, New York, 1958, vol. 5, p. 185.
119. DITCHBURN, R. W., BRADLEY, J. E. S., CANNON, C. G., and MUNDAY, G., *Rocket Exploration of the Upper Atmosphere*, Pergamon Press, New York, 1958, p. 327.
120. CHING, B. K., COOK, G. R., and BECKER, R. A., *J. Quant. Spectrosc. Radiative Transfer* **7**, 323 (1967).
121. NICHOLLS, R. W., *J. Res. Natn. Bur. Stand.* **65** A, 451 (1961).
122. ZARE, R. N., LARSSON, E. O., and BERG, R. A., *J. Molec. Spectrosc.* **15**, 117 (1965).
123. CLARK, K. C., *Phys. Rev.* **87**, 271 (1952).
124. CURTIS, J. P., *Phys. Rev.* **94**, 908 (1954).
125. ASTOIN, N., and GRANIER, J., *Compt. Rend.* **244**, 1350 (1957).
126. HUFFMAN, R. E., TANAKA, Y., and LARRABEE, J. C., *Discuss. Faraday Soc.* **37**, 159 (1964); *J. Chem. Phys.* **39**, 910 (1963).
127. COOK, G. R., and METZGER, P. H., *J. Chem. Phys.* **41**, 321 (1964).
128. COOK, G. R., CHING, B. K., and BECKER, R. A., *Discuss. Faraday Soc.* **37**, 149 (1964).
129. COOK, G. R., and OGAWA, M., *Can. J. Phys.* **43**, 256 (1965).
130. HUDSON, R. D., and CARTER, V. L., *Can. J. Phys.* **47**, 1840 (1969).
131. LAWRENCE, G. M., MICKEY, D. L., and DRESSLER, K., *J. Chem. Phys.* **48**, 1989 (1968).
132. HESSER, J. E., and DRESSLER, K., *J. Chem. Phys.* **45**, 3149 (1966).
133. COLLIN, J. E., and NATALIS, P., *Int. J. Mass. Spectrom. Ion Phys.* **2**, 231 (1969).
134. DE REILHAC, L., and DAMANY, N., *J. Phys.* **32** (C 4), 32 (1971).

135. Baer, P., and Miescher, E., *Nature* **169**, 581 (1952).
136. Wilkinson, P. G., *Can. J. Phys.* **34**, 250 (1956).
137. Joshi, K. C., *Proc. Phys. Soc. London* **87**, 285 (1966).
138. Whittaker, F. L., *Can. J. Phys.* **47**, 1291 (1969).
139. Gilmore, F. R., *J. Quant. Spectrosc. Radiative Transfer* **5**, 369 (1965).
140. Brix, P., and Herzberg, G., *Can. J. Phys.* **32**, 110 (1954).
141. Ditchburn, R. W., and Heddle, D. W. O., *Proc. Roy. Soc. London* A, **226**, 509 (1954).
142. Wilkinson, P. G., and Mulliken, R. S., *Astrophys. J.* **125**, 594 (1957).
143. Ogawa, M., and Chang, H., *Sci. Light, Tokyo* **17**, 45 (1968).
144. Bethke, G. W., *J. Chem. Phys.* **31**, 669 (1959).
145. Halman, M., *J. Chem. Phys.* **44**, 2406 (1966).
146. Watanabe, K., Inn, E. C. Y., and Zelikoff, M., *J. Chem. Phys.* **21**, 1026 (1953).
147. Farmer, A. J. D., Fabian, W., Lewis, B. R., Lokan, K. H., and Haddad, G. N., *J. Quant. Spectrosc. Radiative Transfer* **8**, 1739 (1968).
148. Ackerman, M., Biaume, F., and Nicolet, M., *Can. J. Chem.* **47**, 1834 (1969).
149. Harris, R., Blackledge, M., and Generosa, J., *J. Molec. Spectrosc.* **30**, 506 (1969).
150. Ladenburg, R., and van Vorrhis, C. C., *Phys. Rev.* **43**, 315 (1953).
151. Schneider, E. C., *J. Chem. Phys.* **5**, 106 (1937).
152. Ditchburn, R. W., and Heddle, D. W. O., *Proc. Roy. Soc. London* A, **220**, 61 (1953).
153. Kosinskaya, I. V., and Startsev, G. P., *Optika Spektrosk.* **18**, 735 (1965) [*Optics Spectrosc.* **18**, 416 (1965)].
154. Goldstein, R., and Mastrup, F. N., *J. Opt. Soc. Am.* **56**, 765 (1966).
155. Watanabe, K., Inn, E. C. Y., and Zelikoff, M., *J. Chem. Phys.* **20**, 1969 (1952).
156. Metzger, P. H., and Cook, G. R., *J. Quant. Spectrosc. Radiative Transfer* **4**, 107 (1964).
157. Lombos, B. A., and Sauvageau, P., *Can. Spectrosc.* **13**, 46 (1968).
158. Price, W. C., and Collins, G., *Phys. Rev.* **48**, 714 (1935).
159. Tanaka, Y., *J. Chem. Phys.* **20**, 1728 (1952).
160. Samson, J. A. R., and Cairns, R. B., *J. Opt. Soc. Am.* **56**, 769 (1966).
161. Tanaka, Y., and Takamine, T., *Phys. Rev.* **59**, 771 (1941); *Scient. Pap. Inst. Phys. Chem. Res. Tokyo* **39**, 437 (1942).
162. Huffman, R. E., Larrabee, J. C., and Tanaka, Y., *J. Chem. Phys.* **46**, 2213 (1967).
163. Ogawa, M., (a) *Can. J. Phys.* **46**, 312 (1968); (b) Yamawaki, K. R., *Can. J. Phys.* **47**, 1805 (1969).
164. Yoshino, K., and Tanaka, Y., *J. Chem. Phys.* **48**, 4859 (1968).
165. Codling, K., and Madden, R. P., *J. Chem. Phys.* **42**, 3935 (1965).
166. Weissler, G. L., and Lee, P., *J. Opt. Soc. Am.* **42**, 200 (1952).
167. Lee, P., *J. Opt. Soc. Am.* **45**, 703 (1955).
168. Aboud, A. A., Curtis, J. P., Mercure, R., and Rense, W. A., *J. Opt. Soc. Am.* **45**, 767 (1955).
169. Wainfan, N., Walker, W. C., and Weissler, G. L., *Phys. Rev.* **99**, 542 (1955).
170. Huffman, R. E., Larrabee, J. C., and Tanaka, Y., *J. Chem. Phys.* **40**, 356 (1964).
171. Matsunaga, F. M., and Watanabe, K., *Sci. Light, Tokyo* **16**, 31 (1967).
172. Huffman, R. E., Larrabee, J. C., and Tanaka, Y., *Phys. Rev. Letters* **16**, 1033 (1966).
173. Alberti, F., Ashby, R. A., and Douglas, A. E., *Can. J. Phys.* **46**, 337 (1968).
174. McNeal, R. H., and Cook, G. R., *J. Chem. Phys.* **45**, 3469 (1966).
175. Dibeler, V. H., and Walker, J., *J. Opt. Soc. Am.* **57**, 1007 (1967).
176. Natalis, P., and Collin, J. E., *Chem. Phys. Letters* **2**, 483 (1968).
177. Collin, J. E., and Natalis, P., *Int. J. Mass Spectrom. Ion Phys.* **2**, 231 (1969).
178. Dixon, R. N., and Hull, S. E., *Chem. Phys. Letters* **3**, 367 (1969).
179. Iczkowski, R. P., and Margrave, J. L., *J. Chem. Phys.* **30**, 403 (1959).
180. Cordes, H., and Sponer, H., *Z. Phys.* **63**, 334 (1930).
181. Cordes, H., *Z. Phys.* **97**, 603 (1935).
182. Venkateswarlu, P., *Bull. Am. Phys. Soc.* **14**, 622 (1969); *Can. J. Phys.* **47**, 2525 (1969).

183. HARANATH, P. B. V., and RAO, P. T., *J. Molec. Spectrosc.* **2**, 428 (1958).
184. MULLIKEN, R. S., *Phys. Rev.* **46**, 549 (1934).
185. CORDES, H., and SPONER, H., *Z. Phys.* **79**, 170 (1932).
186. MAEDER, R., and MIESCHER, E., *Nature* **161**, 393 (1948); MAEDER, R., *Helv. phys. Acta* **21**, 411 (1948).
187. TANAKA, Y., and OGAWA, M., *J. Chem. Phys.* **36**, 726 (1962).
188. LAKSHMINARAYANA, G., and NARASIMHAM, N. A., *Current Sci. India* **36**, 533 (1967).
189. BERKOWITZ, J., and CHUPKA, W. A., *J. Chem. Phys.* **50**, 4245 (1969).
190. DRESSLER, K., *Helv. phys. Acta* **28**, 563 (1955).
191. BALFOUR, W. J., and DOUGLAS, A. E., *Can. J. Phys.* **46**, 2277 (1968).
192. CREUTZBERG, F., *Can. J. Phys.* **44**, 1583 (1966).
193. BAUER, S. H., HERZBERG, G., and JOHNS, J. W. C., *J. Molec. Spectrosc.* **13**, 256 (1946).
194. SAFARI, E., *Annls Phys. Paris* **9**, 203 (1954).
195. ROMAND, J., *Annls. Phys. Paris* **4**, 529 (1949).
196. HUEBERT, B. J., and MARTIN, R. M., *J. Phys. Chem.* **72**, 3046 (1968).
197. PRICE, W. C., *Proc. Roy. Soc. London* A, **167**, 216 (1938).
198. BARROW, R. F., and STAMPER, J. G., *Proc. Roy. Soc. London* A, **263**, 259 and 277 (1961).
199. STAMPER, J. G., *Can. J. Phys.* **40**, 1274 and 1279 (1962).
200. KRAUSS, M., WALKER, J. A., and DIBELER, V. H., *J. Res. Natn. Bur. Stand.* **72** A, 281 (1968).
201. FROST, D. C., MCDOWELL, C. A., and VROOM, D. A., *J. Chem. Phys.* **46**, 4255 (1967).
202. LEMPKA, J. H., PASSMORE, T. R., and PRICE, W. C., *Proc. Roy. Soc. London* A, **304**, 53 (1968).
203. NOVIKOV, M. M., and GURVICH, L. V., *Optika Spektrosk.* **23**, 323 (1967) [*Optics Spectrosc.* **23**, 173 (1967)].
204. CHRETIEN, M., *Helv. phys. Acta* **23**, 259 (1950).
205. LEFEBVRE-BRION, H., and MOSER, C. M., *J. Molec. Spectrosc.* **15**, 211 (1965).
206. BERKOWITZ, J., *J. Chem. Phys.* **50**, 3503 (1969); BERKOWITZ, T., and WALTER, T. A., *J. Chem. Phys.* **49**, 1184 (1968).
207. CARROLL, P. K., *Can. J. Phys.* **34**, 85 (1956).
208. HOPFIELD, J. J., and BIRGE, R. T., *Phys. Rev.* **29**, 922 (1927).
209. READ, D. N., *Phys. Rev.* **46**, 571 (1934).
210. TILFORD, S. G., and VANDERSLICE, J., *J. Molec. Spectrosc.* **26**, 419 (1968).
211. HENNING, H. J., *Annln Physik* **13**, 599 (1932).
212. TANAKA, Y., *Scien. Pap. Inst. Phys. Chem. Res. Tokyo* **39**, 447 (1942).
213. TAKAMINE, T., TANAKA, Y., and IWATA, M., *Scient. Pap. Inst. Phys. Chem. Res. Tokyo* **40**, 371 (1943).
214. OGAWA, M., *J. Chem. Phys.* **43**, 2142 (1965).
215. TANAKA, Y., JURSA, A. S., and LE BLANC, F., *J. Chem. Phys.* **26**, 862 (1957).
216. HERZBERG, G., and HUGO, T. J., *Can. J. Phys.* **33**, 757 (1955).
217. TILFORD, S. G., and SIMMONS, J. D., *J. Chem. Phys.* **44**, 4145 (1966); **45**, 2965 (1966).
218. HERZBERG, G., SIMMONS, J. D., BASS, A. M., and TILFORD, S. G., *Can. J. Phys.* **44**, 3039 (1966).
219. KRUPENIE, P. K., and WEISSMAN, S., *J. Chem. Phys.* **43**, 1529 (1965).
220. LEFEBVRE-BRION, H., MOSER, C. M., and NESBET, R. K., *J. Molec. Spectrosc.* **13**, 418 (1964).
221. WATANABE, K., ZELIKOFF, M., and INN, E. C. Y., AFCRC Techn. Rep. 53.23, Geophysical Res. Papers no. 21, 1953.
222. SUN, H., and WEISSLER, G. L., *J. Chem. Phys.* **23**, 1625 (1955).
223. HESSER, J. E., and DRESSLER, K., *Astrophys. J.* **142**, 389 (1965).
224. HUFFMAN, R. E., LARRABEE, J. C., and TANAKA, Y., *J. Chem. Phys.* **40**, 2261 (1964).
225. COOK, G. R., METZGER, P. H., and OGAWA, M., *Can. J. Phys.* **43**, 1706 (1965).
226. CAIRNS, R. B., and SAMSON, J. A. R., *J. Opt. Soc. Am.* **56**, 526 (1966).
227. TANAKA, Y., SEYA, M., and MORI, K., *J. Chem. Phys.* **19**, 979 (1951).
228. SUTCLIFFE, L. H., and WALSH, A. D., *Proc. Phys. Soc.* A, **66**, 209 (1953).

229. Tanaka, Y., *J. Chem. Phys.* **21**, 788 (1953).
230. Ueda, M., *Sci. Light, Tokyo* **3**, 143 (1955).
231. Wilkinson, P. G., and Hauk, N. B., *J. Chem. Phys.* **24**, 528 (1956).
232. Goto, K., *Sci. Light, Tokyo* **5**, 4 (1956).
233. Miescher, E., *J. Quant. Spectrosc. Radiative Transfer* **2**, 421 (1962).
234. Dressler, K., and Miescher, E., *Astrophys. J.* **141**, 1266 (1965).
235. Miescher, E., *J. Molec. Spectrosc.* **20**, 130 (1966).
236. Jungen, C., *Can. J. Phys.* **44**, 3197 (1966).
237. Jungen, C., and Miescher, E., *Can. J. Phys.* **46**, 987 (1968); **47**, 1769 (1969).
238. Roncin, J.-Y., Damany, N., and Romand, J., *J. Molec. Spectrosc.* **22**, 154 (1967).
239. Tanaka, Y., *Scient. Pap. Inst. Phys. Chem. Res. Tokyo* **39**, 456 (1942).
240. Huber, K. P., *Helv. phys. Acta* **34**, 929 (1961).
241. Bethke, G. W., *J. Chem. Phys.* **31**, 662 (1959).
242. Mayence, J., *Ann. Phys.* **7**, 453 (1952).
243. Marmo, F. F., *J. Opt. Soc. Am.* **43**, 1186 (1953).
244. Watanabe, K., Marmo, F. F., and Inn, E. C. Y., *Phys. Rev.* **91**, 1155 (1953).
245. Watanabe, K., *J. Chem. Phys.* **22**, 1564 (1954).
246. Sun, H., and Weissler, G. L., *J. Chem. Phys.* **23**, 1372 (1955).
247. Walker, W. C., and Weissler, G. L., *J. Chem. Phys.* **23**, 1962 (1955).
248. Watanabe, K., Matsunaga, F. M., and Sakai, H., *Appl. Opt.* **6**, 391 (1967).
249. Metzger, P. H., Cook, G. R., and Ogawa, M., *Can. J. Phys.* **45**, 203 (1967).
250. Miescher, E., *Helv. phys. Acta* **29**, 135 (1956).
251. Huber, K. P., *Can. J. Phys.* **46**, 1691 (1968).
252. Rowlinson, H. C., and Barrow, R. F., *J. Chem. Phys.* **21**, 378 (1953).
253. Johns, J. W. C., *Can. J. Phys.* **41**, 209 (1963).
254. Bell, S., *J. Molec. Spectrosc.* **16**, 205 (1965).
255. Beyer, K. D., and Welge, K. H., *Z. Naturforsch.* **22a**, 1161 (1967).
256. Blake, A. J., and Carver, J. H., *J. Chem. Phys.* **47**, 1038 (1967).
257. Dibeler, V. H., Walker, J. A., and Rosenstock, H. M., *J. Res. Natn. Bur. Stand.* **70 A**, 459 (1966).
258. Metzger, P. H., and Cook, G. R., *J. Chem. Phys.* **41**, 642 (1964).
259. de Reilhac, L., and Damany, N., *Spectrochim. Acta* **26 A**, 801 (1970).
260. Weeks, J. L., Meaburn, G. M. A. C., and Gordon, S., *Radiat. Res.* **19**, 559 (1963).
261. Gelebart, F., and Johannin-Gilles, A., *Compt. Rend.* **262 B**, 1509 (1966).
262. Verrall, R. E., and Senior, W. A., *J. Chem. Phys.* **50**, 2746 (1969).
263. Halmann, M., and Platzner, I., *J. Phys. Chem.* **70**, 580 (1966).
264. Onaka, R., and Takahashi, T., *J. Phys. Soc. Japan* **24**, 548 (1968).
265. la Paglia, S. R., *J. Chem. Phys.* **41**, 1427 (1964).
266. Lin, T. F., and Duncan, A. B. F., *J. Chem. Phys.* **48**, 866 (1968).
267. Harada, Y., and Murrell, J. N., *Molec. Phys.* **14**, 153 (1968).
268. Price, W. C., Teegan, J. P., and Walsh, A. D., *Proc. Roy. Soc. London* **A, 201**, 600 (1950).
269. Clark, L. B., and Simpson, W. T., *J. Chem. Phys.* **43**, 3666 (1965).
270. Dibeler, V. H., and Liston, S. K., *J. Chem. Phys.* **49**, 482 (1968).
271. Al-Joboury, M. I., and Turner, D. W., *J. Chem. Soc.* **4434** (1964).
272. Price, W. C., *J. Chem. Phys.* **4**, 147 (1936).
273. Watanabe, K., and Jursa, A. S., *J. Chem. Phys.* **41**, 1650 (1964).
274. Walsh, A. D., *J. Chem. Soc.* 2260 (1953).
275. Mulliken, R. S., *Can. J. Phys.* **36**, 10 (1958).
276. Tanaka, Y., Jursa, A. S., and le Blanc, F. J., *J. Chem. Phys.* **32**, 1199 (1960).
277. Collin, J. E., *J. Chem. Phys.* **57**, 424 (1960); *Mem. Soc. Roy. Sci. Liège* **14**, 1 (1967); Collin, J. E., and Natalis, P., *Int. J. Mass Spectrom. Ion Phys.* **1**, 121 (1968).
278. Turner, D. W., and May, D. P., *J. Chem. Phys.* **46**, 1156 (1967).
279. Schoen, R. I., Boeing Technical Rept. D1-82-0406, 1965.
280. Brundle, C. R., and Turner, D. W., *Int. J. Mass Spectrom. Ion Phys.* **2**, 195 (1969).

281. Bahr, J. L., Blake, A. J., Carver, J. H., and Kumar, V., *J. Quant. Spectrosc. Radiative Transfer* **9**, 1359 (1969).
282. Weissler, G. L., Samson, J. A. R., Ogawa, M., and Cook, G. R., *J. Opt. Soc. Am.* **49**, 338 (1959).
283. Nakata, R. S., Watanabe, K., and Matsunaga, F. M., *Sci. Light, Tokyo* **14**, 54 (1965).
284. Cook, G. R., Metzger, P. H., and Ogawa, M., *J. Chem. Phys.* **44**, 2935 (1966).
285. de Reilhac, L., Damany-Astoin, N., and Romand, J., *Spectrochim. Acta* **25** A, 19 (1969).
286. Peters, D., *J. Chem. Phys.* **45**, 3474 (1966).
287. Newton, A. S., and Sciamanna, *J. Chem. Phys.* **40**, 718 (1964).
288. Price, W. C., and Simpson, D. M., (a) *Proc. Roy. Soc. London* A, **169**, 501 (1939); (b) A, **165**, 272 (1938).
289. Kopp, I., *Can. J. Phys.* **45**, 4011 (1967).
290. Roncin, J.-Y., Damany, N., and Vodar, B., *Chem. Phys. Letters* **3**, 197 (1969).
291. Tanaka, Y., Jursa, A. S., and le Blanc, F. J., *J. Chem. Phys.* **32**, 1205 (1960).
292. Matsunaga, F. M., and Watanabe, K., *J. Chem. Phys.* **46**, 4457 (1967).
293. Cook, G. R., and Ogawa, M., (a) *J. Chem. Phys.* **51**, 647; (b) **51**, 2419 (1969).
294. Al-Joboury, M. I., May, D. P., and Turner, D. W., *J. Chem. Soc.* 6350 (1965).
295. Cook, G. R., Metzger, P. H., and Ogawa, M., *J. Opt. Soc. Am.* **58**, 129 (1968).
296. Weissler, G. L., Samson, J. A. R., Ogawa, M., and Cook, G. R., *J. Opt. Soc. Am.* **49**, 338 (1959).
297. Dibeler, V. H., Walker, J. A., and Liston, S. K., *J. Res. Natn. Bur. Stand.* **71** A, 371 (1967).
298. Judge, D. L., Morse, A. L., and Weissler, G. L., *Proceedings of the Sixth International Conference on Ionization Phenomena in Gases, Paris*, North-Holland, Amsterdam, 1963, vol. 3. p. 373.
299. Dodonova, N. Y., *Optika Spektrosk.* **20**, 493 (1966) [*Optics Spectrosc.* **20**, 271 (1966)].
300. Natalis, P., and Collin, J. E., *Int. J. Mass. Spectrom. Ion Phys.* **2**, 221 (1969).
301. Tanaka, Y., Inn, E. C. Y., and Watanabe, K., *J. Chem. Phys.* **21**, 1651 (1953).
302. Ogawa, M., and Cook, G. R., *J. Chem. Phys.* **28**, 173 (1958).
303. Herron, J. T., and Shiff, H. I., *J. Chem. Phys.* **26**, 542 (1957).
304. Golomb, D., Watanabe, K., and Marmo, F. F., *J. Chem. Phys.* **36**, 958 (1962).
305. Watanabe, K., *J. Chem. Phys.* **26**, 542 (1957).
306. Wehrli, M., *Helv. phys. Acta* **11**, 339 (1938); **13**, 153 (1940).
307. Bell, S., McKenzie, R. D., and Coon, J. B., *J. Molec. Spectrosc.* **20**, 217 (1966).
308. Bell, S., *J. Molec. Spectrosc.* **23**, 98 (1967).
309. Watanabe, K., and Sood, S. P., *Sci. Light, Tokyo* **14**, 36 (1965).
310. Okabe, H., and Lenzi, M., *J. Chem. Phys.* **47**, 5241 (1967).
311. Frost, D. C., McDowell, C. A., and Vroom, D. A., *Can. J. Chem.* **45**, 1343 (1967).
312. Chupka, W. A., and Russel, M. E., *J. Chem. Phys.* **48**, 1527 (1968).
313. Bell, S., Varadarajan, T. S., Walsh, A. D., Warsop, P. A., Lee, J., and Sutcliffe, L., *J. Molec. Spectrosc.* **21**, 42 (1966).
314. Kim, H. H., and Roebber, J. L., *J. Chem. Phys.* **44**, 1709 (1966).
315. Roebber, J. L., Larrabee, J. C., and Huffman, R. E., *J. Chem. Phys.* **46**, 4594 (1967).
316. Lui, T. K., Moe, G., and Duncan, A. B. F., *J. Chem. Phys.* **19**, 71 (1951).
317. Codling, K., *J. Chem. Phys.* **44**, 4401 (1966).
318. Dibeler, V. H., and Walker, J. A., *J. Chem. Phys.* **44**, 4405 (1966).
319. Dibeler, V. H., and Mohler, F. L., *J. Res. Natn. Bur. Stand.* **40**, 25 (1948).
320. Frost, D. C., McDowell, C. A., Sandhu, J. S., and Vroom, D. A., *J. Chem. Phys.* **46**, 2008 (1967).
321. Dibeler, V. H., and Liston, S. K., *J. Chem. Phys.* **48**, 4765 (1968).
322. Berkowitz, J., Chupka, W. A., and Walker, T. A., *J. Chem. Phys.* **50**, 1497 (1969).
323. Dibeler, V. H., and Liston, S. K., *Inorg. Chem.* **7**, 1742 (1968).
324. Noutary, C. J., *J. Res. Natn. Bur. Stand.* **72** A, 479 (1968).
325. Watkins, I. W., *J. Molec. Spectrosc.* **29**, 402 (1969).

Additional bibliography

DIATOMIC MOLECULES

H_2, D_2

ALI, A. W., and ANDERSON, A. D., *H_2 Vacuum u.v. Laser Rate Coefficients*, NRL Report 7282, Washington, 1971.

DALGARNO, A., HERZBERG, G., and STEPHENS, T. L., A new continuous emission spectrum of the hydrogen molecule, *Astrophys. J.* **162**, L49 (1970).

DE HEER, F. J., and CARRIERE, J. D., Emission of the Werner band system and Lyman-α radiation for 0.05-6 eV electrons in H_2, *J. Chem. Phys.* **55**, 3829 (1971).

FINK, E. H., AKINS, D. L., and MOORE, C. B., Relative line intensities in the Lyman bands of HD, *Chem. Phys. Letters* **4**, 283 (1969).

HERZBERG, G., Dissociation energy and ionization potential of molecular hydrogen, *Phys. Rev. Letters* **23**, 1081 (1969).

HERZBERG, G., and JUNGEN, CH., Rydberg series and ionization potential of the H_2 molecule, *J. Molec. Spectrosc.* **41**, 425 (1972).

HODGSON, R. T., and DREYFUS, R. W., Vacuum u.v. laser action observed in H_2 Werner bands: 1161-1240 Å, *Phys. Letters* **26**, 536 (1972); Electron-beam excitation of vacuum u.v. hydrogen laser, *Phys. Letters* **38** A, 213 (1972).

SHAW, G. B., and BERRY, R. S., Photoionization and autoionization of H_2, *J. Chem. Phys.* **56**, 5808 (1972).

TAKEZAWA, S., Absorption spectrum of H_2 in the vacuum u.v. region: I, Rydberg states and ionization energies, *J. Chem. Phys.* **52**, 2575 (1970); *ibid.*, II, Rydberg series converging to the first six vibrational levels of H_2^+ ground state, *J. Chem. Phys.* **52**, 5793 (1970).

TAKEZAWA, S., and TANAKA, Y., Absorption spectrum of HD in the vacuum u. v. region: Rydberg states and ionization energy, *J. Chem. Phys.* **56**, 6125 (1972).

WAYNANT, R. W., SHIPMAN, J. D., ELTON, R. C., and ALI, A. W., Vacuum ultraviolet laser emission from molecular hydrogen, *Appl. Phys. Letters* **17**, 383 (1970).

Rare gases

CASTEX, M.-C., and DAMANY, N., Absorption spectrum of the xenon molecule in the vacuum ultraviolet region, *Chem. Phys. Letters* **13**, 158 (1972).

SANDO, K. M., and DALGARNO, A., The absorption of radiation near 600 Å by helium, *Molec. Phys.* **20**, 103 (1971).

TANAKA, Y., and YOSHINO, K., Absorption spectrum of the He_2 molecule in the 510-611 Å range, *J. Chem. Phys.* **50**, 3087 (1969).

TANAKA, Y., and YOSHINO, K., Absorption spectrum of the Ar molecule in the v.u.v. region, *J. Chem. Phys.*, **53**, 2012 (1970).

TANAKA, Y., and YOSHINO, K., Absorption spectrum of Ne_2 and He Ne molecules in the v.u.v. region, *J. Chem. Phys.*, **57**, 2964 (1972).

N_2

AJELLO, J. M., Emission cross-sections of N_2 in the vacuum ultraviolet by electron impact, *J. Chem. Phys.* **53**, 1156 (1970).

BENESCH, W. M., and SAUM, K. A., The $w^3\Delta_u$ state of molecular nitrogen, *J. Phys.* B, **4**, 732 (1971).

BOURSEY, E., and RONCIN, J.-Y., Experimental deperturbation of the $b\,^1\Pi_u \leftarrow X\,^1\Sigma_g^+$ transition of N_2 in the solid state, pure and trapped in Ne and CF_4 matrices at low temperature, *Phys. Rev. Letters*, **26**, 308 (1971).

CARROLL, P. K., COLLINS, C. P., and YOSHINO, K., The high energy $^1\Sigma_u^+$ states of N_2, *J. Phys.* B, **3**, L127 (1970).

CARTER, V. L., High-resolution N_2 absorption study from 730 to 980 Å, *J. Chem. Phys.* **56**, 4195 (1972).

COMES, F. J., and WEBER, K. W., Wirkungsquerschnitte der Photoabsorption und Photoionisation des N_2-Moleküls, *Z. Naturforsch.* **24** A, 1941 (1969).

COOK, G. R., and OGAWA, M., Oscillator strengths of the Hopfield absorption series of N_2, *J. Chem. Phys.* **53**, 1292 (1970).

HOLLAND, R. F., Excitation of nitrogen by electrons: Lyman-Birge-Hopfield system of N_2, *J. Chem. Phys.* **51**, 3940 (1969).

MILLER, R. E., HIGH resolution emission Lyman-Birge-Hopfield bands of N_2, *J. Opt. Soc. Am.* **60**, 171 (1970).

SAUM, K. A., and BENESCH, W. M., $W^3\Delta_u - X^1\Sigma_g^+$ system of N_2, *Phys. Rev.* A, **2**, 1655 (1970).

TUCKWELL, H. C., Calculation of the photoionization cross-section of N_2 from first threshold to 500 Å, *J. Phys.* B, **3**, 293 (1970).

O_2

ACKERMAN, M., and BIAUMÉ, F., Structure of the Schumann-Runge bands from the 0-0 to the 13-0 band, *J. Molec. Spectrosc.* **35**, 73 (1970).

ACKERMAN, M., BIAUMÉ, F., and KOCKARTS, G., Absorption cross-sections of the Schumann-Runge bands of molecular oxygen, *Planet Space Sci.* **18**, 1639 (1970).

CLARK, I. D., and WAYNE, R. P., The absolute cross-section for photoionization of O_2 ($^1\Delta_g$), *Molec. Phys.* **18**, 523 (1970).

OGAWA, M., Absorption coefficients of O_2 in the metastable state, $a^1\Delta_g$, *J. Chem. Phys.* **53**, 3754 (1970).

OGAWA, M., Absorption cross-sections of O_2 and CO_2 continua in the Schumann and far u.v. regions, *J. Chem. Phys.* **54**, 2550 (1971).

Halogen molecules

GOLE, J. L., and MARGRAVE, J. L., The vacuum ultraviolet spectrum of molecular fluorine, *J. Molec. Spectrosc.* **43**, 65, (1972).

MULLIKEN, R. S., Iodine revisited, *J. Chem. Phys.* **55**, 288 (1971).

WENKATESWARLU, P., Vacuum ultraviolet spectrum of the iodine molecule, *Can. J. Phys.* **48**, 1055 (1970).

Hydrogen halides

GINTER, M. L., and TILFORD, S. G., Electronic spectra and structure of the hydrogen halides: the $b^3\Pi_i$ and $C^1\Pi$ states of HBr and DBr, *J. Molec. Spectrosc.* **34**, 206 (1970); States associated with the $(\sigma^2\pi^3)c\pi$ and $(\sigma^2\pi^3)c\sigma$ configurations of HBr and DBr, *J. Molec. Spectrosc.* **37**, 159 (1971).

TILFORD, S. G., GINTER, M. L., and BASS, A. M., Electronic spectra and structure of the hydrogen halides: the $b^3\Pi_i$ and $C^1\Pi$ states of HI and DI, *J. Molec. Spectrosc.* **34**, 327 (1970).

TILFORD, S. G., GINTER, M. L., and VANDERSLICE, J. T., Electronic spectra and structure of the hydrogen halides: the $b^3\Pi_i$ and $C^1\Pi$ states of HCl and DCl, *J. Molec. Spectrosc.* **33**, 505 (1970).

CO

HERZBERG, G., HUGO, T. J., TILFORD, S. G., and SIMMONS, J. D., Rotational analysis of the forbidden $d^3\Delta_1 \leftarrow X^1\Sigma^+$ absorption band of carbon monoxide, *Can. J. Phys.* **48**, 3004 (1970).

OGAWA, M., and OGAWA, S., Absorption spectrum of CO in the Hopfield helium continuum region, 600-1020 Å, *J. Molec. Spectrosc.* **41**, 393 (1972).

NO

JUNGEN, C., Rydberg series in the NO spectrum: an interpretation of quantum defects and intensities in the *s* and *d* series, *J. Chem. Phys.* **53**, 4168 (1970).

MENTALL, J. E., and MORGAN, H. D., Vacuum ultraviolet excitation cross-sections by electron impact on NO, *J. Chem. Phys.* **56**, 2271 (1972).

MIESCHER, E., Absorption spectrum of the NO molecule: X, The Rydberg complex, its vibrational structure, spin orbit coupling, and interactions with non-Rydberg states, *Can. J. Phys.* **49**, 2350 (1971).

Miscellaneous diatomics

CATON, R. B., and DOUGLAS, A. E., Electronic spectrum of the BF molecule, *Can. J. Phys.* **48**, 432 (1970).

DONOVAN, R. J., HUSAIN, D., and STEVENSON, C. D., Vacuum ultraviolet spectra of transient molecules and radicals: I, CS and S_2, *Trans. Faraday Soc.* **66**, 1 (1970).

DONOVAN, R. J., and LITTLE, D. J., Vacuum ultraviolet spectrum of the SO radical, *Spectrosc. Letters* **4**, 213 (1971).

HERZBERG, G., and JOHNS, J. W. C., New spectra of the CH molecule, *Astrophys. J.* **158**, 399 (1971).

HERZBERG, G., LAGERQVIST, A., and MALMBERG, C., New electronic transitions of the C_2 molecule in absorption in the vacuum ultraviolet region, *Can. J. Phys.* **47**, 2735 (1969).

HERZBERG, G., LAGERQVIST, A., and MCKENZIE, B. J., Absorption spectrum of SiH in the vacuum ultraviolet, *Can. J. Phys.* **47**, 1889 (1969).

HUFFMAN, R. E., Absorption cross-sections of atmospheric gases for use in aeronomy, *Can. J. Chem.* **47**, 1823 (1969).

LAGERQVIST, A., and MALMBERG, C., New absorption systems of Si_2 molecule in the vacuum ultraviolet, *Physica Scripta* **2**, 45 (1970).

LUTZ, B. L., Spectrum of the CN molecule: I, Absorption in the vacuum ultraviolet, *Can. J. Phys.* **48**, 1192 (1970).

MYER, J. A., and SAMSON, J. A. R., Vacuum ultraviolet absorption cross-sections of CO, HCl, and ICN between 1050 and 2100 Å, *J. Chem. Phys.* **52**, 266 (1970).

SASANUMA, M., ISHIGURO, E., MORIOKA, Y., and NAKAMURA, M., Absorption cross sections and ionization efficiencies of several gases in the 300-1000 Å region (O_2, N_2, CO, SF_6), *Third International Conference on Vacuum Ultraviolet Radiation Physics, Tokyo*, 1971, session 1 p. A2-3.

WHITTAKER, F. L., Observation of the $d^1\Sigma^+ - b^1\Sigma^+$ system of NH and ND in the vacuum ultraviolet, *Can. J. Phys.* **47**, 1291 (1969).

POLYATOMIC MOLECULES

BISHOP, D. M., LECLERC, J. C., and TAKAHATA, Y., An investigation of the lower lying excited states of ammonia, *J. Molec. Spectrosc.* **39**, 79 (1971).

BLECHSCHMIDT, D., HAENSEL, R., KOCH, E. E., NIELSEN, U., and SAGAWA, T., The optical spectra of gaseous and solid SF_6 in the extreme ultraviolet and soft X-ray region, *Chem. Phys. Letters* **14**, 33 (1972).

BRUNDLE, C. R., NEUMANN, D., PRICE, W. C., EVANS, D., POTTS, A. W., and STREETS, D. G., Electronic structure of NO_2 studied by photoelectron and vacuum u.v. spectroscopy and Gaussian orbital calculations, *J. Chem. Phys.* **53**, 705 (1970).

CAIRNS, R. B., HARRISON, H., and SCHOEN, R. I., Dissociative photoionization of H_2O, *J. Chem. Phys.* **55**, 4886 (1971).

DUNCAN, A. B. F., and WHITLOCK, R. F., Vacuum ultraviolet absorption spectrum of carbonyl cyanide, *Spectrochim. Acta* **27 A**, 2539 (1971).

DURMAZ, S., MURRELL, J. N., TAYLOR, J. M., and SUFFOLK, R., Calculation of the intensities of the vibrational components of the ammonia ultraviolet absorption bands, *Molec. Phys.* **19**, 533 (1970).

GEDANKEN, A., RAZ, B., EVEN, U., and ELIEZER, I., The vacuum u.v. spectrum of $HgBr_2$, *J. Molec. Spectrosc.* **32**, 287 (1969).

HAYES, W., and BROWN, F. C., Absorption of BF_3 gas in the extreme ultraviolet, *J. Phys.* B, **4**, L85 (1971).

HAYES, W., BROWN, F. C., and KUNZ, A. B., Absorption of SiH_4 in the extreme ultraviolet, *Phys. Rev. Letters* **27**, 774 (1971).

OGAWA, M., and CHANG, H. C., Absorption spectrum of CS_2 in the region from 600 to 1015 Å, *Can. J. Phys.* **48**, 2455 (1970).

OKABE, H., Photodissociation of CS_2 in the vacuum ultraviolet: determination of D_0^0 (SC-S), *J. Chem. Phys.* **56**, 4381 (1972).

ROEBBER, J. L., Vacuum ultraviolet absorption spectrum of carbon suboxide: II, The effect of low temperature and the effect of high pressure on an inert gas on the 1780 Å band, *J. Chem. Phys.* **54**, 4001 (1971).

SHEMANSKY, D. E., CO_2 extinction coefficient 1700-3000 Å, *J. Chem. Phys.* **56**, 1582 (1972).

VINOGRADOV, A. S., and ZIMKINA, T. M., Cross-sections of sulfur hexafluoride photoabsorption in the soft and ultra-soft X-ray range, *Optika Spektrosk.* **32**, 33 (1972) [*Optics Spectrosc.* **32**, 17 (1972)].

WORKMAN, G. L., and DUNCAN, A. B. F., Electronic spectrum of carbonyl fluoride, *J. Chem. Phys.* **52**, 3204 (1970).

CHAPTFR 5

Electronic spectra of paraffins

B. A. Lombos*

Laboratoire des Interactions Moléculaires et des Hautes Pressions du CNRS, Bellevue, 92190 Meudon, France

Contents

* *Present address:* Associate Professor of Engineering, Sir George Williams University, Montreal 107, Quebec, Canada.

Introduction

The transitions of electrons in σ orbitals and the nature of the excited states of saturated hydrocarbons are relatively unknown. Yet the study of the interaction of these molecules with light presents many intriguing problems, and knowledge of their structure is even more fundamental to physics and chemistry than that of, say, aromatic hydrocarbons or co-ordination complexes which have received incomparably more experimental and theoretical attention than have paraffins. The twofold reasons for this are undoubtedly the difficulties encountered in the theoretical handling of electron systems and instrumental difficulties involved in the measurement of the far u.v. absorption spectra. In this chapter a review of the electronic spectra of the alkanes will be presented. Some of the theoretical interpretations and predictions of the transitions will be discussed, and the principal experimental results will be given in order to elucidate the basic problems of the excited states of saturated hydrocarbons.

As is well known, in the Hückel approximation[(1)] the π electrons can be considered as moving independently in the combined field of the nuclei and the σ electrons. For example, 1,3-butadiene presents a four-π electron problem using the Hückel approximation, while n-butane, with the same number of carbon atoms, is a twenty-six-valence electron problem (taking into account the $2s$ and $2p$ carbon and the $1s$ hydrogen electrons but neglecting the $1s$ electrons of carbon), and the molecule is not coplanar. Understandably, the first theoretical investigations of the saturated hydrocarbons involved respectively, an equivalent orbital method by Lennard-Jones and Hall,[(2)] a united atom approach by Franklin,[(3)] a bond orbital technique by Brown,[(4)] and a perturbation treatment by Dewar and Pettit.[(5)] These procedures were based on the strong apparent localization of the σ electrons and on certain, almost additive properties of the σ bonds, in the saturated hydrocarbons. Most of these calculations, however, dealt with bond energies, heats of formation, ionization potentials, and other molecular properties, but not with spectral transitions.

Mulliken[(6)] gave the first theoretical treatment of methane and ethane, discussing the electronic transitions of these molecules in terms of Rydberg-type united-atom orbitals. The molecular orbital method of Hückel was applied to the saturated hydrocarbons by Sandorfy and Daudel.[(7)] In their C approximation they used only the hybridized sp^3 orbitals linking the carbon atoms together. Later, Sandorfy[(8)] included all the hybridized sp^3 orbitals between the C-C and C-H and the 1s hydrogen orbitals in his H approximation. These methods were improved by Yoshizumi[9] and were used for ionization potentials, heats of formation, and chemical reactivity calculations by Fukui *et al.*[10, 11] Klopman[12–14] later obtained calculated values closer to experimental ones for ionization potentials and heats of formation, using parameters based on the heats of formation of certain simple molecules. Hoffman[15] applied a parametrization based on the ionization potentials of the contributing atomic states, and the overlap integrals in which the interactions of the non-neighbouring atoms are also taken into account. A more complete treatment was given by Pople[16] who studied the sources of the non-additivity of certain properties of hydrocarbons, treating the factors causing a delocalization as a perturbation.

The next stage in the evolution of the theoretical treatment of the saturated compounds was the application of the Pariser and Parr method in its original form[17, 18] with or without configuration interaction. It was then a logical step to introduce the self-consistent field theory into Pariser's and Parr's assumption. This has been treated by Pople[16] and by Brickstock and Pople.[19] This approach has been widely used, and Pople has reviewed its applications, mainly to aromatic hydrocarbons, up to 1957.[20] Recently, Klopman[21] gave a similar treatment of diatomic and a few polyatomic molecules, in which all the necessary integrals were deduced directly from atomic spectra and from interatomic bond lengths. Another self-consistent approximation—neglecting completely the differential overlap—was given by Pople *et al.*[22] and Kaufman.[23] Katagiri and Sandorfy[24] used a method of the Pariser-Parr type to examine the electronic transitions of methane, ethane, and propane. These, and some other theoretical contributions concerning this subject, have been reviewed by Daudel.[25]

The second reason for the slow evolution of this field is that the electronic spectra of these molecules fall in the far u.v. where experimental conditions make the measurement of spectra difficult.

The first spectra of methane were obtained by Leifson in 1926;[26] later work on this molecule has been reported by Rose,[27] Duncan and Howe,[28] Price,[29] Watanabe,[30] Wilkinson and Johnston,[31] Ditchburn,[32] Moe and Duncan,[33] and Sun and Weissler.[34] Among the more general recent publications relating to aliphatic hydrocarbons, the following should be noted: Okabe and Becker[35] have studied the spectra of methane, ethane, propane, and n-butane up to 80,000 cm^{-1}; Partridge[36] has investigated the spectrum of polyethylene; Schoen[37] has studied the spectra of ethane, propane, and n-butane between 95,000 and 200,000 cm^{-1}; and Raymonda and Simpson[38, 39] have examined a large number of alkanes with single-beam instrumentation and have interpreted the spectra using the "independent systems" theory. Lombos *et al.*[40, 41] published the first spectra of a number of paraffins, measured with a double-beam arrangement. All these spectra refer to the gas phase.

The spectrum of a film of solid methane was investigated by Dressler and Schnepp[42] and films of solid methane, ethane and propane have been studied by Lombos *et al.*[43]

The electronic spectra of ionic perfluorinated paraffins were also observed. Zobel and Duncan(44) and Stokes and Duncan(45) located the spectrum of perfluoromethane beyond 110,000 cm^{-1}; Belanger *et al.*(46) measured the far u.v. spectra of the first six perfluorinated n-paraffins and of some of the perfluorocycloparaffins.

1. Normal alkanes

1.1. METHANE

Of all the saturated hydrocarbons, methane has received the greatest attention.(26–28, 32, 34, 37, 39, 47–67) The first broad absorption band was found to be centred at about 78,278 cm^{-1}.(41) This is followed by a much stronger absorption toward higher frequencies which, after reaching a maximum near 107,500 cm^{-1},(32) decreases gradually up to 275,000 cm^{-1}.(32, 37, 67)

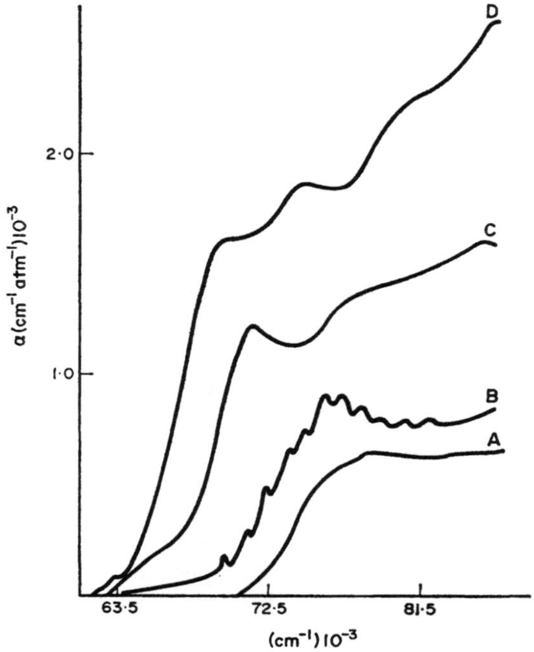

Fig. 1. Spectra of some linear alkanes: (A) methane, (B) ethane, (C) propane, (D) butane.(41) Wave numbers against absorption coefficients.

One can clearly identify a plateau immediately following the shallow peak at 78,278 cm^{-1} (Fig. 1, A).(52, 41) Price(66) and Duncan and Howe(28) attributed the diffuse structure of the methane spectrum to dissociation or predissociation of the excited states since ionization sets in at higher frequencies.(30, 32, 34, 73) However, some slight structure is discernible in the 78,278 cm^{-1} band (see inflections, Fig. 1, A; Table 1).

TABLE 1. ABSORPTION PEAKS IN THE ELECTRONIC SPECTRUM OF METHANE[41, *32]

$1/\lambda$ (cm^{-1})	α (cm^{-1} atm^{-1})	f_{mn}
(70,175)	12	
(71,942)	80	
(76,190)	524	
78,278	600	0.26
(86,022)	619	
107,500*		

Another explanation of the diffuse structure of this band can be found in the discussion of Sponer and Teller[68] who treated the spectra of some molecules by group theory. From the selection rules they deduced that non-degenerate states can only combine with degenerate ones in the case of non-linear molecules. Furthermore, according to the Jahn-Teller theorem,[69] "all non-linear configurations are unstable for an arbitrary degenerate electronic (excited) state ... if the degenerate electrons participate strongly in the binding of the molecules". Thus it can be anticipated that if the absorption arises from the jump of an electron from a non-degenerate ground state to a degenerate excited state, non-totally symmetrical vibrations—corresponding to the ground state—will be simultaneously excited and blurring of the band could result. If, otherwise, the equilibrium positions of the two states are considerably different, the number of transitions could increase up to a point where the spectrum would appear continuous. It is also possible, however, that the excited state does not have an equilibrium configuration, and the molecule dissociates giving rise to a true continuous absorption.

It is not even certain, however, that the 78,278 cm^{-1} band is due to an allowed transition. Its absorption coefficient α is only in the hundreds (cm^{-1}) and stronger bands follow. Ditchburn[32] gave $\alpha \simeq 483$ (cm^{-1} atm^{-1}) for this band; Sun and Weissler[34] determined 537; Moe and Duncan[33] 590; Watanabe[30] 510; and Lombos *et al.*[41] reported 600 (cm^{-1} atm^{-1}). For the oscillator strength f_{mn} of the same band, Moe and Duncan[33] measured 0.342; Wilkinson and Johnston[31] estimated 0.4; while Lombos *et al.*[41] obtained 0.26.†

A point deserving discussion is the appearance of a weak branch at the lower frequency side of the 78,278 cm^{-1} band. It is easily distinguishable in the spectra given by Wilkinson and Johnston,[31] Moe and Duncan,[33] and Sun and Weissler,[34] and its presence has been verified by Lombos *et al.*[41] in work at a pressure of one atmosphere. The question arises whether the weak branch could correspond to a separate forbidden transition. This could be so if the first excited state resulted from the jump of an electron from a triply degenerate f_2 orbital of the ground state to an f_2 orbital of the excited state. In that case, the excited state would

† The conversion factors between the different absorption units are the following: σ (Mb) $= 0.03722\alpha$ (cm^{-1} atm^{-1}); α (cm^{-1} atm^{-1}) $= 0.1028\varepsilon$ (l mole^{-1} cm^{-1}).

split according to $F_2 \times F_2 = A_1 + E + F_2 + F_2$, the jump to F_2 being allowed (from the A_1 ground state) and the others forbidden. The latter may be made allowed by non-totally symmetrical vibrations of suitable symmetry (f_2) which are, indeed, available. The existence of these transitions, even though they are expected to be weak, may also contribute to blurring of the observed bands, and one of them may relate to the weak branch at its low-frequency side.

Katagiri and Sandorfy[24] applied the Pariser and Parr method to saturated hydrocarbons. The molecular orbitals were based on ground state hydrogen ($1s$) and carbon (sp^3) atomic orbitals. The results are shown in Figure 2. In the case of

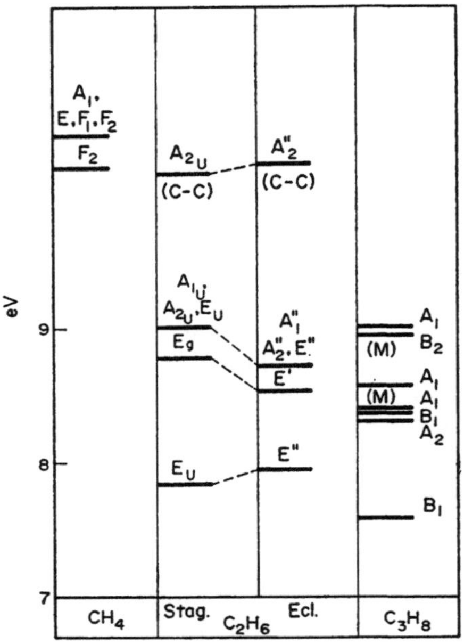

FIG. 2. Energies of the lower singlet-singlet electronic levels of methane, ethane, and propane. C-C means that the orbital from which the transition starts (in absorption) has a high population in the C-C bonds; M (mixed) that it has a fairly large population in the C-C bonds. If the state is non-marked, then all the charge is in C-H bonds in the initial orbital.[24]

methane (T_d symmetry) the lowest energy transition would result from an $a_1 \leftarrow f_2$ jump and not from $f_2 \leftarrow f_2$. This would make the first band $F_2 \leftarrow A_1$, and allowed according to all coordinates. The next higher transition would lead from an A_1 ground state to a degenerate state which would be expected to split as mentioned above (Fig. 2). It should be pointed out that this calculation deals with valence-shell transitions only and not with Rydberg-type transitions. The energy scheme, however, depends in a delicate way on the choice of certain parameters, and a reversal of these states appears quite possible.

Mulliken examined the excited configurations of methane in terms of united-atom orbitals.[6, 70] According to him, the lowest excited orbitals of methane are qualitatively very similar to those of ($3s$) and ($3p$) of an atom from the examination of the nodal surfaces of the LCAO forms of these orbitals. Thus the real molecular orbitals can be much better represented by suitable (3*s*) and (3*p*) atomic orbitals than by the LCAO forms. Omitting carbon (1*s*) orbitals, the first transition would be

$$[s\,a_1]^2\,[p\,f_2]^5\,[3s\,a_1]\ ^1F_2 \leftarrow [s\,a_1]^2\,[p\,f_2]^6\ ^1A_1.$$

This essentially resembles the atomic transition which is strongly allowed. Consequently, Mulliken[6,70] assigned the first intensity maximum of the methane absorption spectrum to the transition $^1A_1 \rightarrow {}^1F_2$ with an excitation energy of 9.9 eV.

Rydberg transitions, in general, are characterized by narrow bands, which is not the case for the methane spectrum, but this is not an absolute criterion. The real orbitals are probably between the Rydberg-type and the valence-shell-type orbitals and are not completely localized in a given bond.

1.2. ETHANE

The majority of the work on ethane[35,37,57,59,81–89] refers to its ionization potential and photolytic behaviour.[96] The detection of vibrational fine structure in the band system of ethane, centred at about 75,838 cm^{-1},[38,40,41,97] is highly significant (Fig. 1, B; Table 2). This shows that the lower excited states of normal-saturated hydrocarbons are not necessarily repulsive or highly predissociated, as is often presumed, i.e. the lifetime of the excited state is not so short as to cause obliteration of all vibrational fine structure.

TABLE 2. ABSORPTION PEAKS IN THE ELECTRONIC SPECTRUM OF ETHANE[41, *37]

	1/λ (cm^{-1})	α (cm^{-1}atm^{-1})	f_{mn}
	(63,494)	23	
	(65,574)	55	
	(67,114)	62	
ν_1	70,269	165	
ν_2	71,347	280	
ν_3	72,464	467	
ν_4	73,594	696	
ν_5	74,677	862	
ν_6	75,838	895	0.3
ν_7	77,095	880	
ν_8	78,290	833	
ν_9	79,536	782	
ν_{10}	80,639	800	
ν_{11}	81,880	816	
	(84,746)	826	
	(86,957)	932	
	120,500*	2150	

Mulliken[6] has discussed the lower excited configurations of ethane. He used CH_3-group orbitals to describe the ground state except for the C-C bond in which two $2p\sigma$ orbitals overlap, building a $[\sigma + \sigma]$-bond orbital. Omitting carbon (1*s*) electrons, the normal state of ethane, assuming D_3 symmetry, may be given as (with the corresponding approximate localization)

$$\underset{CH_3}{[s\,a_1]^2}\;\underset{CH_3}{[s\,a_1]^2}\;\underset{CH_3}{[\pi e]^4}\;\underset{CH_3}{[\pi e]^4}\;\underset{C\text{-}C}{[\sigma + \sigma, a_1]^2},\;{}^1A_1.$$

This electron configuration has, indeed, a strong similarity to that of a halogen molecule, in particular that of F_2. The lowest excited orbital was postulated by Mulliken(6) to be of essentially (3*s*) type; "... we might first think of a (3*s*) orbital for each CH_3 radical. Since these overlap strongly, however, we should then form linear combinations $a(3s \pm 3s)$, of which the form $a(3s + 3s)$ must have the lower energy".(6) The corresponding band would be located near 62,500 cm^{-1} and could be obtained by exciting one $[\sigma + \sigma, a_1]$ and putting it into the $[3s\,a_1]$ orbital thus:

$$\ldots [\sigma + \sigma, a_1]^{-1}\,[3s\,a_1]^{3,1}A_1.$$

Under D_3 symmetry this transition would be of the $A_1 \leftarrow A_1$ type and forbidden. No bands were reported in this region at the time, but recent measurements(40, 41) seem to show the presence of at least one weak band (Fig. 1, B) near 63,494 cm^{-1} (Table 2). This may well be the forbidden band predicted by Mulliken.

The lowest allowed transition may be due to the jump of an electron from a degenerate π-type orbital, which is implied in the C-H σ bonds, to the $[3s\,a_1]$ orbital, giving

$$[\pi e]^{-1}\,[3s\,a_1]^{1,3}E.$$

Thus this transition would be of the allowed type ${}^1E \leftarrow {}^1A_1$, and probably strong since it resembles a $3s \leftarrow 2p$ atomic transition. The well-resolved vibrational structure of the first strong band of the ethane spectrum (Fig. 1, B), suggests such assignment. The absorption maximum of this band is located at 75,838 cm^{-1}, and the estimated oscillator strength f_{mn} of the band is 0.3(41) (Table 2).

According to Katagiri and Sandorfy(24) the first transition of staggered ethane (D_{3d} symmetry) would be ${}^1E_u \leftarrow {}^1A_{1g}$, allowed and polarized perpendicularly to the C-C direction (Fig. 2). The next transition would be $E_g \leftarrow A_{1g}$, which is forbidden. This would be followed by a transition between two doubly degenerate orbitals (e_g to e_u), with the excited state split into $A_{1u} + A_{2u} + E_u$. The transition to A_{1u} is forbidden. Transitions to A_{2u} and E_u are allowed. The former is polarized parallel, the latter perpendicularly, to the C-C direction. The situation is similar for eclipsed ethane (D_{3h} symmetry). The selection rules are the same except for the e'' to e' transition. The resulting excited state would split into $A_1'' + A_2'' + E''$ in this case. However, A_2'' would be only allowed for the latter transition.

As in the case of methane, the order of the transitions may change if somewhat different parameters or if higher approximations are used. Should the first transitions take place between states resulting from a jump between two degenerate orbitals, there could again be weak electronically forbidden transitions, giving bands at the low-frequency side of the first strong band system.

According to both Mulliken[6] and Katagiri and Sandorfy,[24] the electron making the jump giving rise to the first strong absorption band in the case of ethane departs from the C-H bonds, and the transition is of the $E \leftarrow A$ type. Furthermore, Pearson and Innes[97] demonstrated that the polarization of the transition related to this excitation is definitely perpendicular to the C-C bond; consequently, the electronic assignment ${}^1E_u \leftarrow {}^1A_{1g}$ is confirmed. Thus the promotion of the electron associated with this transition is from the C-H bonding system. In all probability there is a third band system at about 83,333 cm^{-1} (Fig. 1, B).

The vibrational fine structure appearing in the first strong band of ethane deserves discussion. The spacing between the observed[40, 41] vibrational bands is approximately 1100-1250 cm^{-1} (Fig. 3, A; Table 3).

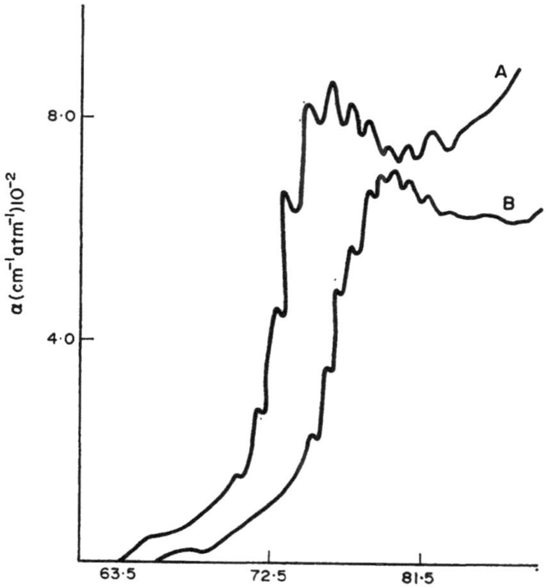

FIG. 3. Spectra of ethane (A) and ethane-d_6 (B).[41] Wave numbers against absorption coefficients.

Under D_{3h} (eclipsed) or D_{3d} (staggered) symmetry ethane has only three totally symmetric vibrations. One of these, a CH stretching motion, ν_1, has a high frequency (2899 cm^{-1}) in the ground state; ν_2, the symmetrical CH bending vibration, is at 1375 cm^{-1}; and ν_3, the C-C stretching, at 993 cm^{-1}; all are Raman-active.[71] An excited-state frequency of 1150 cm^{-1} might well be the ν_2 vibration of the excited state, but it is not impossible that it is ν_3 whose frequency would increase if the electron were taken from an orbital which is anti-bonding in the

TABLE 3. SPACING IN THE VIBRATIONAL FINE STRUCTURE OF ETHANE AND ETHANE-d_6[41]

	C_2H_6 (cm^{-1})	C_2D_6 (cm^{-1})
$\Delta_{1,2}$	1078	814
$\Delta_{2,3}$	1117	902
$\Delta_{3,4}$	1130	887
$\Delta_{4,5}$	1083	924
$\Delta_{5,6}$	1161	860
$\Delta_{6,7}$	1257	994
$\Delta_{7,8}$	1195	941
$\Delta_{8,9}$	1246	752
$\Delta_{9,10}$	1103	1002
$\Delta_{10,11}$	1251	929

C-C link in the ground state. To find an answer to this question, we can examine the spectrum of ethane-d_6[41] (Fig. 3, B). ν_2 has a frequency of 1158 cm^{-1} for C_2D_6 in the ground state; this gives a ratio of 1.187, relative to the ν_2 of C_2H_6. ν_3 is at 993 cm^{-1} for C_2H_6 and at 852 cm^{-1} for C_2D_6, resulting in a ratio of 1.165. Furthermore, according to the normal-coordinate analysis of Schachtschneider and Snyder[72] the two modes are highly mixed for the deuterated species in the ground state. It is noteworthy, however, that the average of the ratios of the corresponding vibrational frequencies for C_2H_6 and C_2D_6 in their excited states is about 1.305.[38, 41] This large isotope effect suggests that the vibration which appears in the fine structure of the spectrum of ethane is mainly a C-H vibration. Thus the deuterated compound did provide at least a partial answer to the problem. It is surprising, however, that ν_2 in the excited state of C_2D_6 is mixed with C-C motion to a much lesser extent than in the ground state, and one cannot be sure that the vibration which is involved would be actually ν_2. It may be a consequence of the distortion of the molecule in the excited state. The blue (hypsochrome) shift of the whole band system is probably connected with this too, and indicates strong vibronic interaction.

Okabe and Becker[85] measured the spectra from methane to n-butane. Raymonda and Simpson[38, 39] and Lombos *et al.*[40, 41] determined the absorption spectra of n-alkanes up to n-nonane and n-octane respectively. There is general agreement between these results.

1.3. PROPANE

A pronounced red shift of the first strong bands, increasing from ethane to propane, can be seen; there are probably three or four electronic band systems in the spectrum of propane. Further lengthening of the chain produces only smaller changes in the frequencies (Figs. 1, 4; Table 4). In all these spectra several electronic transitions are distinguishable to different degrees, depending on overlap conditions for the individual molecules. Katagiri and Sandorfy[24] have given calculated values for the lowest electronic levels of propane (Fig. 2). Re-

TABLE 4. TRENDS IN WAVELENGTH AND INTENSITIES FOR THE FIRST TWO STRONG DIFFUSE BAND SYSTEMS FROM METHANE TO n-OCTANE[41]

Compound	I		II	
	$1/\lambda$ (cm^{-1})	α ($cm^{-1} atm^{-1}$)	$1/\lambda$ (cm^{-1})	α ($cm^{-1} atm^{-1}$)
Methane	78,278	600		
Ethane	75,838	895		
Propane	71,685	1155	77,821	1338
n-Butane	70,922	1586	74,906	1845
n-Pentane	70,671	1781	74,349	2094
n-Hexane	70,175	2010	74,906	2475
n-Heptane	69,930	2154	74,627	2671
n-Octane	70,671	2524	74,627	2977

garding all-trans propane as having a symmetry C_{2v}, the first transition would be $A_1 \rightarrow B_1$, which is allowed along the x axis (perpendicular to the plane of the carbon atoms). The next transitions are as follows: $A_1 \rightarrow A_1$ allowed along the z

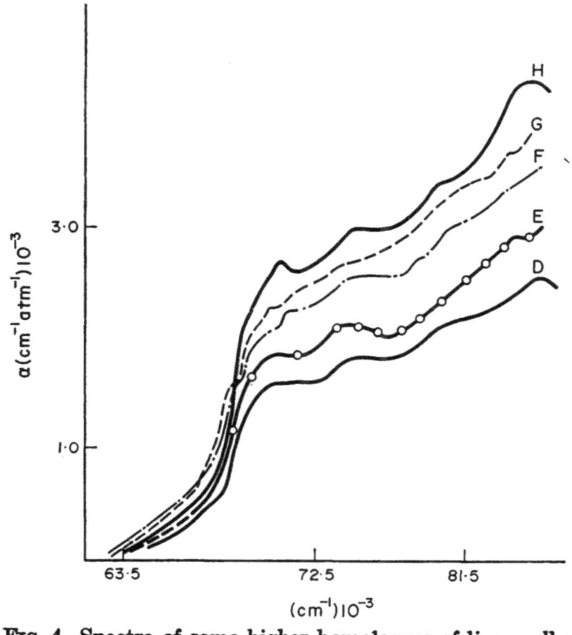

FIG. 4. Spectra of some higher homologues of linear alkanes: (D) butane, (E) pentane, (F) hexane, (G) heptane, (H) octane.[41] Wave numbers against absorption coefficients.

axis (which lies in the molecular plane and bisects the CCC angle); $A_1 \rightarrow B_1$ allowed again for the x axis: $A_1 \rightarrow A_2$ and $A_1 \rightarrow A_2$ forbidden. From these calculations Katagiri and Sandorfy[24] concluded that (a) except for (perhaps) the band system of lowest frequency, the spectra of paraffinic hydrocarbons are likely to contain many overlapping bands; (b) spectral differences between groups may be significant; (c) singlet-triplet transitions are usually less than about 1 eV from the low-frequency side of the related singlet-singlet transition, and they are expected to be very weak; (d) the calculations interpret well the trend toward lower energies, with increasing chain length, of both the first ionization potentials and the first singlet-singlet transitions; (e) the first excitation in methane, ethane, and propane would depart from an orbital where the electronic charge is in the C-H bonds. The lowest C-C type transition, however, shifts considerably to lower frequencies from ethane to propane (Fig. 1), and is likely to become the first band from n-butane. This latter conclusion concerning the first alkanes differs from the results of calculations obtained by Raymonda and Simpson,[38] which will be treated later.

No vibrational structure is present in the spectra of higher homologues (Fig. 4). This does not necessarily mean, however, that dissociation or predissociation are the only causes of these broad bands observed. The existence of many close-lying electronic states, the increasing number of rotational isomers (it was shown in ref. 24 that there may be significant differences in the spectra of these isomers), and the increasing number of totally symmetrical vibrations—due to the low symmetry of the molecule and their size—may originate broad absorption bands.

Raymonda and Simpson[38] treated these spectra with the independent-systems theory. They regarded the total methane spectrum as originating from transition of four identical C-H bonds. Subtracting the spectrum of methane, multiplied by 6/4, and arbitrarily displaced to the blue by 1000 cm^{-1} from the ethane

TABLE 5. ABSORPTION PEAKS IN THE ELECTRONIC SPECTRA OF NORMAL PARAFFINS NEAR 800 Å[37, *52]

Compound	$1/\lambda$ (cm^{-1})	α ($cm^{-1} atm^{-1}$)
Methane*	107,500	1500
Ethane	120,500	2150
Propane	~129,000	3000
Butane	~129,000	4100
Pentane	~129,000	5100

line, they arrived at another symmetrical band, centred at about 75,000 cm^{-1}, and assigned it to a $\sigma \rightarrow \sigma^*$ transition of the C-C bond of ethane. The interpretation of the alkane spectra is based, in their paper, on this fundamental assumption. The treatment of higher homologues, by an analogous procedure, yields absorption seemingly composed of several bands with total oscillator strengths proportional to the number of C-C bonds. Similar results were found by Schoen (Table 5)[37] from examination of the maximum values of the absorption cross-sections of the strong absorption bands of ethane to n-pentane, around 130,000 cm^{-1}. He suggested that the maximum value of the absorption coefficient (α [cm^{-1} atm^{-1}]) for a normal alkane could be estimated by multiplying the total number of bonding electrons by 160 (cm^{-1} atm^{-1}). The application of the independent-systems theory was initiated by those empirical sum rules. The technique of degenerate-perturbation theory was applied to find the first-order energies of excited states. The results of the calculations are given in units of the interaction integral, which represents the electrostatic energy of interaction between transition densities.

In the case of the linear alkanes the transitions to the lowest excited states for each molecule are predicted to be allowed, and transitions to higher-energy states are expected to be weak. This theory, therefore, does not predict the lowest forbidden transition predicted by Mulliken.[6] However, the independent-systems treatment predicts, for higher homologues that each succeeding alkane spectrum will "mesh" outside and inside the transitions of the preceding one. Thus, for example, in the case of propane there are two transitions symmetrically placed on either side of the ethane band predicted at 71,300 and 78,600 cm^{-1}, and no transition of the ethane frequency at 75,838 cm^{-1} (Fig. 1, B, C). The calculations predicted that mainly the C-C bond would be involved in this excited state of ethane.

1.4. BUTANE

Butane has three predicted transitions at 68,900, 75,800, and 79,800 cm^{-1}—two straddling the propane transitions, and the third in the centre at the ethane frequency (Fig. 1, B, C, D). The first two of these most probably correspond to the observed peaks at 70,922 and 74,906 cm^{-1}, and the third might be assigned to the shoulder at 80,808 cm^{-1}.[41]

1.5. n-PENTANE

In the case of n-pentane (Fig. 4, E) transitions at 67,600, 73,300, and 77,900 cm^{-1} are predicted.

1.6. n-HEXANE

The independent-systems approach predicts for n-hexane (Fig. 4, F) absorption peaks at 66,800, 71,500, and 77,900 cm^{-1}; however, a thorough comparison of experiment and calculations is difficult, since the absorption bands overlap severely.

Another interesting result of this treatment is that the absorption edges of these spectra are expected to approach a limit with lengthening of the chain. This is indeed, evident (Table 4) from comparison of the observed spectra. While the topological matching between the observed spectra and the model based on the independent-systems theory is good, the experimental data do not fit the theory quantitatively. The quantitative agreement is, however, much improved when excitations in the C-H bonds are also considered. The "resonance" approach is used, i.e., the C-H bond natural frequency of electronic excitation is taken as equal to the C-C bond excitation frequency, similar to the treatment of Katagiri and Sandorfy.[24] The wave functions are thus derived for the two types of excitation, with coefficients determined mainly by the geometry of the molecule. All the necessary quantities for the calculations were based on the spectra of methane and ethane in the framework of the independent-systems theory.

The calculation also predicts, in the case of linear alkanes, several low-lying excited states in which the excitation is mainly in the C-C bonds. The low-frequency limit, corresponding to the lowest excitation energy for an infinite linear chain, was estimated to be at $\sim 65{,}000\ cm^{-1}$ compared with $\sim 70{,}000\ cm^{-1}$ found experimentally by Lombos *et al.* and with $\sim 70{,}000\ cm^{-1}$ given by Partridge,[36] the latter in the case of infinite n-alkane, i.e., polyethylene.

The resonance calculation also anticipates several transitions originated by the C-H bonds. These transitions are expected to be crowded into a band of energies from 80,500 to 83,000 cm^{-1}, contrary to the predictions of Katagiri and Sandorfy.[24]

It can be argued that the energy of any photon which brings about an electronic (singlet-singlet) transition in a saturated hydrocarbon would be sufficient to dissociate any given bond in these molecules. However, if there is enough interaction between the bonds in the excited state, the excitation energy can be divided between them, precluding dissociation. Similar arguments were put forward by Magee *et al.*,[74, 75] concerning infinite chains, and by Partridge[36] concerning both alkanes and polyethylene. It is often presumed that in paraffins the electrons are localized in one given C-C or C-H bond. This is approximately true in the ground state although it is well known that both the heats of formation[13, 30, 76] and the ionization potentials[77, 78] show a definite trend indicating a certain degree of delocalization.

2. Branched alkanes

The spectra of some branched alkanes were recently measured by Raymonda and Simpson[38] and by Lombos *et al.*[40] The assignments of some distinguishable

TABLE 6. TRENDS IN WAVE NUMBERS FOR THE FIRST TWO STRONG DIFFUSE BAND SYSTEMS OF SOME BRANCHED PARAFFINS
(EXCLUDING THE $\sim 67{,}000\ cm^{-1}$ "CHARGE-TRANSFER" TRANSITIONS)[38, *40]

Compound	$1/\lambda$ (cm^{-1})	
	I	II
Isopentane	72,464*	78,125
2,3-Dimethylbutane	72,000	—
3-Methylpentane	71,500	(76,000)
2-Methylpentane	(72,000)	—
2,4-Dimethylpentane	(72,000)	(76,000)

electronic transitions, using the resonance model of the independent-systems technique, are satisfactory to a certain degree. The results are shown in Figure 5 and Table 6.

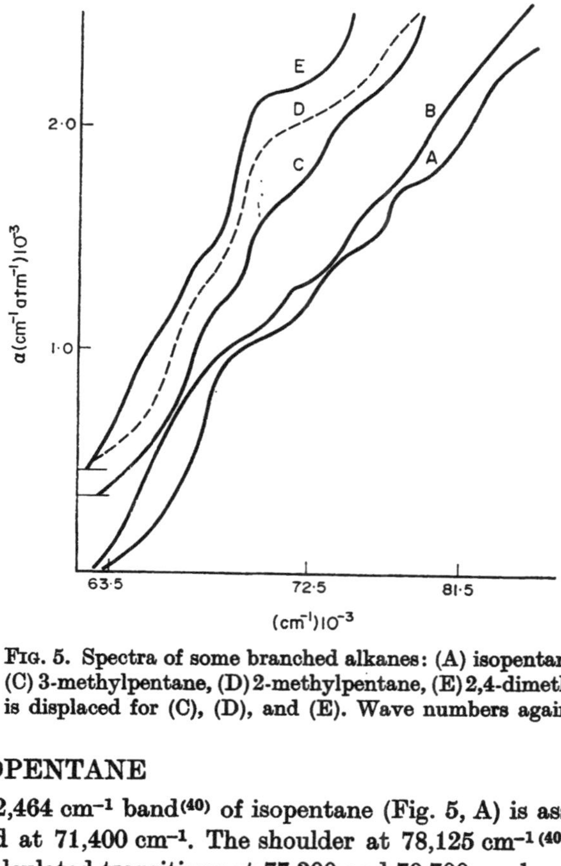

FIG. 5. Spectra of some branched alkanes: (A) isopentane, (B) 2,3-dimethylbutane, (C) 3-methylpentane, (D) 2-methylpentane, (E) 2,4-dimethylpentane.[38] The ordinate is displaced for (C), (D), and (E). Wave numbers against absorption coefficients.

2.1. ISOPENTANE

The 72,464 cm^{-1} band[40] of isopentane (Fig. 5, A) is assigned to the transition predicted at 71,400 cm^{-1}. The shoulder at 78,125 cm^{-1} [40] probably corresponds to the calculated transitions at 77,300 and 79,700 cm^{-1}.

2.2. 2,3-DIMETHYLBUTANE

The peak at 72,000 cm^{-1} in the spectrum at 2,3-dimethylbutane (Fig. 5, B) is assigned to the predicted transition at about 71,200 cm^{-1}. There is also a transition anticipated at 79,000 cm^{-1}, but it might to be hidden under the band.

2.3. 3-METHYLPENTANE

For 3-methylpentane (Fig. 5, C) the absorption band centred at about 71,500 cm^{-1} is associated with the transition predicted at 69,300 cm^{-1} and the shoulder at approximately 76,000 cm^{-1} would correspond to the calculated frequencies of 76,300 and 77,700 cm^{-1}.

2.4. 2-METHYLPENTANE

In the case of the 2-methylpentane (Fig. 5, D) the shoulder at 72,000 cm^{-1} could be associated with the calculated transitions at 71,400 and 73,700 cm^{-1}.

2.5. 2,4-DIMETHYLPENTANE

There is a shoulder at 72,000 cm^{-1} in the spectrum of 2,4-dimethylpentane (Fig. 5, E) which could be assigned to the pair of predicted transitions at 71,300 cm^{-1}.

The tendency toward a slight red-shifting of the bands between isopentane and 2,3-dimethylpentane is correctly predicted by this theory (Fig. 5, B; Table 6). The higher-energy transitions could not be easily assigned, probably because of many overlapping bands.

3. Highly branched alkanes

The spectra of highly branched paraffins will be discussed again on the basis of the work of Raymonda and Simpson.[38] Somewhat better resolved structures

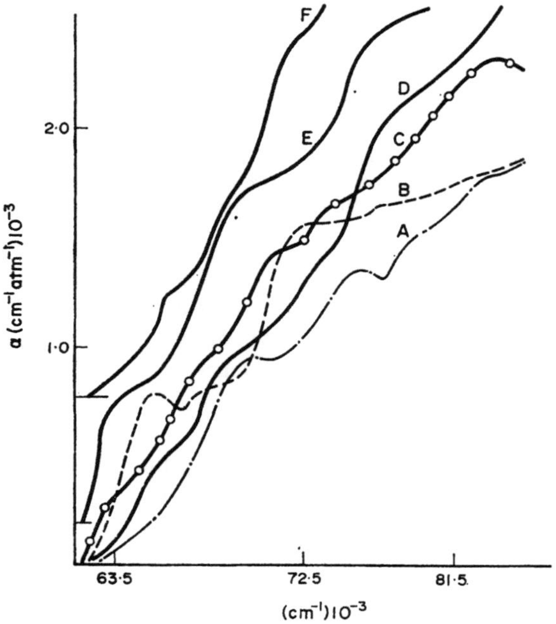

FIG. 6. Spectra of some highly branched alkanes: (A) isobutane, (B) neopentane, (C) 2,2-dimethylbutane, (D) 3-ethylpentane, (E) 2,2,4-trimethylpentane, (F) hexamethylethane.[38] The ordinate is displaced for (E) and (F). Wave numbers against absorption coefficients.

can be seen on some of the spectra reported by Lombos *et al.*[40] who used double-beam instrumentation.[79] The results are shown in Figure 6 and Table 7.[38, 40]

TABLE 7. SOME ABSORPTION PEAKS IN THE ELECTRONIC SPECTRA OF HIGHLY BRANCHED PARAFFINS[38, *40]

Compound	$1/\lambda$ (cm^{-1})	
	I	II
Isobutane	69,204*	74,379*
Neopentane	68,024*	74,074*
2,2-Dimethylbutane	67,500	72,000
3-Ethylpentane	65,000	70,000
2,2,4-Trimethylpentane	67,000	
Hexamethylethane	68,000	75,000

3.1. ISOBUTANE

The lowest calculated transition at 71,200 cm^{-1} is predicted, by the resonance model,[38] to be doubly degenerate for isobutane (Fig. 6, A). This value falls roughly midway between two observed[40] transitions at 69,204 and 74,343 cm^{-1}. In the case of this assignment it is supposed[38] that the spatially degenerate excited state has been split apart in energy by molecular distortion in accord with the Jahn-Teller theorem.[69]

3.2. NEOPENTANE

A triply degenerate excited state at 71,100 cm^{-1} is anticipated for the neopentane (Fig. 6, B) from resonance model calculations. The Jahn-Teller theorem[69] is invoked by Raymonda and Simpson[38] to assign the peaks at 68,024 and 74,074 cm^{-1} [40] to this state. However, it is unlikely that such a large splitting can be explained by the Jahn-Teller effect in this and in the following cases.

3.3. 2,2-DIMETHYLBUTANE

This molecule can be looked upon as neopentane perturbed by an extra methyl group. The calculation[38] predicts that the threefold degeneracy of neopentane has been partially lifted in 2,2-dimethylbutane (Fig. 6, C). The shoulders in the spectrum at 67,500 and 72,000 cm^{-1} can perhaps be associated with the calculated transitions at 68,600 and 71,100 cm^{-1} respectively. The predicted transition at 77,000 cm^{-1} matches exactly a slight shoulder in the spectrum at this frequency.

3.4. 3-ETHYLPENTANE

The lowest calculated[38] transition in the case of 3-ethylpentane (Fig. 6, D) at 67,600 cm^{-1} is between two observed maxima at approximately 65,000 and 70,000 cm^{-1}. It is supposed again that, as in the case of isobutane, the spatially degenerate excited states have been split by molecular distortion in accordance with the Jahn-Teller theorem.[69]

3.5. 2,2,4-TRIMETHYLPENTANE

There are a large number of transitions predicted for 2,2,4-trimethylpentane (Fig. 6, E). Few assignments can be proposed, due to the severe overlapping of the bands. The absorption maximum at 67,000 cm^{-1} may be correlated with the predicted[38] transitions at 68,000 and 67,500 cm^{-1}. This calculated value, as in the case of neopentane and 2,2-dimethylbutane, is the lowest energy anticipated by the resonance model.

3.6. HEXAMETHYLETHANE

For this compound also the calculation[38] predicts a large number of excited states. The band at 68,000 cm^{-1} (Fig. 6, F) probably corresponds to the predicted transition at 67,900 cm^{-1}. The broad absorption centred at 75,000 cm^{-1} may be assigned to the four transitions anticipated in the 70,500-72,000 cm^{-1} range.

One finds, however, in the case of certain compounds, transitions to the red (at about 67,000 cm^{-1}; Figs. 5 and 6) of those predicted by the independent-systems treatment. If one attempts to interpret these as belonging to the independent-systems manifold, the natural way in which the transitions seem assignable is lost. The intensity of these transitions increases with the amount of branching at a given carbon atom. Molecular-orbital calculations of Pariser and Parr[17] indicate that the transitions could arise from charge-transfer around a C-C-C branch point. From some pilot calculations for propane, the authors[38] suggested, therefore, that these bands might be classified as charge-transfer transitions.

4. Cyclic alkanes

These spectra (Fig. 7) were also observed and tentatively assigned by Raymonda and Simpson.[38] Some absorption peaks are collected in Table 8.

TABLE 8. SOME ABSORPTION PEAKS IN THE ELECTRONIC SPECTRA OF CYCLO-PARAFFINS[38]

Compound	$1/\lambda$ (cm^{-1})	
	I	II
Cyclopropane	(63,000)	(69,000)
Cyclobutane	72,000	79,500
Cyclopentane	—	(76,000)
Cyclohexane	70,000	75,000
Cycloheptane	69,500	74,500
Cyclo-octane	68,000	73,000

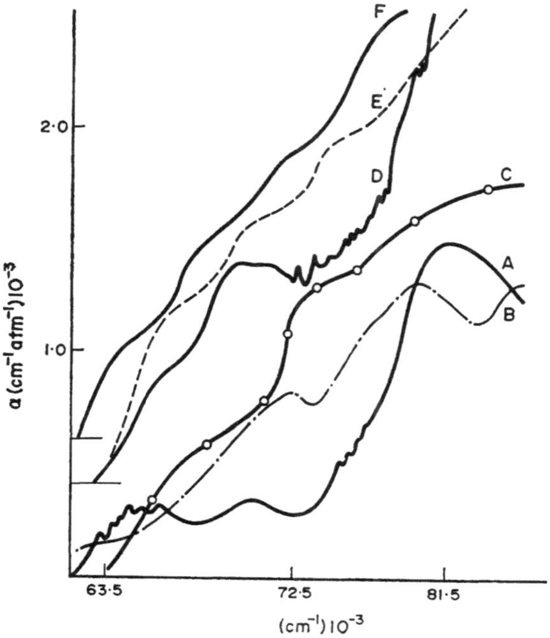

FIG. 7. Spectra of some cyclic alkanes: (A) cyclopropane, (B) cyclobutane, (C) cyclopentane, (D) cyclohexane, (E) cycloheptane, (F) cyclo-octane.[38] The ordinate is displaced for (D), (E), and (F). Wave numbers against absorption coefficients.

4.1. CYCLOPROPANE

It has been shown[38] that, since the spatial geometry of this molecule has to be considered, the independent-systems theory cannot be successfully applied. It is probable that delocalization is important in cyclopropane (Fig. 7, A). A delocalization molecular-orbital calculation has recently been carried out by Brown and Krishna[80] in a manner similar to that of Katagiri and Sandorfy.[24] The electronic configuration of the ground state of cyclopropane has been discussed extensively with emphasis on the orientation and hybridization of bonding orbitals in the molecule.[98, 99] The results are the most easily examined in terms of Walsh's[98] description of the bonding which was used by Brown and Krishna.[80] Six C-C σ electrons were considered. One sp^2 hybrid on each carbon directed to the centre of the ring and one $2p$ on each carbon, taken in the plane of the ring, were used to form the basic orbitals. This calculation predicts a symmetry-forbidden transition at 58,300 cm^{-1} (species A_1' of group D_{3h}), and another forbidden transition (species A_1') at 70,300 cm^{-1}. Two pairs of allowed (species E') transitions are predicted at 70,700 and 93,500 cm^{-1}. A very weak band (not shown in Fig. 7, A in the range of 52,000-57,000 cm^{-1}) would be the A_2' transition. The first band shown in Fig. 7, A, in the 60,000-68,000 cm^{-1} range, would involve the first E' transition, while the large absorption peak at 82,000 cm^{-1} could be assigned to the second E' transition and the peak at 69,000 cm^{-1} to the second A_1' state.

The electronic states of cyclopropane were also treated by a Gaussian orbital-self-consistent-field calculation(100) using contracted Gaussian-type orbitals (GTO). Walsh(98) and Coulson and Moffit(99) have calculated that the internal (sp^2 hybrids) and the external ($2p$ unhybridized) orbitals lead to a particular charge distribution in the plane of the molecule, which would be largest about the centre of the ring and symmetrical outside each edge of it. The GTO calculation suggested also that the $3e'$ molecular orbital is one of the external σ-type variety, while the $3a_1'$ is an internal one, and the $1e''$ could correspond to a π-type orbital related to the out-of-plane C-H bonding. Moreover, the electrons in the $3e'$ molecular orbitals might be considered, in analogy with the unsaturated compounds, as conjugated with the π-type $1e''$ electrons if these are in the plane of the cyclopropane ring.(101–3) The following discussion of the electronic spectrum of cyclopropane is based on this simple picture of bonding, and valence-shell and Rydberg transitions will be considered.

It is a difficult problem, however, to distinguish the Rydberg bands from the valence-shell bands in the electronic spectrum of cyclobutane. In this connection Herzberg(60) pointed out that Rydberg excitations are not expected to be observed in paraffin molecules. However, as the results of the GTO treatment indicate, the originating orbital for the lowest Rydberg series in cyclopropane is $3e'$, which is definitely not a C-H bonding orbital. If the ionization potential of this molecule is taken as 10.53 eV, then Rydberg transitions having $n = 3$, 4, and 5 at 64,530, 74,950, and 79,030 cm^{-1} respectively, are predicted.(100) The two observed (Fig. 7, A) structured bands at 62,000 and 76,000 cm^{-1} seem to fit the expected Rydberg transitions of the GTO calculation. In addition, the vibrational progressions of both of these bands have approximately 490 cm^{-1} spacing.(100) This seems to indicate also that the 62,000 and 76,000 cm^{-1} bands are members of the same Rydberg series. Furthermore, Belanger *et al.*(46) investigated the absorption spectrum of this molecule and found a weak but obvious transition at 80,000 cm^{-1}, which is just the frequency expected for the $n = 5$ member of the Rydberg series mentioned above. If the assignment is correct, the latter band should display a vibrational fine structure, also with 490 cm^{-1} spacing; but this remains to be seen.

To shed more light upon these arguments the spectrum of a solid film of cyclopropane was measured.(100) It is known that, in many cases, the Rydberg bands do not appear in condensed phases.(104) One can see from the comparison of the solid and gas-phase spectra of cyclopropane that the 62,000 cm^{-1} band disappears in the condensed phase—as expected for a Rydberg transition. Therefore the authors(100) assigned the bands at 70,000 and 82,000 cm^{-1} to valence-shell transitions. From consideration of the calculated intensities, the simplest supposition would be that these two valence-shell bands are the strongest and all others are either very weak and buried or lie beyond 91,000 cm^{-1}. Using these assumptions, the 82,000 cm^{-1} band of cyclopropane may be assigned as the doubly degenerate $3e' \rightarrow 1a_2'$, i.e. $^1A_1' \rightarrow {}^1E'$ transition in accord with Brown and Krishna.(80) Clark,(105) however, interpreted this band as a $^1A_1' \rightarrow {}^1A_2''$ out-of-plane polarized excitation. The second strong band is expected to be[100] a $3e' \rightarrow 4e'$, i.e. $^1A_1' \rightarrow {}^1E'$, transition. The calculated excitation energies, however, are about 3 eV too high, though configuration interaction could diminish this discrepancy. All seven of the other transitions have predicted oscillator strengths smaller, by at least a factor of ten, than those of the two strong bands mentioned.

If cyclopropane be considered as a conjugated polyene rather than a paraffin, the two strong $^1A' \rightarrow {}^1E'$ transitions could correspond to the $N \rightarrow V$ band of the straight-chain polyenes, while the higher energy transition could also be formally related to the $^1A_{1g} \rightarrow {}^1E_{1u}$ excitation of benzene, since they both involve in-plane allowed, doubly degenerate transitions between doubly degenerate sets of π^* orbitals.

The observed 490 cm^{-1} vibrational frequency of the excited state was also discussed.[100] The following ground-state vibrational frequencies can be considered:[106] ν_7 (a_2''), the CH_2 rocking motion with a frequency of 854 cm^{-1}; ν_{14} (e''), also a CH_2 rocking vibration at the frequency of 741 cm^{-1}; and $\nu_{11}(e'')$, a CH_2 wagging motion having a frequency of 866 cm^{-1}. The first two vibrations show an isotope effect of 1.4 on going from C_3H_6 to C_3D_6[100] in the ground state, but the third has a ratio of only 1.20. The excited state frequencies in C_3D_6, however, could not be determined for these 62,000 and 76,000 cm^{-1} bands, so that the ambiguity of the vibrational assignment cannot yet be resolved.

4.2. CYCLOBUTANE

The independent-systems theory predicted allowed doubly degenerate pair transitions for each of the other cyclic-alkanes. The pair at 77,600 cm^{-1} in the case of cyclobutane (Fig. 7, B) could correspond to the observed transitions at 72,000 and 79,500 cm^{-1}, split by the Jahn-Teller effect, according to Raymonda and Simpson.[38]

4.3. CYCLOPENTANE

For cyclopentane (Fig. 7, C) the predicted value seems to coincide with a definite shoulder at 76,000 cm^{-1}, and no Jahn-Teller splitting can be seen.

4.4. CYCLOHEXANE

A doubly degenerate pair of states, at approximately 74,000 cm^{-1}, has been predicted by the calculations.[38] The observed transitions at 70,000 and 75,000 cm^{-1} (Fig. 7, D) have been assigned to this theoretical value taking into consideration the Jahn-Teller effect.[38]

4.5. CYCLOHEPTANE

In the case of the cycloheptane (Fig. 7, E) the inflections at 69,500 and 74,500 cm^{-1} were correlated with the predicted doubly degenerate transition at 72,000 cm^{-1} using the Jahn-Teller theorem again.[38]

4.6. CYCLO-OCTANE

The two shoulders at 68,000 and 73,000 cm^{-1}, found on the experimental absorption curve of cyclo-octane (Fig. 7, F), were associated with the anticipated doubly degenerate transitions at 71,000 cm^{-1}. Raymonda and Simpson(38) invoke tentatively the Jahn-Teller effect for this assignment also, which is again very questionable.

The C-H bands are predicted, by the resonance treatment, to be in the range of 81,000-81,900 cm^{-1}. These absorptions are spread out over broad bands, as in the case of the branched and linear alkanes, resembling the methane spectrum. Here, as in the branched alkanes, ionization processes are setting-in. The spectroscopic first ionization potentials for cyclopentane and cyclohexane(73) are at 84,773 and 79,530 cm^{-1} respectively. Further details about the decomposition of excited molecules, including data on all processes which occur beyond the frequency of 96,000 cm^{-1}, can be found in the excellent work of Steiner *et al.*(88)

5. Liquid alkanes

Liquid paraffins were mainly investigated as transparent media suitable for use as solvents.

The measured(90–95) transmission limits of *n-pentane, n-hexane, n-heptane, n-decane, n-dodecane, 3-methylhexane, 2,2-dimethylpentane, 2,2-dimethylbutane, 2,2,3-trimethylbutane, 2,2,4-trimethylpentane, 2,2,3-trimethylpentane, 2,2,3,3-tetramethylpentane, cyclopentane, cyclohexane, methylcyclopentane, and methylcyclohexane,* were found to be between 65,000 and 60,000 cm^{-1}.

6. Perfluoro-alkanes

The far u.v. spectra of the perfluoro-n-paraffins up to hexane, were measured from 50,000 to 86,000 cm^{-1} by Belanger *et al.*(46) Figure 8 and Table 9 show the

TABLE 9. COMPARISON OF THE WAVE NUMBERS AND MOLECULAR EXTINCTION COEFFICIENTS OF THE FIRST STRONG BANDS OF n-PARAFFINS AND PERFLUORO-n-PARAFFINS(46, *41)

Compound	$1/\lambda$ (cm^{-1})	α ($cm^{-1} atm^{-1}$)
C_3F_8	84,034	179
C_3H_8*	71,685	1155
C_4F_{10}	79,840	288
C_4H_{10}*	70,922	1586
C_5F_{12}	74,074	413
C_5H_{12}*	70,671	1781
C_6F_{14}	71,174	775
C_6H_{14}*	70,175	2010

results. The first observation is the great relative transparency of these molecules in this part of the v.u.v.

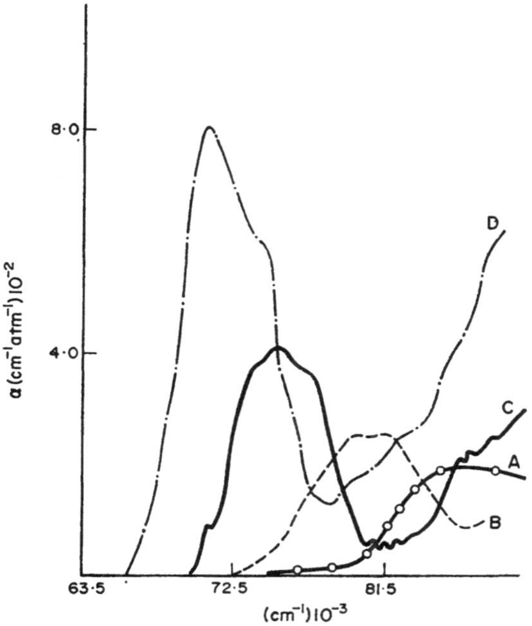

FIG. 8. Spectra of perfluoro-normal: (A) propane, (B) butane, (C) pentane, (D) hexane.[46] Wave numbers against absorption coefficients.

6.1. CARBONTETRAFLUORIDE

Carbontetrafluoride has no absorption between 50,000 and 86,000 cm^{-1} under pressures varying from 0.1 to 0.5 torr.[46] Zobel and Duncan[107] investigated its spectrum, and only one electronic transition, with an intensity maximum at 109,890 cm^{-1}, was found; this has also been confirmed by Cook and Ching.[108] According to Stokes and Duncan,[109] the transitions of CF_4 can be interpreted in a way similar to Mulliken's united-atom treatment[6] of CH_4 with the $2p\sigma$ electrons of fluorine taking the place of the hydrogen $1s$ electrons. The lone pairs of the fluorine seem to have little influence on the orbitals involved in the spectral transitions discussed. In the part of the spectrum examined, (50,000-86,000 cm^{-1}), no evidence for $\sigma \leftarrow n$ transitions were found.[46]

6.2. PERFLUOROETHANE

Perfluoroethane exhibits a weak band near 77,000 cm^{-1} ($\alpha \simeq 0.5$ cm^{-1} atm^{-1}) which might belong either to the molecule itself[46] or to impurities.

6.3. PERFLUOROPROPANE

Perfluoropropane enters the region investigated by Belanger *et al.* with a well-defined peak at 84,034 cm^{-1} (Fig. 8, A; Table 9). This band shifts gradually to lower frequencies with increasing carbon-chain length (Fig. 8, B, C, D), thus exhibiting a behaviour similar to that of the first strong band of the corresponding paraffins themselves. The frequencies of the perfluoro compounds are, however, higher and the intensities of the absorptions are lower (Table 9). Second and third bands occur for the higher homologues of the series, but these show less analogy with the related paraffin spectra.

Furthermore, there are differences in the nature of the red-shift observed for the alkanes and the perfluoro-paraffins. In the case of the alkanes the shifts are large from methane to ethane and from ethane to propane, and they tend to converge to a limit at about 70,000 cm^{-1} (Table 4). The higher members of the perfluoro-paraffins shift by nearly equal amounts and do not reach their convergence limit for perfluoro-normal-hexane (Table 9). In Mulliken's united-atom treatment[6] the excited orbital would be again $(3s + 3s + \cdots)$, and the shift with increasing number of carbon atoms may then be due to lowering of this orbital. The departure orbital is probably of C-F character for CF_4 and C_2F_6, but it may become C-C for the higher members. The independent-systems technique,[38] as in the case of the paraffins, would certainly favour the latter hypothesis. We do not know for sure if the first ionization of the n-paraffins leaves the hole mainly in the C-C or mainly in the C-H bonds. Most calculations favour C-C, but Al-Joboury and Turner's recent photoelectron data[110] favour C-H, at least in the case of ethane. The first (adiabatic) ionization potential of methane is 12.99 eV[110] and that of CF_4 is probably closer to 15 eV.[111] Davies[112] quotes 15.9 eV for the adiabatic value of ionization potential of CF_4. The difference is considerable. A similar lowering of the energy of the highest filled-orbital of the ground state is probably the main cause of the general displacement of the spectra of the perfluoro-paraffins to higher frequencies with respect to the paraffins. Unfortunately we have no data on the ionization potentials of the homologues of perfluoro-paraffins to verify this supposition.

The lower intensities are probably related to impoverishment of the electrons of the C-C chains due to the high electron affinity of the fluorine atom.

From the parallelism of the spectra of n-paraffins and perfluoro-n-alkanes one could conclude that the excitation process is similar, at least for the first singlet-singlet transition.

7. Perfluoro-cyclic alkanes

The spectra of perfluorocyclo-propane, -butane, and -pentane have been measured by Belanger *et al.*[46] Some of the absorption peaks are collected in Table 10.

7.1. PERFLUOROCYCLOPROPANE

Four band systems with vibrational fine structure can be distinguished, with the maxima at 49,853, 66,000, 75,930, and 79,554 cm^{-1}. All four structured

TABLE 10. SOME ABSORPTION PEAKS IN THE ELECTRONIC SPECTRA OF PERFLUORO-CYCLO-PARAFFINS[46]

Compound	I		II	
	$1/\lambda$ (cm^{-1})	α (cm^{-1}atm^{-1})	$1/\lambda$ (cm^{-1})	α (atm^{-1}cm^{-1})
Perfluorocyclopropane	49,853	54.8	66,000	92.5
Perfluorocyclobutane	57,637	4.4	72,464	15.0
Perfluorocyclopentane	64,516	40.0	77,220	53.6

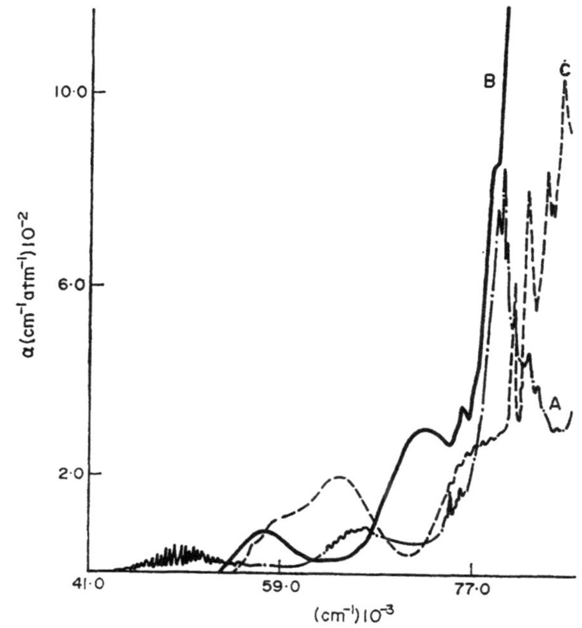

FIG. 9. Spectra of (A) perfluorocyclopropane, (B) perfluorocyclobutane, (C) perfluorocyclopentane. The ordinate is extended twenty times for (B) and five times for (C).[46] Wave numbers against absorption coefficients.

bands (Fig. 9, A) seem to contain vibrational progressions, having approximately 370 cm^{-1} spacing. The first and the second transitions, however, at 49,853 and 66,000 cm^{-1}, reveal some vibrations with approximately 410 and 610 cm^{-1} respectively. The assignment of the electronic transitions would thus be an even more difficult problem than in the case of cyclopropane. The GTO treatment would suggest again both Rydberg and valence-shell excitations. This might be accepted for the 75,930 and 79,554 cm^{-1} transitions. In addition, the first two transitions show a combination of vibrations with the frequencies above mentioned, and thus could not be taken to belong in the same Rydberg series. Unfortunately, no condensed phase spectra are available to clarify these ambiguities.

7.2. PERFLUOROCYCLOBUTANE

The first two transitions were measured[46] at 57,637 and 72,464 cm^{-1} (Fig. 9, B). It is interesting to note the large red-shift of the first band, compared with the spectrum of cyclobutane (Tables 8 and 10). The intensity of the absorption of this band is very low, indicating a similar behaviour to that demonstrated in the case of the linear paraffins. Impoverishment in electrons of the C-C ring, due to the high electron affinity of the fluorine atom, seems to be most prevalent in the four-member ring.

7.3. PERFLUOROCYCLOPENTANE

Two electronic transitions can be distinguished at 64,516 and 77,220 cm^{-1} in the spectrum[46] (Fig. 9, C). The second band exhibits a vibrational fine structure with a spacing of approximately 750 cm^{-1}. The bathochromic shift of the spectrum is contrary again, to the displacements observed in the case of the open-chain paraffins. The absorption intensities are, however, lower than in the case of cyclopentane,[38] in accordance with the general trend observed between the corresponding paraffins and the perfluoro-paraffins.

8. Conclusion

The electronic spectra of saturated alkanes exhibit a number of interesting and basic features. However, further theoretical and experimental work is certainly highly desirable since many assignments are ambiguous or are not possible on account of the small amount of good resolution experimental data and theoretical complexity.

The weak band found in the 61,350 and 63,500 cm^{-1} region in the case of the linear alkanes may be related to the forbidden $^1A_1 \leftarrow {}^1A_1$ transition which Mulliken predicted in the case of ethane.

The first allowed electronic transition of ethane was confirmed to be correctly assigned to $^1E_u \leftarrow {}^1A_{1g}$ indicated by the polarization of the transition.[113] The promotion of the electron in this excitation would thus be from the C-H bonding system rather than from the C-C, which is in accord with the conclusion of ionization potential studies.[73]

The relatively well-resolved vibrational structure of the first strong band system of ethane and of the bands of cyclopropane, perfluorocyclopropane, and perfluorocyclopentane shows that caution should be exercised in attributing the diffuse character of most of the other bands entirely to dissociation or predissociation. It should be emphasized that, while in photochemistry it is assumed that the lifetime of the ^1E_u state is less than 10^{-14} s, the observed line-width for C_2D_6, for example, indicates a lifetime closer to 10^{-11} s.[113]

The systematic shifting and intensity increase of the strong bands, corresponding to higher electronic transitions, bear a certain analogy to those known for conjugated chains.

The large hypsochromic shift of the spectra of the perfluoro-paraffins is partly due to lowering of the energy of the highest filled-orbital of the ground state. The lower intensities, related to the impoverishment of the electrons of the C-C chains by the high electron affinity of the fluorine atom, indicates the usefulness of the linear perfluoro-alkanes as transparent solvents in the far u.v.

References

1. Hückel, E., *Z. Phys.* **70**, 204 (1931).
2. Lennard-Jones, J., and Hall, G. G., *Trans. Faraday Soc.* **48**, 581 (1952).
3. Franklin, J. L., *J. Chem. Phys.* **22**, 1304 (1954).
4. Brown, R. D., *J. Chem. Soc.* **21**, 2615 (1953).
5. Dewar, M. J. S., and Pettit, R., *J. Chem. Soc.* **22**, 1625 (1954).
6. Mulliken, R. S., *J. Chem. Phys.* **3**, 517 (1935).
7. Sandorfy, C., and Daudel, R., *Compt. Rend.* **238**, 93 (1954).
8. Sandorfy, C., *Can. J. Chem.* **33**, 1337 (1955).
9. Yoshizumi, H., *Trans. Faraday Soc.* **53**, 125 (1957).
10. Fukui, K., Kato, H., and Yonezawa, T., *Bull. Chem. Soc. Japan* **33**, 1201 (1960).
11. Fukui, K., Kato, H., and Yonezawa, T., *Bull. Chem. Soc. Japan* **34**, 1111 (1961).
12. Klopman, G., *Sur la structure electronique des molécules saturées*, Cyanamid European Res. Inst., Geneva, Switzerland, 1962.
13. Klopman, G., *Tetrahedron* **19**, suppl. 2, 111 (1963).
14. Klopman, G., *Helv. chim. Acta* **45**, 711 (1962).
15. Hoffman, R., *J. Chem. Phys.* **39**, 1397 (1963).
16. Pople, J. A., *Trans. Faraday Soc.* **49**, 1375 (1953).
17. Pariser, R., and Parr, R. G., *J. Chem. Phys.* **21**, 466, 762 (1953).
18. Pariser, R., and Parr, R. G., *J. Chem. Phys.* **21**, 568 (1953).
19. Brickstock, A., and Pople, J. A., *Trans. Faraday Soc.* **50**, 901 (1954).
20. Pople, J. A., *J. Phys. Chem.* **61**, 6 (1957).
21. Klopman, G., *J. Am. Chem. Soc.* **86**, 1463, 4550 (1964).
22. Pople, J. A., Santry, D. P., and Segal, G. A., *J. Chem. Phys.* **43**, 129, 136 (1965).
23. Kaufman, J., *J. Chem. Phys.* **43**, 152 (1965).
24. Katagiri, S., and Sandorfy, C., *Theor. chim. Acta Berlin* **4**, 203 (1966).
25. Daudel, R., *Structure electronique des molécules*, Gauthier-Villards, 1962, p. 210.
26. Leifson, S. W., *Astrophys. J.* **63**, 87 (1926).
27. Rose, A., *Z. Phys.* **81**, 751 (1933).
28. Duncan, A. B. F., and Howe, J. P., *J. Chem. Phys.* **2**, 851 (1934).
29. Price, W. C., *Phys. Rev.* **45**, 529, 843 (1934).
30. Watanabe, K., *J. Chem. Phys.* **22**, 1564 (1954).
31. Wilkinson, R. G., and Johnston, H. L., *J. Chem. Phys.* **18**, 190 (1950).
32. Ditchburn, R. W., *Proc. Roy. Soc. London* A, **229**, 44 (1955).
33. Moe, G., and Duncan, A. B. F., *J. Am. Chem. Soc.* **74**, 3140 (1952).
34. Sun, H., and Weissler, G. L., *J. Chem. Phys.* **23**, 1160 (1955).
35. Okabe, H., and Becker, D. A., *J. Chem. Phys.* **39**, 2549 (1963).
36. Partridge, R. H., *J. Chem. Phys.* **45**, 1685 (1966).
37. Schoen, R. I., *J. Chem. Phys.* **37**, 2032 (1962).
38. Raymonda, J. W., and Simpson, W. T., *J. Chem. Phys.* **47**, 430 (1967).
39. Raymonda, J. W., and Simpson, W. T., *Symposium on Molecular Structure and Spectroscopy*, Ohio State University, Columbus, 1965.
40. Lombos, B. A., Sauvageau, P., and Sandorfy, C., *Chem. Phys. Letters* **1**, 42, 221 (1967).
41. Lombos, B. A., Sauvageau, P., and Sandorfy, C., *J. Molec. Spectrosc.* **24**, 253 (1967).

42. Dressler, K., and Schnepp, O., *J. Chem. Phys.* **33**, 270 (1960).
43. Lombos, B. A., Sauvageau, P., and Sandorfy, C., *Chem. Phys. Letters* **1**, 382 (1967).
44. Zobel, C. R., and Duncan, A. B. F., *J. Am. Chem. Soc.* **77**, 2611 (1955).
45. Stokes, S., and Duncan, A. B. F., *J. Am. Chem. Soc.* **80**, 6177 (1958).
46. Belanger, G., Sauvageau, P., and Sandorfy, C., *Chem. Phys. Letters* **3**, 649 (1969).
47. Scheibe, G., and Grieneisen, H., *Z. phys. Chem.* B 25, 52 (1934).
48. Hese, H., and Rose, A., *Z. Phys.* **81**, 745 (1935).
49. Leighton, P. A., and Steiner, A. B., *J. Am. Chem. Soc.* **58**, 1823 (1936).
50. Groth, W., *Z. Elektrochem.* **45**, 262 (1939).
51. Watanabe, K., Inn, E. C. Y., and Zelikoff, M., *J. Chem. Phys.* **21**, 1026 (1953).
52. Weissler, G. L., and Sun, H., *Phys. Rev.* **92**, 533 (1953).
53. Sun, H., and Weissler, G. L., *J. Chem. Phys.* **23**, 1160 (1955).
54. Watanabe, K., *J. Chem. Phys.* **26**, 542 (1957).
55. Mahan, B. H., and Mandal, R., *J. Chem. Phys.* **37**, 207 (1962).
56. Back, R. A., and Walker, D. C., *J. Chem. Phys.* **37**, 2348 (1962).
57. Walker, C. D., and Back, R. A., *J. Chem. Phys.* **38**, 1526 (1963).
58. Magee, E. M., *J. Chem. Phys.* **39**, 855 (1963).
59. Metzger, P. H., and Cook, G. R., *J. Chem. Phys.* **41**, 642 (1964).
60. Herzberg, G., *Molecular Spectra and Molecular Structure*, vol. III, Electronic Spectra and Electronic Structure of Polyatomic Molecules, D. van Nostrand Co., Inc., 1966.
61. Auloos, P. J., and Lias, S. G., *J. Chem. Phys.* **38**, 2207 (1963).
62. Lassettre, E. N., and Francis, S. A., *J. Chem. Phys.* **40**, 1208 (1964).
63. Bowman, C. R., and Miller, W. D., *J. Chem. Phys.* **42**, 681 (1965).
64. Laufer, A. H., and McNesby, J. R., *Can. J. Chem.* **43**, 3487 (1965).
65. Rustgi, O. P., *J. Opt. Soc. Am.* **54**, 464 (1964).
66. Price, W. C., *J. Chem. Phys.* **3**, 256 (1935).
67. Wainfan, N., Walker, W. C., and Weissler, L., *Phys. Rev.* **99**, 542 (1955).
68. Sponer, H., and Teller, E., *Rev. Mod. Phys.* **13**, 75 (1941).
69. Jahn, H. A., and Teller, E., *Proc. Roy. Soc. London* A **161**, 220 (1937).
70. Mulliken, R. S., *J. Am. Chem. Soc.* **86**, 3183 (1964).
71. Herzberg, G., *Infrared and Raman Spectra of Polyatomic Molecules*, van Nostrand, New York, 1945, p. 343.
72. Schachtschneider, J. H., and Snyder, R. G., *Spectrochim. Acta* **19**, 117 (1963).
73. Turner, D. W., *Adv. Phys. Organic Chem.* **4**, 67 (1966).
74. Magee, J. L., and Funabashi, K., *J. Chem. Phys.* **34**, 1715 (1961).
75. Lorquet, J. C., El Komoss, S. G., and Magee, J. L., *J. Chem. Phys.* **37**, 1991 (1962).
76. Rossini, F. D., Pitzer, K. S., Arnett, R. L., Braun, R. M., and Pimentel, G. C., *Selected Values of Physical and Thermodynamic Properties of Hydrocarbons and Related Compounds*, Carnagie Press, Pittsburgh, Pennsylvania, 1953.
77. Price, W. C., Bralsford, R., Harris, P. V., and Ridley, R. G., *Spectrochim. Acta* **14**, 45 (1959).
78. Honig, R. E., *J. Chem. Phys.* **16**, 105 (1948).
79. Lombos, B. A., and Sauvageau, P., *Can. Spectrosc.* **46** (1968).
80. Brown, R. D., and Krishna, V. G., *J. Chem. Phys.* **45**, 1482 (1966).
81. Scheibe, G., and Grieneisen, H., *Z. phys. Chem.* **52**, 1325 (1934).
82. Price, W. C., *Phys. Rev.* **47**, 444 (1935).
83. Klevens, H. B., and Platt, J. R., *J. Am. Chem. Soc.* **69**, 3055 (1947).
84. Isaacs, L. D., Price, W. C., and Ridley, G. D., *The Threshold of Space*, Proceedings of the conference on Chemical Aeronomy, Cambridge, Mass., 1956, p. 143.
85. Wijnen, M. M. J., *J. Chem. Phys.* **24**, 851 (1956).
86. Watanabe, K., *J. Chem. Phys.* **26**, 542 (1957).
87. Watanabe, K., Nakayama, T., and Mottl, J., *Final Report on Ionization Potentials of Molecules*, Hawaii, 1959.
88. Steiner, B., Giese, C. F., and Inghram, H. G., *J. Chem. Phys.* **34**, 189 (1961).

89. Hampson, R. F., jr, McNesby, J. R., Akimoto, H., and Tanaka, I., *J. Chem. Phys.* **40**, 1099 (1964).
90. Klevens, H. B., and Platt, J. R., *J. Am. Chem. Soc.* **69**, 3055 (1947).
91. Pickett, L. W., Muntz, M., and McPherson, M., *J. Am. Chem. Soc.* **73**, 4862 (1951).
92. Platt, J. R., Rusoff, I., and Klevens, H. B., *J. Chem. Phys.* **11**, 535 (1943).
93. Jones, L. C., and Taylor, L. W., *Analyt. Chem.* **27**, 228 (1955).
94. Turner, D. W., *Chem. Ind.*, **21**, 626 (1958).
95. Potts, W. J., *J. Chem. Phys.* **20**, 809 (1952).
96. Scheerr, M. D., McNesby, J., and Klein, R., *J. Chem. Phys.* **36**, 3504 (1962).
97. Pearson, E. F., and Innes, K. K., *J. Molec. Spectrosc.* **30**, 232 (1969).
98. Walsh, A. D., *Nature* **159**, 167, 712 (1947).
99. Coulson, C. A., and Moffit, W. E., *Phil. Mag.* **40**, 1, (1949).
100. Basch, H., Robin, M. B., Kuebler, N. A., Baker, C., and Turner, D. W., *J. Chem. Phys.* **51**, 52 (1969).
101. Eastman, R. H., and Freeman, S. K., *J. Am. Chem. Soc.* **77**, 6642 (1955).
102. Kosower, E. M., and Ito, M., *Proc. Chem. Soc.* 25 (1962).
103. Heathcock, C. H., and Poutter, S. R., *J. Am. Chem. Soc.* **90**, 3766 (1968).
104. Robin, M. B., Hart, R. R., and Kuebler, N. A., *J. Chem. Phys.* **44**, 1803 (1966).
105. Clark, D. T., *Theoret. chim. Acta* **10**, 111 (1968).
106. Gunthard, H., Lord, R. C., and McCubbin, T. K., jr, *J. Chem. Phys.* **25**, 768 (1956).
107. Zobel, C. R., and Duncan, A. B. F., *J. Am. Chem. Soc.* **77**, 2611 (1955).
108. Cook, G. R., and Ching, B. K., *J. Chem. Phys.* **43**, 1794 (1965).
109. Stokes, S., and Duncan, A. B. F., *J. Am. Chem. Soc.* **80**, 6177 (1958).
110. Al-Joboury, M. I., and Turner, D. W., *J. Chem. Soc.* **B**, 373 (1967).
111. Kiser, R. W., and Hobrock, D. L., *J. Am. Chem. Soc.* **87**, 922 (1965).
112. Davies, D. W., *Chem. Phys. Letters* **2**, 173 (1968).

Additional bibliography

Brongersma, H. H., and Oosterhoff, L. J., Singlet-triplet excitation of saturated hydrocarbons by low-energy electron impact, *Chem. Phys. Letters* **3**, 437 (1969).

Brundle, C. R., and Robin, M. B., Electronic energies and electronic structures of the fluoromethanes, *J. Chem. Phys.* **53**, 2196 (1970).

Doucet, J., Sauvageau, P., and Sandorfy, C., Vacuum ultraviolet and photoelectron spectra of fluoro-chloro derivatives of methane, *J. Chem. Phys.* **58**, 3708 (1973).

Edwards, L., and Raymonda, J. W., Assignments of electronic transitions in the fluoromethanes, *J. Am. Chem. Soc.* **91** (22) 5937 (1969).

Koch, E. E., and Skibowski, M., Optical absorption of gaseous methane, ethane, propane, and butane and reflection of solid methane and ethane in the vacuum ultraviolet, *Chem. Phys. Letters* **9**, 429 (1971).

Narayan, B., Electronic spectra of paraffin hydrocarbons, *Molec. Phys.* **23**, 281 (1972).

Raymonda, J. W., Rydberg states in cyclic alkanes, *J. Chem. Phys.* **56**, 3912 (1972).

CHAPTER 6

Spectra of condensed gases

J.-Y. RONCIN

Laboratoire des Interactions Moléculaires et des Hautes Pressions du CNRS, Bellevue, 92190 Meudon, France

Contents

1. Introduction

This chapter will review the far u.v. absorption of molecular solids. The emphasis will be mainly on the spectra of substances trapped in a matrix, while including certain pure substances which retain their characteristic discrete spectra in the condensed state. Trapping in a matrix consists of isolating a substance (atom, radical, or molecule) inside a chemically neutral and optically transparent rigid medium which is referred to as a matrix.

During the last 24 years the study in the far u.v. of solid substances, either pure or trapped in a matrix, has been pursued in four principal research centres:

(i) At the High Pressure Laboratory (Bellevue), as early as 1950, Vodar, Romand, and Granier-Mayence[1–4] began the pioneering work which was resumed from 1958 onwards by Damany-Astoin, Roncin *et al.*[5,6].

(ii) As a part of the research programme on trapped radicals, initiated in 1956-9 under the direction of H. Broida at the National Bureau of Standards (Washington, D.C.),[7] Schnepp and Dressler made experiments at wavelengths down to 1200 Å and at temperatures down to 4.2 K.[8a]

(iii) During 1961—64, at the Institute of optics (Rochester) Baldini made a very complete study of solid rare gases (except neon) down to about 900 Å.[9,10]

(iv) A group formed in Israel by Schnepp and Jortner has been working since 1964 on pure and trapped molecules.[11, 12]

For many molecules, studies of the condensed phase in the far u.v. afford the distinct advantage of easy characterization of electronic transitions, since the spectral analyses are considerably simplified by the absence both of rotational structure and of transitions from the excited vibrational levels of the ground state ("hot bands").

From a more general point of view the methods has a twofold interest. (i) Such studies may be regarded as an extension of the observation of pressure effects on spectral lines under high pressure of foreign gas as outlined by Vodar.[13] However, until recently this sort of comparison was very limited as there were few spectroscopic data under high pressure in the far u.v. due to lack of suitable transparent window materials. (ii) It permits the stabilization of free radicals in the solid for a duration long enough for observation. Unfortunately, the absorption of the parent molecule, which is itself not always known, may mask the free radical spectrum. Hence the method is restricted to certain favourable cases of a strongly allowed transition of the radical or of disappearance of the overlying absorption of the parent molecule in the solid state. For this reason, only three unstable species (H, P, and CH_3) have so far been successfully detected in the spectral range concerned.

2. Experimental methods

The experimental techniques utilized are conditioned by the fact that almost all solids of macroscopic thickness absorb completely in the v.u.v. Therefore the molecular solids considered are almost invariably obtained by the condensation of a gaseous beam on a cooled substrate; the necessary thickness of the active substance may vary from a few tens of angstroms to some microns, depending on the intensity of the transition. For this reason studies of solidified liquids are practically limited to 1700 Å, with organic solvents.[14]

However, Raz and Jortner were able to obtain transparent liquids of rare gases in a 1 cm cell sealed with LiF windows. By slowly cooling the liquid under a pressure of 4-5 atm they even got an optically transparent 1 cm thick solid sample.[15] In any case the only solids—apart from solid hydrogen—which are still transparent in the far u.v. (above their resonance lines) are those of the rare gases; this necessitates cooling of the sample to very low temperature (liquid H_2: 20-14 K; liquid He: 4.2-1.5 K).

Two distinct types of technique are used for the investigations:

(i) The classical photographic method (Fig. 1a) in which the sample is deposited on the cooled transparent substrate interposed between the source and the spectrograph: this is widely employed, but the spectral domain is limited to 1050 Å by the transparency of an LiF substrate.

(ii) The photoelectric method, in which the sample is condensed on a cooled support covered by a fluorescent deposit and placed between the monochromator and the photoelectric receptor (Fig. 1b). The u.v. radiation, traversing only the sample, is converted into visible light by the fluorescent deposit; the limitation in wavelength is thereby eliminated, but possible interactions of the sample with the fluorescent layer are ignored.

The gaseous components of the mixtures under study are either mixed in advance or are evaporated separately if their masses are too different in order to prevent confusion from effects of differential evaporation. The standard technique

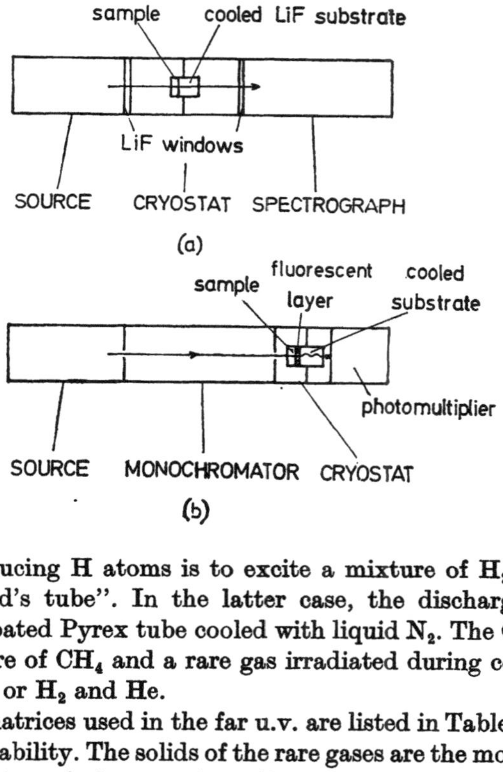

FIG. 1. Block diagram of (a) the photographic method and (b) of the photoelectric method.

for producing H atoms is to excite a mixture of H_2 and A or of H_2O and A in a "Wood's tube". In the latter case, the discharge products pass through a teflon-coated Pyrex tube cooled with liquid N_2. The CH_3 radical is produced from a mixture of CH_4 and a rare gas irradiated during condensation by a lamp filled with Kr or H_2 and He.

The matrices used in the far u.v. are listed in Table 1 with their respective limits of applicability. The solids of the rare gases are the most interesting since they offer a gradation of the trapping effect, and their crystal structure is particularly simple—f.c.c. with undoubtedly a small proportion of the h.c.p. phase.[16] These solids offer highly symmetric centres for trapping (O_h for the substitutional site of the f.c.c. phase*). Moreover, removal of an atom from the crystal lattice of a solidified rare gas leaves sufficient room so that the low-lying transitions of small molecules at least are not profoundly modified. The space available at a lattice site may be specified as follows. The size of an atom or a molecule can be represented by its Lennard-Jones diameter σ which is the abscissa of the turning

* The deposits are supposed to be at least polycrystalline as corroborated by the features of the spectra discussed in detail later on, i.e. crystal field splitting for the pure solid, and sharp structure for the trapped molecule.

TABLE 1. SHORT WAVELENGTH LIMITS OF UTILITY OF SOLID MATRICES IN THE FAR ULTRAVIOLET

	λ (Å)
Organic liquids	1600
N_2	1500
Xe	1500
Kr	1250
H_2-D_2	1100
A	1070
CF_4	930
Ne	800

point, that is to say, the point at which the repulsive term ($1/r^{12}$) in the intermolecular Lennard-Jones potential

$$V(r) = 4\varepsilon\left[\left(\frac{\sigma}{r}\right)^{12} - \left(\frac{\sigma}{r}\right)^{6}\right] \qquad \text{(Fig. 2)}$$

balances the attractive one ($-1/r^6$).

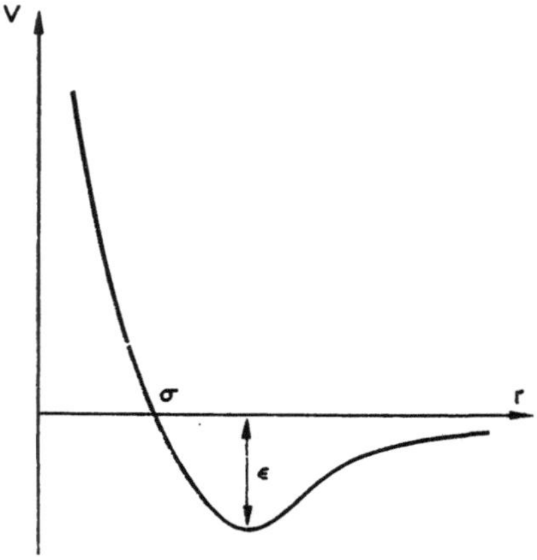

FIG. 2.
Lennard-Jones intermolecular potential.

TABLE 2. LENNARD-JONES DIAMETERS σ AND DIMENSIONS OF LATTICE-SITES (Å)

	$\sigma^{(17)}$	$a_0^{(16)}$	$\frac{a_0}{\sqrt{2}}$	$\varrho_s = a_0\sqrt{2} - \sigma$	$\varrho_i = a_0 - \sigma$
Ne	2.79	4.46	3.15	3.51	1.67
Ar	3.42	5.31	3.75	4.08	1.89
Kr	3.61	5.64	3.99	4.37	2.03
Xe	4.06	6.13	4.33	4.60	2.07
N_2	3.68	5.67	4.01	4.34	1.99
D_2	2.95	5.07	3.59	4.22	2.12
H_2	2.97	5.29	3.74	4.51	2.32

Table 2 gives σ[17] and the lattice parameter a_0[16] of most of the matrices listed in Table 1. The quantity ϱ_s is a rough measure of the space available at a substitutional site; ϱ_i corresponds to the larger of the two interstitial sites (Fig. 3).

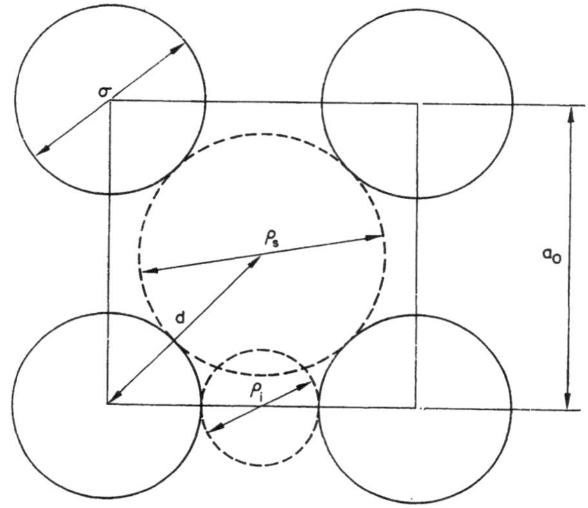

FIG. 3. Substitutional and interstitial (octahedral) sites of a f.c.c. lattice.

A quantitative description is obtained by taking account of the spatial distribution of the matrix particles, which actually gives rise to a potential of the form

$$V_{A-M}(r) = 4\varepsilon_{A-M}\left[12.13\left(\frac{\sigma_{A-M}}{r}\right)^{12} - 14.45\left(\frac{\sigma_{A-M}}{r}\right)^{6}\right]^{(18)}$$

with $\sigma_{A-M} = \frac{1}{2}(\sigma_A + \sigma_M)$ and $\varepsilon_{A-M} = \sqrt{(\varepsilon_A \varepsilon_M)}$, where M stands for matrix and A for active species. This gives $V_{A-M}(a_0/\sqrt{2})$ for the displacement of the ground state of A. From the experimental shift, viz.

$$\nu_{\text{solid}} - \nu_{\text{gas}} = \Delta\nu = V_{A^*-M}\left(\frac{a_0}{\sqrt{2}}\right) - V_{A-M}\left(\frac{a_0}{\sqrt{2}}\right),$$

the value of $V_{A^*-M}(a_0/\sqrt{2})$ is obtained for the excited species A^*. Approximate values of σ_{A^*} and ε_{A^*} can then be derived in such a way that the results are compatible in more than two matrices.[19]

Several researches on the structure of condensed rare gases[20, 21] have revealed that the h.c.p. phase is more frequent than was previously thought, especially when impurities are present. However, the lattice parameters are not yet known, so that it is not possible to modify the values of ϱ. However, the differences may not affect the calculations seriously, and only reasoning based on group theory may be upset. The structure of solid H_2 is still more controversial, but the f.c.c. phase seems to predominate for H_2 as well as for D_2 when the solid is formed by condensation of a gaseous beam as is the case here.[22]

3. Pure and trapped atoms (Xe, Kr, Ar, Ne, Hg, P, H)

The only stable atoms giving rise to molecular solids with resonance lines in the far u.v. are those of the rare gases. Each of these solids may be studied either pure or trapped in a matrix of another lighter rare gas with its resonance lines at shorter wavelengths; that is to say, Xe in Kr, Ar, and Ne; Kr in Ar and Ne; and, finally, Ar in Ne. Other atoms can only be studied in a matrix since at low temperature they associate: for example Hg, P, and H.

Pure solidified rare gases. The spectra of pure solidified Xe, Kr, and Ar are quite well known down to 900 Å.[9, 23–26] Apart from the two intense bands corresponding to the resonance doublet of the free atom, the spectra of annealed deposits,[9, 26] clearly show a hydrogenic series associated with the first doublet component of xenon

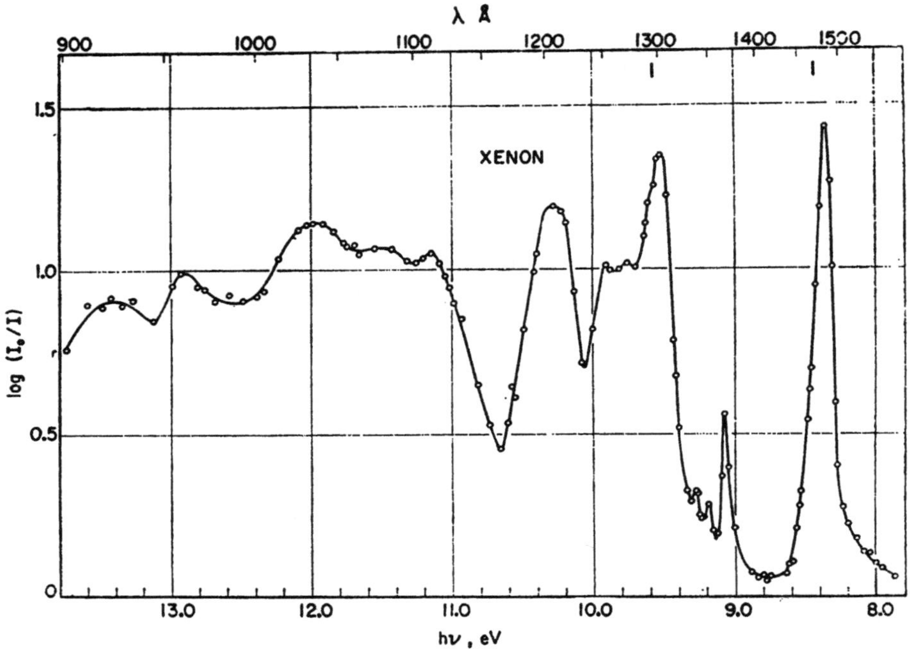

FIG. 4. Absorption spectrum of pure solid xenon at 21 K, annealed at 53 K.[9]

(Fig. 4) as well as of krypton (Fig. 5). This series is interpreted according to Wannier's exciton model and its limit furnishes the energy gap at the centre Γ of the Brillouin zone. In the spectra of unannealed deposits only the first two series members are detectable. In all cases the series associated with the second component of the doublet is doubtful. The spectra of solid argon present the beginning of the two series whose $n = 2$ members have been determined by electron energy loss experiments[27] and interpreted by Moorjani[28] on the same model as for xenon.

However the applicability of this model to molecular solids has been criticized by Dressler.[8b]

The structures observed at higher energies (> 10 eV) in the spectra of solid xenon have been ascribed by Phillips to metastable excitons at L and X points of the Brillouin zone.[29] The theoretical calculations of Jain[30] and Herman-

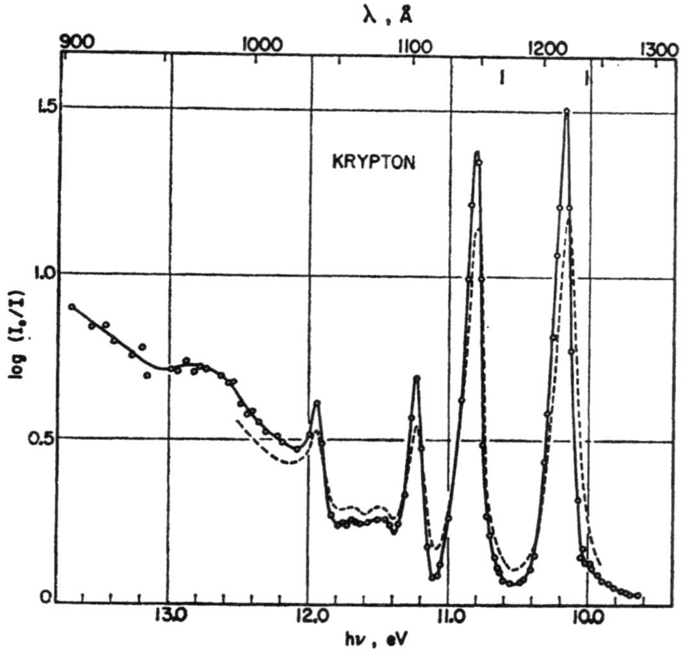

FIG. 5. Absorption spectrum of pure solid krypton at 20 K (annealed at 44 K) (solid line) and at 40 K (dashed line).[9]

son[31] support this hypothesis. Nonetheless, it seems reasonable that at about 10-11 eV superposition of transitions of atomic character occurs.[25, 32–34]

Besides these results on absorption spectra there are, to our knowledge, only three papers on the luminescence of solidified rare gases.[35, 36a, 36b]

Trapped atoms. The displacements of the resonance lines of Xe, Kr, Ar, Hg, P, and H trapped in a matrix are assembled in Table 3.

Xenon

The spectrum of xenon trapped in argon presents a hydrogenic series more clearly resolved than in the case of pure xenon. This series is associated with the first member (1470 Å) of the resonance doublet [$\Gamma(3/2)$ 1*s*, 2*s*, 3*s*, and 4*s* on Fig. 6].[25, 37] Baldini and Knox have offered an explanation in terms of effective mass-impurity states converging towards the conduction band of solid argon. At short wavelengths the assignment of the observed structures is very uncertain;

TABLE 3. SHIFT (cm^{-1}) OF RESONANCE LINES OF TRAPPED ATOMS AT 4.2 K (* EXCEPT, AT 20 K) IN EACH MATRIX (λ, σ, and ϱ_s in Å)

Free Atom			Matrix				
			Xe	Kr	D_2[a]	Ar	Ne
	σ	λ \ ϱ_s	4.60	4.37	4.22	4.08	3.51
Xe	4.06	1470		4650*	5210	6330*	5210
		1296			6290	7770*	13820
Kr	3.61	1236			4820	6130*	4760
		1165				6450*	5320
Ar	3.42	1067					7070
		1048					7030
Hg	2.90	1850	−1240	610		2055	
P		1775	−3345*	−1605*		−265*	
H		1216				2340	

[a] Although $\sigma_{D_2} < \sigma_{Ar}$ (Table 2) it seems well justified to put D_2 in between Kr and Ar as the shifts in Kr, D_2, and Ar are increasing toward the blue while the value of ϱ_s decreases, that is to say when the repulsive forces become stronger. Except for the neon matrix this rule is usually satisfied.(60) In particular ϱ_s of N_2 and Kr are 4.34 and 4.37 Å and the respective shifts +795 and +760 cm^{-1} for $^3P \leftarrow {}^1S$ of Hg trapped in Kr and N_2 at 20 K.(76)

the hydrogenic series associated with the second component (1296 Å) of the doublet overlaps with the bands probably due to perturbation of the atomic transitions of higher energy (Fig. 6). The analysis of Xe-Ne spectra is still more doubtful and

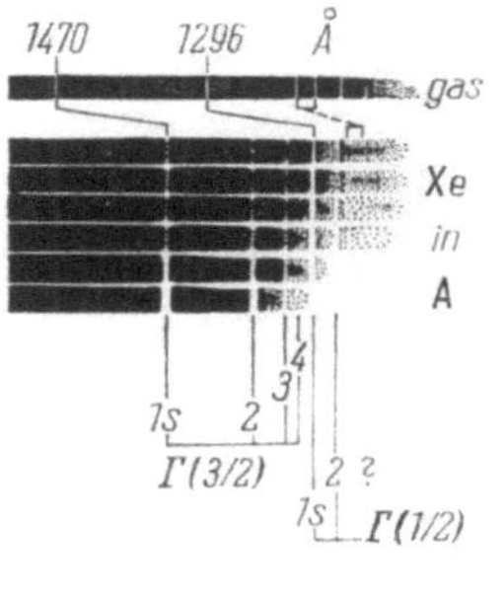

FIG. 6. Absorption spectra of xenon free and trapped in an argon matrix.

needs an extension of the measurements below 1050 Å. In Xe-Kr the intrinsic absorption of the Kr matrix interrupts the first series at $n = 2$. On the other hand, in a matrix of deuterium the first three terms are clearly observable.(38)

Even in the cases of a sample of Xe-Ar or Xe-Kr, 1 cm thick, $\Gamma(3/2)2s$ is still observed in the solid and in the liquid phase as well. This indicates that Wannier-type excitons also exist in disordered systems.(15)

For Xe-Ar it is possible to compare the spectrum of the solid with that obtained for xenon perturbed by argon under pressure up to a density of 500 amagats;* the displacement of the 1470 Å line increases under these conditions more rapidly than the square of the density.[39]

Krypton

The spectral data on krypton trapped in Ar, Ne, and D_2 are even more fragmentary since the resonance lines fall very near the LiF cut off.[10,23,38]

Argon

Baldini succeeded in recording as far as 900 Å the spectrum of argon trapped in neon.[10] However, the structures corresponding to the resonance doublet are obscured by parasitic absorption which might be due to condensation of hydrogen from the source.[38]

Mercury

The band corresponding to the resonance line $^1P \leftarrow {}^1S$ (1850 Å) of Hg has been observed by Roncin and Damany-Astoin in argon and in isopentane at 20 K[40] and by Duley in Xe, Kr, and Ar at 4.2 K.[41] At 4.2 K the band is split into three components, the structure being much less pronounced for Hg-Ar than for Hg-Kr and Hg-Xe; in these two latter cases the structure persists up to the melting point of the matrix. The triplet structure suggests that the symmetry site is not higher than D_{2h}, which would be the case for an interstitial site in a local h.c.p. environment. Crystal field parameters have been derived by Duley for Hg in Ar, Kr, and Xe.[42]

Although the shift of the 1850 Å line perturbed by compressed argon is only known up to a density of 400 amagats,[43] the shift in the solid appears to fit a reasonably extrapolated curve at the density of solid argon as in the case of xenon in argon. The effects observed in a solid matrix may thus be compared with the pressure effect at very high density. In the two cases of gaseous mercury and xenon, the line broadens with increasing pressure and satellites appear which also broaden till they merge with the parent line itself. Proceeding to the solid, the corresponding bands become again relatively sharp since the intermolecular distance is fixed in the solid. The perturbations of the ground and excited states, and consequently the observed shift, then have fixed values. Splitting of the bands may even appear in some cases depending on the site symmetry in the solid.

Phosphorus

When the photographic method is used, the irradiation from the source can give rise to photodecomposition. In the case of PH_3 trapped in a matrix, an atomic-like absorption is actually observed corresponding to the resonance line ... $(3p)^2 4s\ {}^4P \leftarrow \cdots (3p)^3\ {}^4S$ of phosphorus.[67] The band is strongly red-shifted and very clearly split into three components in Xe, Kr, and Ar just as for mercury.

* The density in amagat is the ratio of the density under the experimental conditions to the density in standard conditions.

Atomic hydrogen

The Lyman α line ($2p \leftarrow 1s$) of hydrogen has been observed in an Ar matrix with both methods of investigation.[44,45] The success of this experiment was favoured by two major factors: the long lifetime of hydrogen and the fact that the band, even though shifted towards shorter wavelengths, remains above the absorption edge of solid H_2. Keil and Gold calculated theoretically the energy of the $2p \leftarrow 1s$ transition of hydrogen in a substitutional site of solid argon using the tight-binding approximation and taking into account several contributing factors.[46] The large half-width (1300 cm^{-1}) might be due to the mobility of H atoms at 4.2 K as well as the possibility of several trapping sites for this small atom, as indicated by e.s.r. studies.[47] In spite of many efforts, Baldini[44] and Roncin[48] failed to get a significant result in attempts at trapping hydrogen in a Ne matrix, contrary to the expectation that the smaller site in the Ne lattice would stabilize the H atoms better. However, it is quite reasonable to suppose that in the narrow site of neon the $2p$ level undergoes such a large perturbation that the line is too broad to be detectable experimentally.

For the observed transitions of trapped rare gases, the displacements are towards higher energies (Table 3). A qualitative discussion[48] has shown that this is compatible with the nature of the Rydberg atomic transitions. A blue shift means that the repulsive forces are more important for the excited state than for the ground state. This is easily justifiable since the excited orbital of the optical electron is farther out from the core during the transition, and therefore the electron approaches the neighbouring atoms of the matrix sufficiently to experience repulsive forces.

Moreover, the shifts ν_{solid}-ν_{gas} increase in magnitude as the site dimension decreases—except for the anomaly that on going to neon the blue shift is smaller than in argon. However, the observed shift results from a balance between the displacement of the excited and ground states and, in the case of a Ne matrix, even the ground state of xenon or krypton is itself affected by the repulsive forces (Table 2). This apparently anomalous behaviour is quantitatively well accounted for by the calculation outlined in § 2.[19]

4. Pure and trapped molecules

The early work on pure molecules like NO, O_2, etc., did not seem very promising as the discrete band spectrum of the free molecule was lost in the condensed state. Such a profound alteration arises essentially from the electronic structure of these molecules which loose their individuality in the solid state by formation of polymers. This is not the case with molecules like CO, N_2, N_2O, which possess electronic configurations forming closed shells. The spectra of non-aggregated pure solids may be explained by Davydov's theory of the interaction of light with molecular solids for which the intermolecular forces are weak compared with intramolecular forces.[9,50] According to this theory, for a crystal containing more than one molecule per unit cell, the number of components in the spectrum of the solid follows from the correlation between the molecular group and the unit cell group (factor group); transitions in the solid are then subjected to the selection rules inside the factor group.

On the other hand, a molecule isolated in a rigid matrix always preserves its individuality through avoidance of resonance interactions between identical molecules. Thus it is easy to study the influence of a neutral environment on each molecular transition characterized by its Rydberg or non-Rydberg nature. From the shift and broadening it is possible to deduce qualitative and quantitative information on the extent of the excited orbital provided the molecular size be known for the ground state.[6, 19, 51] Thus confining the molecule in a matrix appears, in conjunction with high-pressure experiments, as an elegant method for probing molecular electronic structure.

4.1. DIATOMIC MOLECULES (D_2, CO, N_2, NO, O_2, I_2)

D_2

The spectrum of solid deuterium at 4.2 K consists of a broad absorption band between 1100 and 900 Å with a maximum at 940 Å and a shoulder at 1030 Å.[38] The absence of a clear vibrational progression makes it difficult to assign the observed structures.

CO

Owing to the closed shell configuration ... $(\pi\, 2p)^4(\sigma\, 2p)^2$ of CO, the spectrum of the pure solid remains essentially that of the free molecule. Thus the vibrational band system of the fourth positive group $A\,^1\Pi \leftarrow X\,^1\Sigma^+$ has been clearly

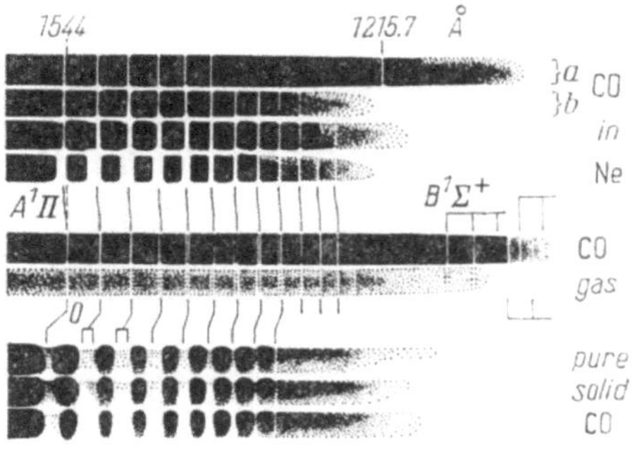

Fig. 7. Absorption spectra of CO free, pure solid, and trapped in a neon matrix.

identified by Dressler,[8a] Roncin and Damany,[52] and by Brith and Schnepp.[11] This very intense transition is observed even with a deposit a few tens of angstroms thick. The most intense bands are most strongly shifted (−800 to −300 cm⁻¹). The splitting, clearly seen for the most intense bands, is classified as Davydov-type (Fig. 7); it arises from the coupling between molecules occupying translationally non-equivalent sites and its magnitude (at most 400 cm^{-1}) is approximately proportional to the Franck-Condon factor.[53]

Brith and Schnepp have also discerned several weak bands corresponding to the forbidden transition $d\ {}^3\Delta \leftarrow X\ {}^1\Sigma^+$ displaced by about $-400\ cm^{-1}$. Furthermore, Hexter observed, in spectra of annealed solid CO, the splitting of the first few bands of the forbidden transition $a\ {}^3\Pi \leftarrow X\ {}^1\Sigma^+$ (Cameron system); in this case, the splitting ($\sim 110\ cm^{-1}$) is due either to a triplet exciton (spin-orbit splitting) or to a vibronic interaction in the crystal so that the observed splitting is in fact a lattice frequency of solid CO.[54]

The spectra obtained in argon by Dressler[8a] and in all of the rare gas matrices by Roncin *et al.*[5] show that in various matrices the resonance interaction between CO molecules vanishes so that the splitting disappears and the bands become sharp. Further, the shift is practically constant for all vibrational terms, indicating the absence of deformation of the potential energy curve of the excited state but rather a simple vertical translation (Fig. 7).

N_2

Being isoelectronic with CO, N_2 also keeps its free-molecule character in the condensed state. However, as the transitions of N_2 are dipole forbidden above 1000 Å, their intensities are so weak that a relatively thick layer (a few tenths of a micron) is required to record the bands. Four to eleven bands of the $a\ {}^1\Pi_g \leftarrow X\ {}^1\Sigma_g^+$ progression (Lyman-Birge-Hopfield bands) have been observed in the pure solid.[8a, 11, 52] Besides this system, the solid spectrum exhibits another system—sharper, more intense, and observable up to $v' = 17$,[52] the limitation near 1050 Å being imposed by the LiF cut-off (Fig. 8). This is the $w\ {}^1\Delta_u \leftarrow X\ {}^1\Sigma_g^+$ transition which is enhanced by a factor 10^2 to 10^3 in the solid owing to change of the selection rules in going from the molecular group to the factor group. This enhancement explains why seventeen bands are observed in the solid, and only eleven in the gas.

In spite of the relatively great thickness (10-50 μ) of the layers required for observation of the bands in a matrix, Roncin *et al.* have successfully recorded both progressions in the spectrum of N_2 isolated in neon at 1.5 K.[5] The bands are very sharp and shifted very little as in the case of CO-Ne. Further, the intensity ratio of the two systems is similar to that of the gas since $w \leftarrow X$ remains dipole-forbidden, as $a \leftarrow X$, in the matrix. The conservation of this ratio has made it possible to deduce the molecular orientation of N_2 in its site, along the body diagonal, from symmetry considerations.[6]

NO

No definite conclusions could be drawn regarding Rydberg transitions of CO and N_2 in the solid due to their energy levels being too high; the NO molecule affords, in contrast, a particularly favourable case since no less than twenty-five Rydberg states overlap five valence states between 2300 and 1300 Å.

Because of its odd number of electrons, the behaviour of NO is completely different from that of CO and N_2; in fact, in the pure solid spectrum only a broad absorption band ascribed to N_2O_2 has been observed at 2100 Å[2] with no other structure at shorter wavelengths.[5] In contrast, on isolating NO in a matrix (Kr,

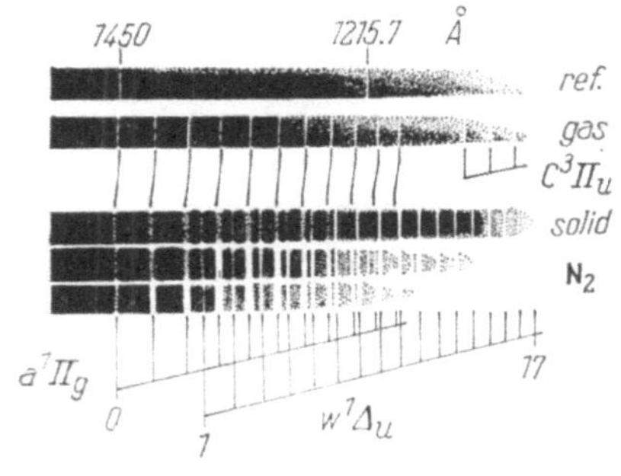

FIG. 8. Absorption spectra of gaseous and solid nitrogen.

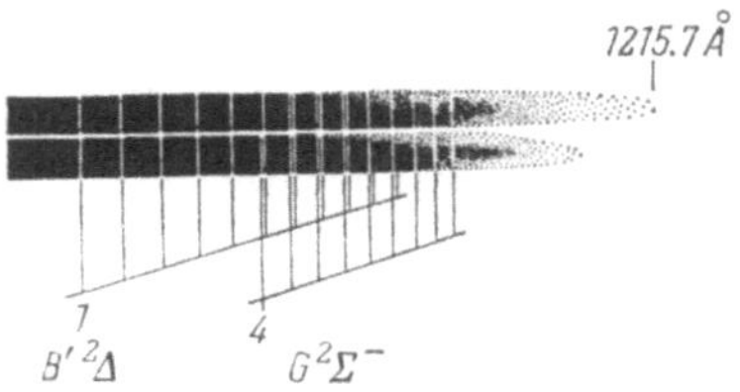

FIG. 9. Absorption spectrum of nitric oxide trapped in a neon matrix.

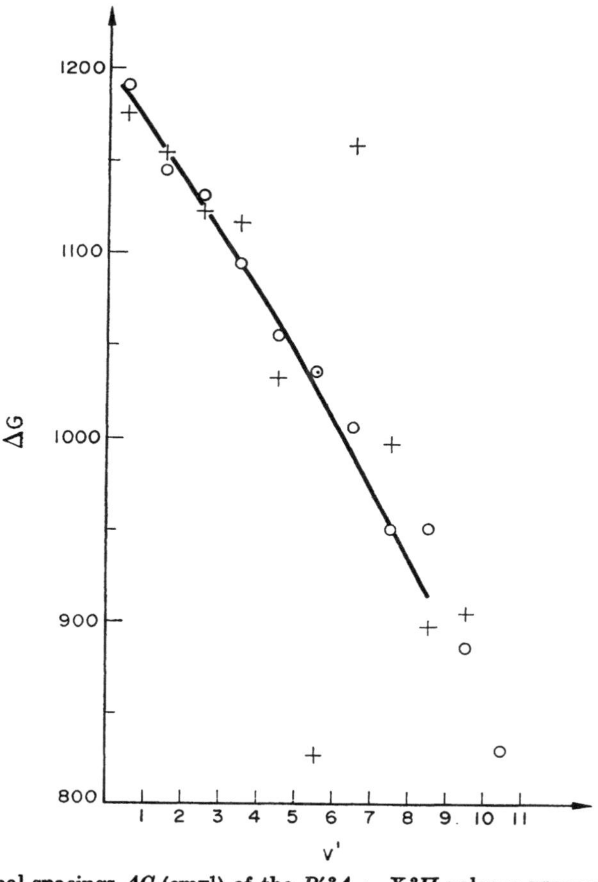

FIG. 10. Vibrational spacings ΔG (cm^{-1}) of the $B'^2\Delta \leftarrow X^2\Pi$ valence progression of NO. +, Experimental values in the free molecule. Continuous line—theoretical deperturbed values.[56], o, Experimental deperturbed values in the Ne matrix.[48]

Ar, and Ne) Roncin *et al.* have found forty bands readily assigned to the three valence transitions $B\,^2\Pi, B'\,^2\Delta$, and $G\,^2\Sigma^- \leftarrow X\,^2\Pi$.[5] In NO-Ne the sharpness of the bands was limited only by the resolution of the spectrograph ($\sim$ 0.6 Å). The NO molecule provides the best-known example of homogeneous interactions which give rise to very irregular vibrational spacings in the $B\,^2\Pi$ and $B'\,^2\Delta \leftarrow X\,^2\Pi$ progressions; this corresponds to crossing of potential energy curves of the valence states and the Rydberg states of the same symmetry. The remarkable feature of the trapped NO spectrum is that the progression become perfectly regular as there seem to be no Rydberg states to cause perturbations anymore. Furthermore, the measured spacings in a matrix fit very closely the deperturbed values calculated by Lagerqvist and Miescher for $B\,^2\Pi$[55] and Jungen for $B'\,^2\Delta$[56] (Figs. 9 and 10).

For all three molecules CO, N_2, and NO whose Lennard-Jones diameter σ is very close to ϱ_s of neon (Tables 2 and 4), the small red shift (-50 to -100 cm^{-1}) indicates a small change of molecular dimension from the ground to the excited states under consideration (Table 5). This description is consistent with the nature of a valence transition consisting of a simple rearrangement of electrons inside a purely molecular orbital of the same principal quantum number. On the contrary,

TABLE 4. MEAN SHIFTS (cm^{-1}) OF VIBRATIONAL PROGRESSIONS FOR SEVERAL MOLECULES TRAPPED IN RARE GAS MATRICES AT 1.5 K (except * at 4 K and † at 20 K) (σ and ϱ_s in Å)

Free molecule			Matrix				
			Xe	Kr	N_2	Ar	Ne
	σ	transition \ ϱ_s	4.60	4.37	4.34	4.08	3.51
CO	3.59	$A\,^1\Pi \leftarrow X\,^1\Sigma^+$	−1660	−1000		−590	−65
N_2	3.68	$a\,^1\Pi_g \leftarrow X\,^1\Sigma_g^+$ $w\,^1\Delta_u \leftarrow X\,^1\Sigma_g^+$					−60 −100
NO	3.47	$B\,^2\Pi \leftarrow X\,^2\Pi$ $B'\,^2\Delta \leftarrow X\,^2\Pi$ $G\,^2\Sigma^- \leftarrow X\,^2\Pi$		−350 −690 −970		−185* −500* −605*	−70 −50 −70
O_2	3.43	$B\,^3\Sigma_u^- \leftarrow X\,^3\Sigma_g^-$	−1139	−185	−280	+175	+390
N_2O	3.88	$\tilde{B} \leftarrow \tilde{X}\,^1\Sigma^+$ $\tilde{E} \leftarrow \tilde{X}\,^1\Sigma^+$ (Rydberg)		+2000†		+2550† +4300†	+2960 +5200
C_2H_4	4.23	$\pi^* \leftarrow \pi$ (Rydberg)		+4040†			
C_2H_2	4.22	$\tilde{C} \leftarrow \tilde{X}$ (Rydberg)		+4160†		+6940†	
C_6H_6	5.27	$\tilde{C}\,^1E_{1u} \leftarrow \tilde{X}\,^1A_{1g}$ $\tilde{D}\,^1E_{1u} \leftarrow X\,^1A_{1g}$ (Rydberg)	−2350 +1610†	−1610 +3150†	−1540	−1130 +6000†	

a Rydberg transition is characterized by a change in the principal quantum number analogous to transitions in atoms, so that the optical electron, moving farther out from the core, is subjected to strong repulsive forces which profoundly

TABLE 5. LENNARD-JONES PARAMETERS FOR SOME EXCITED STATES OF TRAPPED SPECIES[19]

A	Ground state	σ_A (Å)	$\varepsilon_A \times \frac{1}{k}$ (K)	Excited state	σ_{A^*} (Å)	$\varepsilon_{A^*} \times \frac{1}{k}$ (K)
Xe	^1S	4.06	229	^3P	4.8	210
				^1P	5.0	185
Kr	^1S	3.61	190	^3P	4.65	225
Hg	^1S	2.90	851	^1P	4.25	825
CO	$X\,^1\Sigma^+$	3.59	110	$A\,^1\Pi$	3.75	275
NO	$X\,^2\Pi$	3.47	119	$B\,^2\Pi$	3.70	175
				$B'\,^2\Delta$	3.70	230
				$G\,^2\Sigma^-$	3.70	275
O_2	$X\,^3\Sigma_g^-$	3.43	113	$B\,^3\Sigma_u^-$	4.05	260
N_2O	$\tilde{X}\,^1\Sigma^+$	3.88	220	$\tilde{B}$	4.75	115

modify the transition. The disappearance of Rydberg transitions in the spectrum of trapped NO is supported by the important broadening and shift of these transitions in the spectrum of NO perturbed by a moderate pressure of foreign gas. The nearby valence transitions, on the other hand, are still relatively sharp and unshifted at 150 atm.[57]

O_2

In the spectrum of solid O_2 at 20 K, Romand and Granier-Mayence found no trace of the Schumann-Runge bands ($B\,^3\Sigma_u^- \leftarrow X\,^3\Sigma_g^-$) but only a broad continuous absorption band with its maximum below 1500 Å.[3] The disappearance of the discrete bands suggests that the continuum is due to polymerization like O_4 rather than dissociation of O_2. The external shell $\pi_g 2p$ of O_2 is, in fact, incomplete, and such an aggregation is probable in the condensed state.

On isolating O_2 in a matrix of Ar or N_2, the Schumann-Runge bands were found by Bass and Broida, starting from a liquid cooled between 90 and 60 K[58] and by Schnepp and Dressler by condensing gaseous mixtures at 4.2 K.[59] The data indicate that the $B\,^3\Sigma_u^-$ state of both $^{16}O_2$ and $^{18}O_2$ has a narrower potential well in a matrix than in the free molecule. The deformation may be explained here by the large internuclear distance in the excited state: $r_e = 1.6$ Å and $r_{max} \sim 2$ Å for $v' = 6$, while for the preceding molecules $r_{max} < 2$ Å even for large v'. By comparing the previous results to their spectra in krypton and xenon, Boursey *et al.* were able to make a more consistent assignment of the vibrational terms in each matrix.[60]

I_2

Between 1700 and 1300 Å the spectrum of I_2 trapped in argon exhibits about ten diffuse bands which do not appear to correspond with the spectrum of the gas which is, however, not well known in this region. In neon the structures disappear on account of the strong perturbation of the relatively large molecule I_2 (σ = 4.98 Å) in the narrow site of neon.[61] The study of I_2-Kr, in which the bands might become sharper, would assist the correct assignment of the bands. Study of the analogous molecules Br_2, Cl_2, and F_2 would be equally worth while.

Other diatomic molecules such as hydrogen halides, have not been investigated, as their transitions are of Rydberg character and are expected to disappear in the solid. This deserves verification even though negative results are, generally, not gratifying.

4.2. TRIATOMIC MOLECULES (N_2O, OCS, CS_2, H_2O, NO_2)

N_2O

The broad band at 1830 Å in the spectrum of the gas gives rise to a broad absorption (half-width ~ 3000 cm^{-1}) at about 1780 Å in the pure solid at 63 K.[4] In the region 1500-1400 Å (transition $\tilde{B} \leftarrow \tilde{X}\,^1\Sigma^+$), the spectra look very different, depending on the temperature at which the layers are deposited:[51] at 63 K there is a progression of diffuse bands shifted by +2500 cm^{-1}, while at 14 K no structure appears. The bands persist if the deposit, formed at 63 K, is cooled to 14 K, while if the deposit is grown at 14 K and warmed up, the bands appear around 30 K; it would be necessary to know the low-temperature crystal structure to make a correct hypothesis on the symmetry species of the $\tilde{B}$ state.

In a matrix (Kr, Ar, and Ne) the bands are much more pronounced and numerous. Taking into account that $\sigma_{N_2O} \sim \varrho_s$ of argon, the large blue shift (+2000 to 3000 cm^{-1}) suggests that the optical electron moves in an excited orbital which is unusually extended for a valence state. Analogous conclusions hold for the two broad bands at 1450 and 1284 Å and, *a fortiori*, for the two sharp bands at 1178 and 1152 Å, which are still more blue-shifted (Table 4). The band at 1178 Å is, in fact, the $n = 2$ member ($\tilde{E} \leftarrow \tilde{X}$) of the first Rydberg series of N_2O.

OCS

Like N_2O, the spectra of trapped OCS (in Kr, Ar, and Ne) exhibit two types of bands. Between 1800 and 1600 Å a vibrational progression appears as in the gas, the spacing of which corresponds to the bonding frequency ν = 520 cm^{-1}.[62] For this transition, the molecule is bent in its excited state as foreseen from Walsh's diagram. At shorter wavelengths, more or less isolated bands are not easily assignable to definite transitions of the free molecule, whose spectrum has not yet been analysed above 1400 Å. These isolated bands are sharp in krypton while in neon they are very indistinct; this may be explained by the extension of the excited orbitals of the already large OCS molecule. In contrast, the first progression is much less sensitive to trapping since the valence transition is accompanied by a very small change of molecular size.

CS_2

The continuous absorption of pure, solid CS_2 has no relation to the spectrum of the gas, while, in matrices of Kr, Ar, and Ne, many structures appear.[62] Unfortunately, very little can be said because of the lack of knowledge about the spectrum of the gas, in which transitions are not yet identified. However, even in the two favourable matrices of argon and neon no discrete structure is observed below the starting point of Rydberg transitions of the gas, although solid films of (CS_2-Ar) and (CS_2-Ne) are quite transparent below this limit and long wavelength bands are well developed. As in the case of OCS, the short-wavelength bands are very indistinct in neon.

For CS_2, as for OCS, many unassigned bands of the trapped-molecule spectrum may be due to photodecomposition products.

H_2O

In spectra of ice at $T \geq 77$ K, the broad band at 1660 Å seems to be shifted by about 9900 cm^{-1}.[48, 63, 64] However, this considerable shift is consistent with the large change in geometry accompanying the transition of a lone pair electron to an antibonding orbital. Rydberg transitions below 1250 Å are not seen,[48] but further experiments below 1050 Å are necessary to search for transitions beyond this limit.

NO_2

The absence of any discrete or continuous absorption in the spectrum of NO_2 in argon at 20 K[61] might confirm the Rydberg character assigned to the $\tilde{E}\,^2\Sigma_u^+ \leftarrow \tilde{X}\,^2A_1$ transition, although experimental inadequacies cannot be excluded.

4.3. OTHER MOLECULES (CH_3, NH_3, PH_3, CH_4, C_2H_6, C_3H_8, C_2H_4, C_2H_2, C_6H_6, $C_6H_5CH_3$, $C_6H_4(CH_3)_2$, CH_3I, C_2H_5I, C_2H_5Br)

CH_3

The v.u.v. photolysis of methane in argon at 14 K leads to the formation of a sufficient concentration of the methyl radical for observation near 1500 Å of a band assigned to overlapping of $\tilde{C}\,^2E''$ and $\tilde{D}\,^2A_1' \leftarrow \tilde{X}\,^2A_2''$ transitions. This is supported by simultaneous observation in the infrared of the out-of-plane deformation fundamental for CH_3 and CD_3.[65]

NH_3, PH_3

For some deposits of solid ammonia at 175 K, Dressler and Schnepp observed, above 1800 Å, nine faint bands corresponding to the lowest transition of the free molecule.[63] The large shift ($\sim$ 7000 cm^{-1}) of the onset of absorption is explained in the same way as for ice.

In the spectra of trapped NH_3, two systems of bands appear. The first system consists of progression of diffuse bands above 1600 Å where every second band in the spectrum of the gas is missing in the matrix. Dressler was able to ascribe the A_2'' species to the excited state on the basis of rotational selection rules.[66] The same behaviour is observed for the second system lying between 1600 and 1300 Å,[67] in which region several incompletely analysed progressions overlap in the spectrum of the gas. However, the bands are too diffuse to allow a one-to-one correspondence. In any case the long- and short-wavelength bands of the solid correspond to transitions of the gas which are known to be the $n = 3$ and $n = 4$ members of the first Rydberg series of NH_3.

In PH_3, a series of even more diffuse bands has been observed near 1400 Å in krypton and argon, with the absence of every second band.[67]

CH_4

Using a double-beam spectrophotometer, Lombos *et al.* observed a shift of -4200 cm^{-1} for the main peak in the absorption spectrum of solid methane at 80 K.[68] This red shift would be consistent with the compact spherical shape of the molecule and with a transition of a bonding electron to an atomic 3*s* orbital.

C_2H_6

In the spectrum of solid ethane, the weak bands at about 1575 Å undergo a slight red shift whereas the prominent progression near 1300 Å gives rise to two intense blue-shifted maxima with faint vibrational structure. The appearance of the two maxima may indicate a (large) Davydov splitting of 5300 cm^{-1}.[68]

C_3H_8

The weak bands at about 1600 Å exhibit a small red shift, but the strong bands below 1400 Å shift towards the blue by $+4500$ cm^{-1}.[68]

C_2H_4

In the spectrum of ethylene-d_4 trapped in krypton at 20 K, Katz and Jortner have observed three vibrational terms of the lowest Rydberg transition $\pi^* \leftarrow \pi$ of ethylene near 1600 Å.[69] The shift of $+4000$ cm^{-1} is consistent with the nature of the excited state.

C_2H_2

In the spectra of C_2H_2 in krypton and argon at 20 K, Jortner *et al.* observed the three first members of the lowest Rydberg transition of acetylene ($\tilde{C} \leftarrow \tilde{X}$).[69, 70] The extension of the Rydberg orbital explains the large blue shift of $+4200$ cm^{-1} in krypton, and $+6800$ cm^{-1} in argon.

C_6H_6

Whereas the third transition ($\tilde{C}\,{}^1E_{1u} \leftarrow \tilde{X}\,{}^1A_{1g}$) and the first Rydberg transition ($\tilde{D}\,{}^1E_{1u} \leftarrow \tilde{X}$) overlap in the spectrum of gaseous benzene, the former is red-shifted in the pure solid[71] and in various matrices,[12, 72, 73] while the Rydberg transition, starting at 1789 Å in the gas, is blue-shifted in Xe, Kr, and Ar.[12, 73] All the bands, particularly the vibrational terms of $\tilde{C} \leftarrow \tilde{X}$, are resolved better in krypton, where there is more room for the large benzene molecule.

$C_6H_5CH_3$ and $C_6(H_4CH_3)_2$

As in the case of benzene in a hydrocarbon matrix frozen from the liquid, the corresponding bands of toluene (1850 Å) and paraxylene (1910 Å) are equally red-shifted by −1250 and −600 cm^{-1} respectively and a faint structure appears in the broad continuum.[71] The structure has been resolved for toluene, toluene-d_8, *p* and *m* xylene trapped in a Kr matrix.[74]

CH_3I, C_2H_5I, and C_2H_5Br

Spectra of pure solid alkyl halides exhibit only a broad continuous absorption in the region of the sharp Rydberg bands,[72, 75] in agreement with the strong pressure effect observed at only 140 atm.[57]

5. Conclusions

It is now possible to draw general conclusions from the admittedly incomplete results so far obtained, thereby contributing to our knowledge of the electronic structures of atoms and molecules.

The uniqueness of the band progressions indicates that all trapping sites are equivalent, the dilution of the order of 1/100 to 1/1000 being sufficient to assure isolation of the active substance. Actually, it has been shown theoretically that a trapped species is well isolated only if the concentration is 0.001 or smaller.[77] However, in the case of a valence transition of a small diatomic molecule, which fits approximately the matrix site, the environment is not sensibly modified by the replacement of one surrounding matrix atom by another active molecule. On the another hand, in the case of a Rydberg-type transition, the perturbation arising from a small proportion of neighbouring trapping sites is masked by the usually large perturbation caused by the environment. That is why the observed spectra are not greatly affected by the relatively large concentration, ranging from 0.02 to 0.01, which is commonly used. As the bands are sharp for many transitions, especially when the matrix site fits exactly the dimension of the active species in both excited and ground states, the site is very likely to be highly symmetric. Furthermore, the weakness of the perturbation experienced by low-lying molecular transitions rules out the possibility of interstitial sites for which the space available is so small that large perturbations would arise from strong repulsive forces. Consequently, the trapping takes place most probably at a substitutional site of the matrix lattice.

The u.v. transitions of atoms, for which a change of the principal quantum number takes place, are generally blue-shifted owing to promotion of the optical electron to an orbital sufficiently large to allow repulsive forces to play a role. The red shift observed in some cases results from a fortuitous balance between the displacement of the excited and ground states when the intermolecular distance is close to the value corresponding to the minimum of the intermolecular potential curve; this is well accounted for by calculation.[19]

Behaviour of molecules varies considerably according to the nature of the excited states, even though the corresponding transitions may fall in the same spectral region for the free molecule, differing only in vibrational spacings. For small diatomic molecules the distinction is very clear between valence transitions and atom-like Rydberg transitions. Nitric oxide offers a particularly striking example of a trapped molecule which is practically free in a Ne matrix as far as valence transitions are concerned, but whose Rydberg transitions are completely inhibited. For polyatomic molecules the difference is less clear between the two types of transitions; the valence transitions may even be blue-shifted, showing a relatively large excited orbital, while Rydberg-like and low-lying Rydberg transitions may become detectable, even though largely blue-shifted, if the excited orbital is not exceedingly large. Consequently, although molecular solids are mainly bound by weak van der Waals forces ($V = -1/r^6$), repulsive forces may prevail when the transition involves excitation of an electron to a large orbital, e.g. a Rydberg orbital. In that case, the tight-binding approximation generally fails unless several corrections be made as for H-Ar. However, matrix isolation has not yet proved to be a powerful means of discrimination between different states of polyatomic molecules as it does not prevent linear triatomic molecules from bending and pyramidal molecules from becoming planar.

One important conclusion is that the features of a matrix-isolated spectrum depend more on the nature and dimension of the excited orbital than on the energy of the transition. Further, assuming the simple model of a Lennard-Jones potential for intermolecular interaction, it has been possible to acquire a quantitative idea of the dimension of the excited species. In Table 5 are assembled the intermolecular-potential parameters σ^* and ε^* for several atoms and molecules, computed along the lines described in § 2.(19) These empirical data have been found to be in agreement with the theoretical van der Waals constants calculated for an excited impurity interacting with a matrix atom in the ground state. Heinrichs achieved this calculation by applying his general theory of the dielectric constant to deep impurity states in solid rare gases.(78) Information about excited atoms and molecules derived from matrix shifts can also be taken as a starting point to explain the pressure effects on spectral lines (shifts, asymmetrical broadening, and satellite lines). Many spectra of trapped species are actually closely related to their spectra under a pressure of foreign gas. However, the data deduced from matrix shifts should be regarded only as tentative; in fact, if the fit is satisfactory near the turning point, a Lennard-Jones model for the excited state is ineffective in explaining many features appearing under pressure.(79)

It has been shown that strong perturbations are exhibited, in the solid state by the first Rydberg transitions which converge towards the first ionization potential. Hence the ionization process seems to be strongly modified. Moreover, since the matrix possesses closed electronic shells, the optical electron has no tendency to leave the active species and go into the matrix lattice. Future photoconductivity measurements should therefore be attempted on the pure solid matrix before studying trapped species.

In the near future several avenues are open for the continuation of such studies. Beside the extension of investigations below 1050 Å, it is possible to study a large number of simple molecules (and perhaps some radicals) by trapping them in all available matrices and at various temperatures in order to derive σ^* and ε^* in a wide variety of cases. In all cases of very sharp structures it would be interesting to use higher resolution in order to specify the perturbation of vibrational levels to study the contribution of lattice vibrations and to search for any rotational structure of the excited state.

By any standards, the matrix-isolation method has proved fruitful in providing new qualitative and quantitative information about intermolecular interactions, not directly accessible, for electronically excited atoms and molecules.

References

1. Romand, J., and Vodar, B., *Compt. Rend.* **233**, 930 (1951).
2. Romand, J., and Granier-Mayence, J., *Compt. Rend.* **235**, 1023 (1952).
3. Granier-Mayence, J., and Romand, J., *Compt. Rend.* **238**, 1148 (1953).
4. Romand, J., and Granier-Mayence, J., *J. Phys. Radium* **15**, 62 (1954).
5. Roncin, J.-Y., Damany, N., and Romand, J., *J. Molec. Spectrosc.* **22**, 154 (1967).
6. Roncin, J.-Y., *J. Molec. Spectrosc.* **26**, 105 (1968).
7. *Fourth International Symposium on Free Radical Stabilization, 31 August-2 September 1959, Washington D.C.*, Nat. Bur. Stand. pulication.
8. Dressler, K.: (a) *J. Quant. Spectrosc. Radiative Transfer* **2**, 683 (1962) (last review on the subject; (b) *Tenth European Conference on Molecular Spectroscopy, Liège, September 29—October 3, 1969.*
9. Baldini, G., *Phys. Rev.* **128**, 1562 (1962).
10. Baldini, G., *Phys. Rev.* **137**, A 508 (1965).
11. Brith, M., and Schnepp, O., *Molec. Phys.* **9**, 473 (1965).
12. Katz, B., Brith, M., Ron, A., Scharf, B., and Jortner, J., *Chem. Phys. Letters* **2**, 189 (1968).
13. Vodar, B., *Proc. Roy. Soc. London* A, **255**, 44 (1960).
14. Potts, W. J., jr., *J. Chem. Phys.* **21**, 191 (1953).
15. Raz, B., and Jortner, J., *Chem. Phys. Letters* **4**, 511 (1970).
16. Pollack, G. L., *Rev. Mod. Phys.* **36**, 748 (1964).
17. Hirschfelder, J. O., Curtiss, C. F., and Bird, R. B., *Molecular Theory of Gases and Liquids*, Wiley, 1964, p. 1110.
18. Kihara, T., and Koba, S., *J. Phys. Soc. Japan* **7**, 348 (1952).
19. Roncin, J.-Y., *Chem. Phys. Letters* **3**, 408 (1969).
20. Pollack, G. L., *J. Appl. Phys.* **38**, 4867 (1967).
21. Kovalenko, S. I., and Bagrov, N. N., *Fizika tverd. Tela* **9**, 3032 [*Soviet Phys. Solid State* **9**, 2396 (1968)].
22. Curzon, A. E., and Maxall, I. J., *Br. J. Appl. Phys.* **16**, 1301 (1965).
23. Dressler, K., *J. Opt. Soc. Am.* **50**, 50 (1960).
24. Roncin, J.-Y., Chandrasekharan, V., and Damany, N., *Compt. Rend.* **258**, 2513 (1964).
25. Roncin, J.-Y., and Moorjani, K., *Phys. Status Solidi* **23**, Kl (1967).
26. Steinberger, I. T., and Schnepp, O., *Solid State Commun.* **5**, 417 (1967).
27. Bostanjoglo, O., and Schmidt, L., *Phys. Letters* **22**, 130 (1966).
28. Moorjani, K., *Phys. Letters* **23**, 652 (1967); *J. Phys.* **28**, C 3, 87 (1967).
29. Phillips, J. C., *Phys. Rev.* **136**, A 1714 (1964).
30. Jain, K. P., *Phys. Rev.* **139**, A 544 (1965).

31. Hermanson, J., *Phys. Rev.* **150**, 660 (1966); **166**, 893 (1968).
32. Beaglehole, D., *Phys. Rev. Letters* **15**, 551 (1965).
33. O'Brien, J. F., and Teegarden, J., *Phys. Rev. Letters* **17**, 919 (1966).
34. Reilly, M. H., *J. Phys. Chem. Solids* **28**, 2067 (1968).
35. Jortner, J., Meyer, L., Rice, S. A., and Wilson, E. G., *J. Chem. Phys.* **42**, 4250 (1965).
36. (a) Basov, N. G., Bogdanovitch, O. V., Danilytchev, V. A., Deviatkov, A. G., Kachnikov, G. N., and Lantsov, N. P., *Zh. éksp. teor. Fiz. Pis'ma* **7**, 404 (1968) [*JETP Letters* **7**, 317 (1968)]; (b) Basov, N. G., Bogdanovitch, O. V., Danilytchev, V. A., Kaschnikov, G. N., Lantsov, N. P., and Khodkevitch, D. D., *J. Lumin*, **1**, 900 (1969).
37. Baldini, G., and Knox, R. S., *Phys. Rev. Letters* **11**, 127 (1963).
38. Baldini, G., *Japan J. Appl. Phys.* **4**, suppl 1, 613 (1965).
39. Robin, S., *J. Chim. Phys.* **49**, 1 (1952).
40. Roncin, J.-Y., and Damany-Astoin, N., *Compt. Rend.* **252**, 835 (1961).
41. Duley, W. W., *Phys. Letters* **19**, 361 (1965).
42. Duley, W. W., *Proc. Phys. Soc.* **88**, 1049 (1966).
43. Robin, S., and Robin, S., *Rev. Opt.* **37** (4) 161 (1958).
44. Baldini, G., *Phys. Rev.* **136**, A 248 (1964).
45. Roncin, J.-Y., Damany, N., and Vodar, B., *Compt. Rend.* **260**, 96 (1965).
46. Keil, T. H., and Gold, A., *Phys. Rev.* **136**, A 252 (1964).
47. Cochran, E. L., Bowers, V. A., Foner, S. N., and Jen, C. K., *Phys. Rev. Letters* **2**, 43 (1959).
48. Roncin, J.-Y., Thesis, Paris 1967, (available from the author on request).
49. Davydov, A. S., *Theory of Molecular Excitons*, McGraw-Hill, 1962.
50. Craig, D. P., and Walsmley, S. H., *Physics and Chemistry of the Organic Solid State*, Wiley, New York, 1963, vol. 1, p. 585.
51. Sibleyras, M., Roncin, J.-Y., and Damany, N., *Compt. Rend.* **266**, B 175 (1968).
52. Roncin, J.-Y., and Damany, N., *Compt. Rend.* **260**, 6069 (1965).
53. Fox, D., and Hexter, R. M., *J. Chem. Phys.* **41**, 1125 (1964).
54. Hexter, R. M., *J. Chem. Phys.* **46**, 2300 (1967).
55. Lagerqvist, A., and Miescher, E., *Helv. phys. Acta* **31**, 221 (1958).
56. Jungen, Ch. *Can. J. Phys.* **44**, 3197 (1966).
57. Robin, M. B., and Kuebler, N. A., *Twenty-fourth Symposium on Molecular Structure and Spectroscopy, Columbus, September 2-6, 1969*.
58. Bass, A. M., and Broida, H. P., *J. Molec. Spectrosc.* **12**, 221 (1964).
59. Schnepp, O., and Dressler, K., *J. Chem. Phys.* **42**, 2482 (1965).
60. Boursey, E., Roncin, J.-Y., and Damany, N., *Chem. Phys. Letters* **5**, 584 (1970).
61. Roncin, J.-Y., and Damany, N., unpublished results.
62. Roncin, J.-Y., Damany, N., and Vodar, B., *Chem. Phys. Letters* **3**, 197 (1969).
63. Dressler, K., and Schnepp, O., *J. Chem. Phys.* **33**, 270 (1960).
64. Onaka, R., and Takahashi, T., *J. Phys. Soc. Japan* **24**, 548 (1968).
65. Milligan, D. F., and Jacox, M. E., *J. Chem. Phys.* **47**, 5146 (1967).
66. Dressler, K., *J. Chem. Phys.* **35**, 165 (1961).
67. Vodar, B., Eur. Res. Off. of US Army Contract Nb DAJA, 3769 C 0756.
68. Lombos, B. A., Sauvageau, P., and Sandorfy, C., *Chem. Phys. Letters* **1**, 382 (1967).
69. Katz, B., and Jortner, J., *Chem. Phys. Letters* **2**, 437 (1968).
70. Pysh, E. S., Rice, S. A., and Jortner, J., *J. Chem. Phys.* **43**, 2997 (1965).
71. Potts, W. J., jr., *J. Chem. Phys.* **23**, 73 (1955).
72. Roncin, J.-Y., Ch'en, S. Y., Granier, J., and Damany-Astoin, N., *Spectrochim. Acta* **18**, 907 (1962).
73. Katz, B., Brith, M., Sharf, B., and Jortner, J., *J. Chem. Phys.* **50**, 5195 (1969); **52**, 88 (1970).
74. Brith, M., Katz, B., Sharf, B., and Jortner, J., *Tenth European Congress of Molecular Spectroscopy, Liège, September 29-October 3, 1969*.
75. Astoin, N., Granier, J., and Cordier, M., *J. Phys. Rad.* **19**, 506 (1958).
76. Brewer, L., Meyer, B., and Brabson, G. D., *J. Chem. Phys.* **43**, 3973 (1965).

77. HARVEY, K. B., SHURVELL, H. F., and HENDERSON, J. R., *Can. J. Chem.* **42**, 911 (1964).
78. HEINRICHS, J., *Phys. Rev.* B, **2**, 518 (1970).
79. RONCIN, J. Y., *J. Quant. Spectrosc. Radiative Transfer* **11**, 1151 (1971).

Additional bibliography

BAJAJ, K. K., Optical absorption due to substitutional hydrogen atom impurity in solid rare gases, *Solid State Commun.* **8**, 1665 (1970).

BOURSEY, E., and RONCIN, J.-Y., Experimental deperturbation of the $b^1\Pi_u \leftarrow X^1\Sigma_g^+$ transition of N_2 in the solid state, pure and trapped in Ne and CF_4 matrices at low temperature, *Phys. Rev. Letters* **26**, 308 (1971).

BOURSEY, E., RONCIN, J.-Y., DAMANY, H., and DAMANY, N., Double-beam monochromator for absorption studies of condensed gases from 2000 to 500 Å, *Rev. Phys. Appl.* **6**, 11 (1971).

BOURSEY, E., RONCIN, J.-Y., and DAMANY, H., Fundamental absorption spectrum of solid neon, *Phys. Rev. Letters* **25**, 1279 (1970).

BUXTON, R. A. H., and DULEY, W. W., Reflection spectra of solid CO_2 and CO_2-O_2 mixtures, from 4 to 22 eV, *J. Chem. Phys.* **56**, 1783 (1972).

BUXTON, R. A. H., and DULEY, W. W., Reflection spectrum of solid nitrogen in the v.u.v., *Phys. Rev. Letters* **25**, 801 (1970).

CHESCHNOVSKY, O., RAZ, B., and JORTNER, J., Vacuum u.v. α-induced luminescence phenomena in rare gas liquids and solid alloys, *Proceeding of the Third International Conference on Vacuum Ultraviolet Radiation Physics, Tokyo, 1971*, session 1aA2-5.

GEDANKEN, A., RAZ, B., and JORTNER, J., Vacuum u.v. spectroscopy of deep lying impurity states in solid CF_4, *Chem. Phys. Letters* **14**, 172 (1972).

GEDANKEN, A., RAZ, B., and JORTNER, J., Trapped exciton in solid hydrogen, *Chem. Phys. Letters* **14**, 326 (1972).

HAENSEL, R., KEITEL, G., KOCH, E. E., KOSUCH, N., and SKIBOWSKI, M., Optical excitation of solid Ne in the v.u.v., *Phys. Rev. Letters* **25**, 1281 (1970).

HAENSEL, R., KEITEL, G., KOCH, E. E., SKIBOWSKI, M., and SCHREIBER, P., Reflection spectrum of solid Kr and Xe in the v.u.v., *Optics Commun.* **2**, 59 (1970).

HAENSEL, R., KEITEL, G., KUNZ, C., and SCHREIBER, P., Line shapes in soft X-ray absorption spectra of solidified rare gases, *Phys. Rev. Letters* **25**, 208 (1970).

HAENSEL, R., KEITEL, G., SCHREIBER, P., and KUNZ, C., Optical absorption of solid Kr and Xe in the far u.v., *Phys. Rev.* **188**, 1375 (1969).

HAENSEL, R., KOSUCH, N., NIELSEN, U., SONNTAG, B., and RÖSSLER, U., Optical properties of dilute solid rare gas mixtures in the extreme ultraviolet, *Phys. Rev.* **B 7** (1973).

KOCH, E. E., and SKIBOWSKI, M., Electronic excitation of solid CO_2 in the extreme u.v., *Chem. Phys. Letters* **14**, 37 (1972).

MCCARTY, M., JR., Mercury $^3P_1$—$^1S_0$ transition in solid rare gases, *J. Chem. Phys.* **52**, 4973 (1970).

MEYER, B., *Low Temperature Spectroscopy*, Am. Elsevier, New York, 1971.

MIRON, E., RAZ, B., and JORTNER, J., A comment on the *V* excited states of the ethylene molecule, *Chem. Phys. Letters* **6**, 563 (1970).

MORRIS, G. C., and ANGUS, J. G., Do antiresonances in C_6H_6-rare gas solids exist?, *J. Chem. Phys.* **54**, 3233 (1971).

RÖSSLER, U., Interpretation of far u.v. spectra in solid rare gases from calculated density of states, *Phys. Status Solidi* b, **45**, 483 (1971).

SHARBER, S. R., and WEBBER, S. E., Optical constants of simple molecular crystals: I, Results for CO and O_2: II, Results for Kr and Xe, *J. Chem. Phys.* **55**, 3977, 3985 (1971).

STEINBERGER, I. T., ATLURI, C., and SCHNEPP, O., Optical constants of solid Xe in the v.u.v. region, *J. Chem. Phys.* **52**, 2723 (1970).

ZUMOFEN, G., Application of dipole-wave summations for $\mathbf{K} \cdot \mathbf{R} \simeq 1$ to the u.v. absorption spectrum of solid CO, *Helv. phys. Acta* **44**, 579 (1971).

CHAPTER 7

Photoelectron spectroscopy: Principles and instrumentation

S. LEACH
Laboratoire de Photophysique Moléculaire, Faculté des Sciences, 91405 Orsay, France

Contents

Introduction

Vacuum ultraviolet light sources and monochromators are important tools in a number of fields outside that of optical spectroscopy. Among recent developments, photoelectron spectroscopy is of particular interest. It can provide important information on energy levels and electron configurations of atomic and molecular ions. In the latter case some understanding of the internal dynamics and the geometrical structure can be obtained when vibrational structure is resolved. In conjunction with other techniques, photoelectron spectroscopy can be used to clarify photoionization and autoionization processes. Promising applications have been made in problems related to catalysis and solid-state structure.

Photoelectron spectroscopy is essentially concerned with the process of photoionization and the study of the structural properties of ions through an examination of the kinetic energies and angular distributions of photoelectrons ejected from neutral species. The maximum u.v. photon energy used so far is 40.8 eV, corresponding to the 303.78 Å resonance emission of ionized helium He II.

A considerable amount of photoelectron spectroscopy at higher energies, using X-ray sources in the 1-23 keV range, has been carried out in recent years, mainly by

Siegbahn *et al.*[1] Siegbahn calls his technique ESCA (electron spectroscopy for chemical analysis), thus emphasizing its physicochemical and analytical applications.

Photoelectron spectroscopy of negative ions has also been studied using low energy sources in the visible and near u.v. regions.[2] Such experiments are usually classified as photodetachment experiments.

The principles of the photoelectron spectroscopic method are essentially the same over the whole energy range from the X-ray to the visible spectral region. The present review is restricted to the principles of photoelectron spectroscopy and the instrumentation used for investigating the ionization of neutral species with far u.v. sources.

1. Principles of photoelectron spectroscopy

1.1. THE PHOTOIONIZATION PROCESS

In the single particle approximation the electronic structure of an atomic or molecular species is characterized by an electron configuration $S(j)$ based on a set of atomic or molecular orbitals $j_1, j_2, j_3, \ldots$, having definite energies. Ionization corresponds to removal of a bound electron, with simultaneous reorganization of the orbitals and energies of the remaining electrons, leaving the newly formed ion in the reorganized electron configuration $S'(j)$. The ionization potential I is the energy expended in removing the expelled electron to infinity.

Study of ionization processes is usually carried out by collision experiments between a neutral species and photons or charged particles. In this connection it should be pointed out that for ionization by electron impact the time of interaction with the neutral species may be shorter than the time for complete rearrangement of the orbitals.[3] For example, with electrons of energies in the 10 eV range the interaction time is of the order of 1.6×10^{-16} s. This is much shorter than the interaction time for a 10 eV photon which is calculated to be of the order of 2×10^{-14} s, assuming 50 cycles of the radiation field. If the orbital reorganization time is of the order of 10^{-15} s, as is probable in saturated hydrocarbons,[3] ionization by impact with electrons of 10 eV kinetic energy will be a partially diabatic process from the viewpoint of the electron orbitals, whereas with 10 eV photons it will be an electronic adiabatic process. The ionization potential measured by electron impact will be higher than by photoionization under these circumstances and will correspond to energies of localized orbitals.

In the following discussion, concerning photoionization, the adiabatic case is assumed.

Let us consider a neutral stable molecular species M which initially is in a discrete quantum state n described by a total wave function Ψ_n. Within the limits of the validity of the Born-Oppenheimer approximation we may separate Ψ_n into a product of electronic e, vibrational v, and rotational r wave functions:

$$\Psi_n = \Theta_e \Theta_v \Theta_r, \tag{1}$$

and separate the internal energy of the system E_n into a sum of electronic, vibrational, and rotational energies:

$$E_n = E_e + E_v + E_r. \tag{2}$$

On ionization, in which we assume that a discrete state of the ion is formed, and we exclude ion pair production and ion fragmentation, the new wave function Ψ_f describing the system $M^+ + e^-$ will have as components the total wave function Ψ_k of the kth discrete state of the ion and Ψ_c, the wave function corresponding to a continuum state of the electron removed from the jth orbital of the neutral species: $\Psi_f = \Psi_k \Psi_c$.

Interaction between the electrons in a particular electronic configuration can often give rise to several electronic states having different energies. An ionization potential will therefore be characteristic of the formation of the ion M^+ in a particular electronic state e'.

For a molecular species, the ion will be formed not only in a particular electronic state but also in well-defined vibrational v' and rotational r' levels, so that we can characterize the ionization potential as

$$I = E[S'(j); e', v', r'] - E[S(j); e, v, r]. \tag{3}$$

If the molecule, initially in the energy state $E[S(j); e, v, r]$ suffers a collision with a photon of energy $h\nu \geqq I$, photoionization will occur with a cross-section σ_I, leading to ejection of an electron:

$$h\nu + M[S(j); e, v, r] \rightarrow M^+[S'(j); e', v', r'] + e^-. \tag{4}$$

In practice, the recoil energy of the ion is negligible so that the kinetic energy of the photoelectron is given by

$$\tfrac{1}{2} mv^2 = h\nu - I, \tag{5}$$

where m is the mass of the electron and v its velocity.

The above discussion naturally applies also to atoms, but in this case we have to consider only electronic configurations, states, and energies.

With the relatively low photon energies used in far u.v. photoelectron spectroscopy, v is small compared to the velocity of light, so that the electron will be photoejected with an angular distribution $F(\theta)$ given, for *linearly polarized light*,[4,5] by

$$F(\theta) = \left(\frac{\sigma_{total}}{4\pi}\right)\left[1 + \frac{\alpha}{2}(3\cos^2\theta - 1)\right], \tag{6}$$

where σ_{total} is the total cross-section, θ the angle between the direction of the ejected electron and the electric vector of the polarized incident radiation and α an asymmetry parameter which is a function of the energy of the ejected electron and depends also on the nature of the outgoing partial waves, and hence on $S'(j)$ and $S(j)$.

For *unpolarized light*, the angular distribution is given by

$$F(\omega) = \left(\frac{\sigma_{\text{total}}}{4\pi}\right)\left[1 - \frac{\alpha}{4}(3\cos^2\omega - 1)\right], \tag{7}$$

where ω is the angle between the direction of the ejected electron and the direction of propagation of the light beam.

These formulae are valid for single-photon, electric dipole transitions.

One electron is emitted for every singly charged ion formed, so that the ion appearance potentials can be determined by measurement of the kinetic energy $\frac{1}{2}mv^2$ of the photoejected electrons. The relative ion yields can be obtained from electron current intensities with appropriate corrections, where necessary, for anisotropy in the angular distribution of the photoelectrons.

In low energy photoelectron spectroscopy, the limited photon energy restricts the information obtained mainly to outer electrons, whereas with high-energy techniques one can eject deep lying core electrons. However, experimentally, the energy resolution of the order of 0.01 eV obtainable with u.v. photons cannot be matched in high energy photoelectron spectroscopy where the best resolution obtainable is of the order of 1 eV.

1.2. PHOTOIONIZATION CROSS-SECTIONS

The cross-section σ_k for photoionization producing a singly charged ion in a particular energy level E_k is ideally a step function of the incident radiation energy $h\nu$.

For atoms one observes two kinds of general behaviour for the variation of σ_k with $h\nu$. The first type is illustrated by helium in Fig. 1, which is based on the

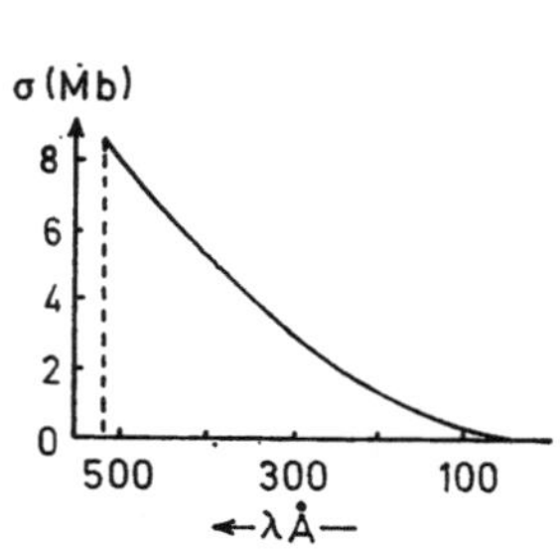

FIG. 1. Photoionization cross-section function for helium.[6]

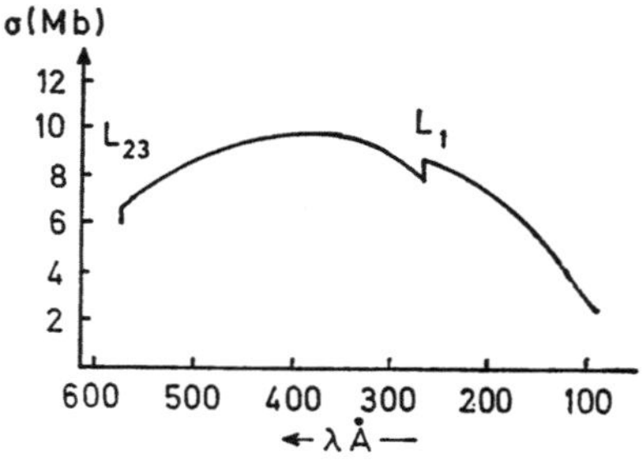

FIG. 2. Photoionization cross-section function for neon.[7]

data of Lowry *et al.*[6] There is a sharp rise of σ_k at threshold followed by a slow regular decrease to higher photon energies.

The photoionization data for neon[7] in Fig. 2 shows the maximum which is observed in the second type of cross-section function.

The threshold for production of the $^2P_{3/2}$ and $^2P_{1/2}$ states of $2s^22p^5$ Ne II is in the neighbourhood of 575 Å, with a cross-section of 6.3 Mb [1 Megabarn (Mb) $= 10^{-18}$ cm^2]. As photon energies increase, σ first increases to a maximum of 10 Mb at 375 Å and then decreases to a value of 7.8 Mb at about 260 Å, the threshold wavelength for production of the excited $2s\,2p^6\,{}^2S_{1/2}$ state of Ne II.

The two types of photoionization cross-section functions are related to the presence or absence of a node in the bound state radial wave function of the ejected electron.[8]

When a new ionization level $E_{k'}$ is reached with increasing photon energy, there should be a new threshold step in the σ vs. $h\nu$ curve. Above this new threshold the $\sigma_{k'}$ function will be superposed on the σ_k function. This is clearly seen in Fig. 2 where the jump at the $2s\,2p^6\,{}^2S_{1/2}$ level of Ne II is of the order of 0.7 Mb.

Discontinuities at the photoionization thresholds above the lowest ionization potential are not always observed. Indeed, in more recent work, Samson[9] found no evidence for a discontinuity at 260 Å for neon. Fano and Cooper[10] have shown that interference effects related to autoionization processes can make the photoionization cross-section function increase, decrease, or remain virtually unchanged at thresholds corresponding to higher ionization potentials. In this section we will neglect autoionization effects and reserve their discussion to § 1.3.

For a molecule, $E_{k'}$ may be a vibrational or rotational as well as an electronic level of the ion.

In the absence of the interference effects mentioned above, the energy level structure of the ion can be studied by measuring σ_I as a function of $h\nu$. This may be done either directly in the integral form as illustrated in Figs. 1 and 2, or in the differential form $d\sigma_I/d\nu$ as a function of $h\nu$ which reveals the energy levels as a series of peaks in the energy spectrum.

Photoionization cross-sections are generally determined by measuring either the current of ions or the current of electrons produced. Absolute cross-section values are difficult to obtain since this requires much additional information. Most of the published data on photoionization cross-sections are relative measurements.

In photoelectron spectroscopy, measurements are made at a fixed frequency ν of the incident radiation. The total photoelectron current is therefore proportional to the total photoionization cross-section $\sigma_I(\nu)$ at this frequency. From the above discussion it is clear that $\sigma_I(\nu)$ is the sum of the cross-sections for ionization to all E_k levels for which photoionization is possible, i.e. to those levels whose energies lie between $E_k = h\nu$ and the first ionization potential E_0.

We may therefore write

$$\sigma_I(\nu) = \sum_k \sigma_k(\nu).$$

The photoelectrons ejected on formation of the ion in its kth level will have kinetic energy $w_k = (1/2)mv_k^2$. Measurement of the spectrum of kinetic energies of photoelectrons produced at a fixed incident radiation frequency will therefore correspond to scanning the spectrum of energy levels of the ion over the energy range $h\nu - E_0$, where E_0 is the first ionization potential.

Since, as we have seen, the individual σ_k functions vary with photon energy, the relative intensities in the energy spectrum so obtained will not correspond exactly to the relative photoionization probabilities at threshold. However, the σ_k functions are often approximately constant within a few electron volts of the threshold (see Figs. 1 and 2) so that true relative photoionization probabilities are reasonably approximated to in photoelectron spectroscopy. Nevertheless, caution must be taken in ensuring that this condition is valid in particular cases.

1.3. AUTOIONIZATION

The relatively simple behaviour of the photoionization and photoelectron cross-section functions described previously can be considerably more complex in the presence of autoionization processes.

Autoionization essentially involves an isoenergetic radiationless transition, of probability P_{if}, from an initial bound zero-order atomic or molecular excited electronic state i of a neutral species to a zero-order final state f which lies in the ionization continuum of a lower-lying electronic state k, leading to formation of the ion M^+ in state k and an electron whose kinetic energy is equal to the difference in energy between states f and k (Fig. 3).

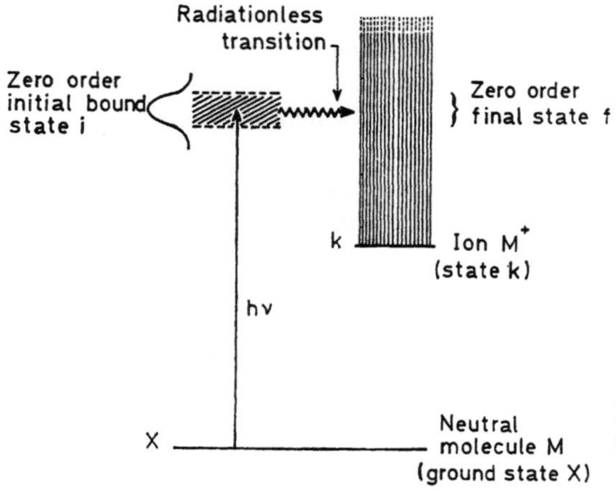

FIG. 3. Autoionization as a radiationless transition.

The angular momenta and symmetries of states i and f must be related by the Kronig rules for radiationless transitions applied to autoionization[11] for P_{if} to be non-zero.

If P_{if} is much smaller than the Bohr frequency of the excited neutral state i, then the autoionization probability is given by a Fermi golden rule expression

$$P_{if} = \frac{2\pi}{\hbar} \beta^2, \tag{8}$$

where β is a matrix element:

$$\beta = \langle \Psi_f \mid V \mid \psi_i \rangle \tag{9}$$

in which V is the interaction operator coupling initial and final states. The radiationless transition may be induced electronically and/or, for molecules, vibrationally or through rotation.[12]

The Born-Oppenheimer approximation allows us to decompose β into the product of an electronic part $\beta(e)$ and a Franck-Condon factor involving the vibrational wave functions χ_v:

$$\beta = \beta(e)\ (\langle \chi_{vk} | \chi_{vi} \rangle)^2. \tag{10}$$

The autoionizing level will have an energy half-width given by $\hbar\tau^{-1}$, where τ is the lifetime of this level. The domain of isoenergetic interaction between states i and f will therefore depend on τ^{-1}. Assuming only first-order relaxation processes, the value of τ is determined by

$$\frac{1}{\tau} = P_{if} + \sum_q k_q, \tag{11}$$

where k_q is the rate of the qth radiative or non-radiative relaxation process leading to depopulation of level i other than by autoionization. For free atoms, $\sum_q k_q$ will generally be of the order of 10^9 or 10^{10} s^{-1}.

Normally the energy width of a single ground state level X will be much less than $\hbar\tau^{-1}$ so that the line shape for photon absorption $i \leftarrow X$ will be broad to an extent that is proportional to τ^{-1}.

Since the photoexcitation function is expressed by the absorption line shape it is clear that autoionization is essentially an optical resonance phenomenon when only one excited neutral state i is involved.

The true wave function Φ_a of the state a reached by photon absorption from the true ground state o, whose wave function is Φ_o, will be a linear combination of ψ_i with the continuum states Ψ_f. The true wave functions Φ_a and Φ_o are eigenfunctions of the molecular Hamiltonian, whereas ψ_i, ψ_k, Ψ_f, corresponding to the zero-order states i, k, f, are eigenfunctions of a Hamiltonian which excludes coupling terms between electronic states. Since in general, the zero-order ground state X is far removed energetically from other electronic states, the true ground state o will be quasi-identical with X.

If more than one zero-order ion state lies below the autoionizing level, then these will each contribute to the true state a.

The absorption cross-section will be proportional to K_{ao}^2, the square of the total probability amplitude K_{ao} for the optical transition $a \leftarrow o$.

Assuming only one ion level k below the autoionizing level, K_{ao} can be decomposed into a sum of terms corresponding to the zero-order absorption processes

$i \leftarrow X$ and $f \leftarrow X$:

$$K_{ao} = K_{iX} + K_{fX}. \tag{12}$$

The total transition probability is then

$$K_{ao}^2 = K_{iX}^2 + K_{fX}^2 + 2Re\,|K_{iX}K_{fX}|. \tag{13}$$

The last term on the right-hand side of eq. (13) expresses the interference effects between absorption to the directly ionizing zero-order state f and to the zero-order autoionizing state i.

For photon energies resonant with the autoionizing level i, the absorption cross-section can have a complicated form[10, 13]. If the $a \leftarrow o$ transition has a much smaller oscillator strength than the zero-order transition $f \leftarrow X$, there can be a marked decrease in transition intensity in the resonance region. In other cases the absorption line profile can exhibit an asymmetric resonance peak and an anti-resonance minimum. It should be noted that the $f \leftarrow X$ transition in question is to a band of continuum states of half-width $\hbar\tau^{-1}$ to match the width of the auto-ionization level.

The photoionization cross-section will also be very complicated in this region since it will also depend on the amount of mixing between the zero-order states, one of which leads to direct ionization, the other to autoionization. Increase and decrease of the photoionization cross-section will be mirrored in the photon absorption cross-section.

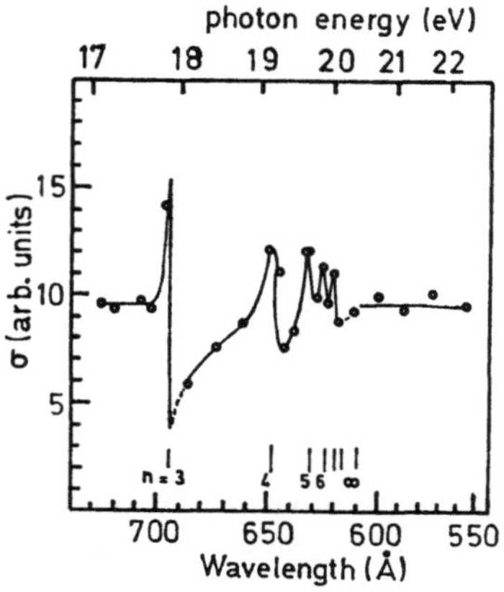

FIG. 4. Photoionization cross-section function for atomic nitrogen (after Comes[14]).

The contribution to the photoionization cross-section in the neighbourhood of autoionizing state energies and the concomitant peak asymmetry and antireso-nance minima are illustrated in Fig. 4 which shows the results of Comes[14] on the photoionization cross-section function of atomic nitrogen.

The photoionization threshold for the production of the $2s^2 2p^{2\,3}P$ state of N II is at 852.2 Å. The first excited state of the ion to which an optical dipole transition is allowed from the $2s^2 2p^3 {}^4S$ state of the neutral N I atom is the $2s 2p^3 {}^5S$ state of N II at 609.3 Å. Figure 4 shows the portion of the photoionization cross-section function between 750 and 550 Å. Asymmetrical autoionization peaks and anti-resonance minima are observed in the 700-600 Å region. They correspond to autoionization into the essentially ${}^3P$ continuum from $n = 3, 4, \ldots$ of the $2s 2p^3({}^5S)np {}^4P$ Rydberg series levels leading to the $2s 2p^3 {}^5S$ state of the ion.

From the previous discussion it is clear that the width of an autoionization peak in a photoionization spectrum is, *grosso modo*, a function of τ^{-1}. In particular, it will be narrow if the corresponding autoionization rate is relatively slow, assuming autoionization to be the dominant relaxation process of the autoionizing state.

Let us now consider the role of autoionization in photoelectron spectroscopy. Most photoelectron spectroscopy has been carried out at a fixed photon energy $h\nu$. Under these conditions, the true wave function of the final state reached on absorption of a monochromatic photon will have significant components of every ion state in the energy interval $h\nu - I_0$ and its associated continuum electron state. The component of a particular autoionization state will be negligible unless $h\nu$ is close to such a state or if the mixing between the neutral state and the ion state is so strong that the autoionization state acquires an extensive continuum character with significant oscillator strength at $h\nu$.

If we first assume that a particular autoionization state i provides a negligible component to the true wave function at the $h\nu$ level, then no photoelectrons will be produced from the autoionizing state. Photoelectron spectroscopy under these conditions will scan the spectrum of electron kinetic energies relative to the ions produced only by direct photoionization, and will reveal only the energy levels of the ion. It must be remembered, however, that when a particular level in the continuum interacts with an autoionization level, the energies of these levels will not be those of the zero-order states but will correspond to the perturbed continuum and autoionization levels after this interaction has been taken into account.

When state i contributes significantly to the true wave function of the absorption final state, photoelectrons will be produced not only by direct ionization but also via autoionization. A photoelectron spectrum taken under these conditions will have a distorted structure as compared with the previous case.

As an illustration for atoms, we may cite the results of Samson and Cairns(15) on the photoelectron spectra of xenon as a function of photon energy. The ionization processes studied were:

$$\mathrm{Xe}\,(5s^2 5p^6\,{}^1S_0) + h\nu \rightarrow \mathrm{Xe}^+\,(5s^2 5p^5\,{}^2P_{3/2}) + e^-, \tag{14a}$$

$$\mathrm{Xe}\,(5s^2 5p^6\,{}^1S_0) + h\nu \rightarrow \mathrm{Xe}^+\,(5s^2 5p^5\,{}^2P_{1/2}) + e^-, \tag{14b}$$

whose thresholds are respectively at 1022.1 and 922.8 Å. The next ionization threshold is at 529.9 Å for the production of $\mathrm{Xe}^+\,(5s\,5p^6\,{}^2S_{1/2})$.

Samson and Cairns were able to determine R, the ratio of the cross-sections for production of ions in the ${}^2P_{3/2}$ and ${}^2P_{1/2}$ ion states, as a function of photon energy.

They found that R was constant, with a value of 1.60 for about 28 wavelengths in the range 922.8-460 Å, except near some xenon absorption resonances. At 558.6 Å, $R = 1.46$; at 543.9 Å, $R = 0.77$; at 521.8 Å, $R = 1.49$. Calculation of the specific cross-sections showed that $\sigma_{3/2}$ decreased at these wavelengths, no change in $\sigma_{1/2}$ being observed at 558.6 and 521.8 Å, but an increase was found at 543.9 Å.

Distortion of the photoelectron spectrum at photon energies close to auto-ionization levels has also been observed for molecules.

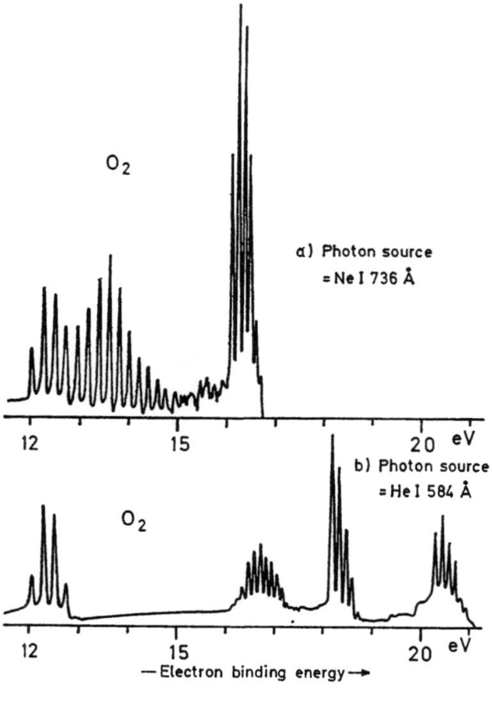

FIG. 5. Photoelectron spectra of O_2 excited by (a) Ne I 736 Å (16.85 eV) photons, and (b) He I 584 Å (21.21 eV) photons.[16]

Price[16] studied the photoelectron spectrum of O_2 at two photon energies, He I 584 Å (21.21 eV) and Ne I 736 Å (16.85 eV). With the helium source, the spectrum showed four or five vibrational bands in the 12 eV region, corresponding to excitation to vibrational levels of the $X\,^2\Pi_g$ state of O_2^+, whose ionization potential is 12.059 ± 0.001 eV.[17, 18] A further set of vibrational bands exist, corresponding around 16.1 eV to the $a\,^4\Pi_u$ level and around 16.8 eV possibly to the $A\,^2\Pi_u$ levels of O_2^+.[19] Higher electronic states with associated vibrational structure are observed near 18 and 20 eV; these probably correspond to the $b\,^4\Sigma_g^-$ and the $B\,^2\Sigma_g^-$ states of O_2^+ respectively,[20] (Fig. 5).

The photoelectron spectrum taken with the neon source shows a considerable development of the vibrational structure of the $X\,^2\Pi_g$ state above 13 eV where there exist a series of Rydberg levels whose limit is the $A\,^2\Pi_u$ state of O_2^+.[21] Just above

16 eV there is a hundred-fold enhancement of the vibrational states of $a\,^4\Pi_u$ which probably arises from autoionization of Rydberg levels of O_2 whose limit is the $b\,^4\Sigma_g^-$ state of O_2^+. Similar results were obtained by Collin and Natalis[22] who observed the development of the $X\,^2\Pi_g$ vibrational levels up to $v = 20$ with a neon source.

In our previous discussion we have considered the autoionization process following photon absorption of a species M in the o ground state as involving a single autoionizing state R_1 according to the scheme:

$$h\nu + M(o) \rightarrow M(R_1), \tag{15a}$$

$$M(R_1) \rightsquigarrow M^+ + e^-. \tag{15b}$$

Berry[12] has considered the possibility that in some cases autoionization may proceed via a series of intermediate Rydberg states. We represent this process as follows for the case of three Rydberg states:

$$h\nu + M(o) \rightarrow M(R_3), \tag{15c}$$

$$M(R_3) \rightsquigarrow M(R_2), \tag{15d}$$

$$M(R_2) \rightsquigarrow M(R_1), \tag{15e}$$

$$M(R_1) \rightsquigarrow M^+ + e^-. \tag{15f}$$

The Rydberg states are coupled together. In terms of the zero-order states it can be considered that steps (15d) and (15e) represent radiationless transitions between Rydberg states of the neutral molecule. Such radiationless inter-Rydberg transitions could be particularly important for molecules since, in this case, states may also be coupled via vibrational and rotational motion.

The results of Price on O_2 show considerable enhancement of the vibrational structure of the $X\,^2\Pi_g$ state of the ion near 13 eV. The $v = 8$ maximum lies some 3 eV below the energy of the neon photons used.

The complex structure of the absorption spectrum of O_2 in the 700-1000 Å region can be analysed, at least partially, into various Rydberg electronic transitions whose vibrational progressions may overlap.[21] It may therefore be possible that autoionization in the 13 eV region proceeds via radiationless transitions between a set of highly excited states of the neutral molecule, though other interpretations are possible.

Vibrationally induced autoionization effects in the photoelectron spectra of H_2 have been observed. It is instructive to compare data obtained at 21.21 eV (Fig. 6) and at threshold (Fig. 7). Figure 6 shows the photoelectron spectrum obtained by Siegbahn *et al.*[1] with a helium 584 Å (21.21 eV) line source. Well-developed vibrational structure was observed up to $v = 13$ of the $X\,^2\Sigma_g^+$ state of H_2^+ (I.P. = 15.42556 eV[23]).

In a similar study, Turner[24] observed vibrational structure up to $v = 15$, i.e. to about 0.1 eV below the dissociation limit of the $X\,^2\Sigma_g^+$ state. The peak electron currents at each vibrational level were generally within a factor of two

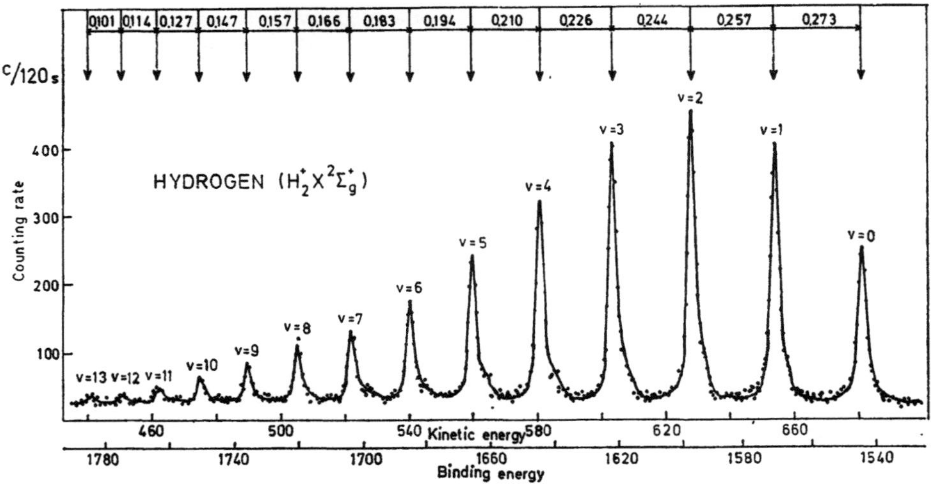

Fig. 6. Photoelectron spectrum of H_2 excited by He I 584 Å (21.21 eV) photons.[1]

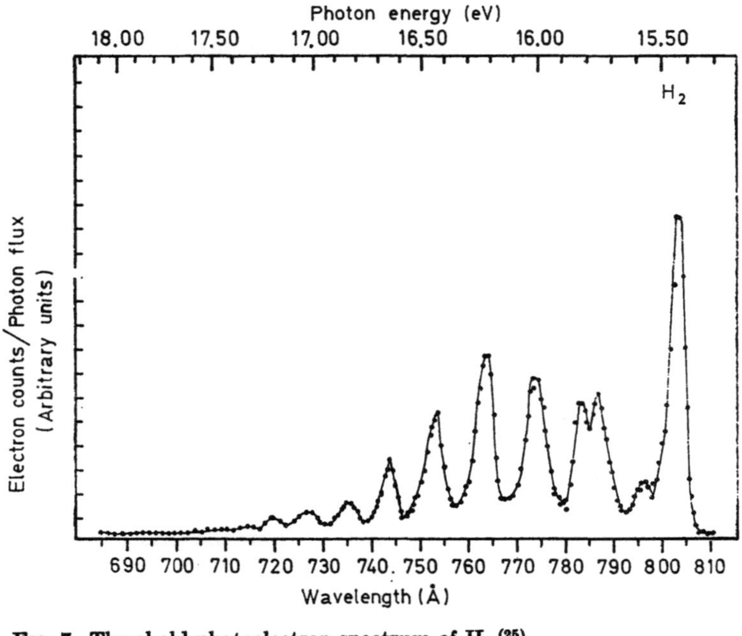

Fig. 7. Threshold photoelectron spectrum of H_2.[25]

of the values calculated from Franck-Condon factors, normalized to $X\,^2\Sigma_g^+ v = 0$, for the vibronic transitions

$$H_2(X\,^1\Sigma_g^+;\, v'' = 0) + h\nu \rightarrow H_2^+(X\,^2\Sigma_g^+;\, v' = 0, 1, 2, \ldots).$$

Villarejo[25] has measured the photoelectron spectrum of H_2 at varying photon energies; initial kinetic energies of the photoelectrons were fixed approximately equal to zero since in this case threshold photoelectrons were sought. Figure 7 shows the threshold photoelectron spectrum in H_2 under experimental conditions where the analysis energy and acceleration energy were both equal to 6 eV and the energy analyser band width was approximately 0.7 eV.

The observed peaks correspond to formation of H_2^+ $(X\,^2\Sigma_g^+)$ vibrational levels $v = 0, 1, 2, 3, \ldots$ The peaks at 797 and 784 Å are extraneous and were interpreted by Villarejo as arising from photoelectrons produced by autoionization from vibrational levels of the $4p\sigma B''\,^1\Sigma_u^+$ and/or $4p\pi D'\,^1\Pi_u$ Rydberg states of H_2. The relative peak heights of the $H_2^+(X\,^2\Sigma_g^+)$ state vibrational levels $v = 0$, 1 and 2 did not agree with the $H_2(X\,^1\Sigma_g^+) \rightarrow H_2^+(X\,^2\Sigma_g^+)$ Franck-Condon factors normalized to $v = 3$, whereas relatively good agreement was found for the $v = 4$, 5 and 6 levels. This indicates that the $v = 0$, 1 and 2 peaks are influenced by contributions from autoionizing Rydberg levels of H_2, but that levels $v = 3$, 4, 5 and 6 are formed principally by direct photoionization.

Theoretical calculations of the rates of autoionization from various vibrational levels of a number of H_2 Rydberg states have been carried out.[12, 26–28] The quantitative values for autoionization rates differ according to the theoretical model used, but the results of the various authors support the vibrational propensity rule proposed by Berry[12] for the inter-Rydberg state and autoionization transitions in H_2: $\Delta v = \pm 1$ radiationless transitions should be more probable than $\Delta v = \pm 2$ and still more so than $\Delta v = \pm 3$, etc. Some of the recent photoelectron spectroscopy results of Berkowitz and Chupka[29] appear to contradict this propensity rule. The disagreement may possibly indicate that coupling between Rydberg states is involved in some of the vibrationally induced autoionizations in H_2.

One other point should be mentioned regarding autoionization as related to photoelectron spectroscopy. The *angular distribution* of photoelectrons ejected via an autoionization process may differ from that via direct photoionization depending on whether the outgoing partial waves differ for the two processes.[12, 30] Studies of photoelectron angular distributions with photon energies near ionization thresholds could thus provide useful information concerning autoionization states and their nature.[5, 31]

1.4. PHOTOIONIZATION AND PHOTOELECTRON EFFICIENCIES

The efficiency of photoionization is an important factor in determining electron currents in photoelectron spectroscopy. The photoionization efficiency γ_I is defined as the number of ions produced per photon initially absorbed. For atoms, γ_I

will be generally equal to unity at wavelengths shorter than the first ionization threshold but could conceivably be less than unity in regions where the ionization continuum interacts with superexcited neutral states.

Let us consider, for a molecule AB, the various possible decay channels of the final state reached on absorption of monochromatic photons $h\nu$. These will be considered as independent competitive processes, although if there is very strong interaction between these channels there may exist important cross-terms (interference effects).

The primary act of photon absorption may lead to the following processes having specific cross-sections σ. It can form a stable ion state with a cross-section σ_k

$$AB + h\nu \xrightarrow{\sigma_k} AB^+ + e^- \tag{16}$$

or it may lead to ion fragmentation

$$AB + h\nu \xrightarrow{\sigma_g} A^+ + B + e^- \tag{17}$$

or to ion pair production

$$AB + h\nu \xrightarrow{\sigma_p} A^+ + B^-. \tag{18}$$

The primary act can also form a superexcited neutral state i:

$$AB + h\nu \xrightarrow{\sigma_i} AB^*. \tag{19}$$

This superexcited state may decay by autoionization with rates k_{if}, k'_{if}, and k''_{if} for the three following processes:

$$AB^* \overset{k_{if}}{\rightsquigarrow} AB^+ + e^-, \tag{20}$$

$$AB^* \overset{k'_{if}}{\rightsquigarrow} A^+ + B + e^-, \tag{21}$$

$$AB^* \overset{k''_{if}}{\rightsquigarrow} A^+ + B^-. \tag{22}$$

Non-ionic decay of the superexcited state is also possible (assuming only first order processes):

Dissociation: $AB^* \rightsquigarrow A + B.$ (23)

Predissociation: $AB^* \rightsquigarrow AB^{**} \rightsquigarrow A + B.$ (24)

Fluorescence: $AB^* \rightarrow AB + h\nu.$ (25)

Internal conversion: $AB^* \rightsquigarrow AB.$ (26)

Let us consider k_n to be the total probability for non-ionic radiative and non-radiative decay processes.[23–26]

The total photon absorption cross-section σ_a will be given by

$$\sigma_a = \sigma_k + \sigma_g + \sigma_p + \sigma_i \tag{27}$$

and the total ionization cross-section σ_I by

$$\sigma_I = \sigma_k + \sigma_g + \sigma_p + \sigma_i \left(\frac{k_{if} + k'_{if} + k''_{if}}{k_{if} + k'_{if} + k''_{if} + k_n} \right). \tag{28}$$

For conditions where the ion pair production processes (18) and (22) may be neglected (i.e. assuming $\sigma_p = 0$ and $k''_{if} = 0$), the photoionization efficiency γ_I will simply be given by the ratio σ_I/σ_a.

$$\gamma_I = \frac{\sigma_I}{\sigma_a} = \frac{\sigma_k + \sigma_g + \sigma_i \left(\frac{k_{if} + k'_{if}}{k_{if} + k'_{if} + k_n} \right)}{\sigma_k + \sigma_g + \sigma_i}. \tag{29}$$

Under these circumstances, γ_I will be equal to the photoelectron yield.

However, if ion pair production processes (18) and (22) are significant, the photoionization efficiency as defined earlier is no longer given by σ_I/σ_a. Since these processes do not produce photoelectrons, the ratio σ_I/σ_a will be greater than the photoelectron yield γ_e under these conditions, the latter being given by

$$\gamma_e = \sigma_e/\sigma_a, \tag{30}$$

where σ_a is given by eqn. (27), and σ_e is the sum

$$\sigma_e = \sigma_k + \sigma_g + \sigma_i \left(\frac{k_{if} + k'_{if}}{k_{if} + k'_{if} + k''_{if} + k_n} \right). \tag{31}$$

Photoionization efficiencies less than unity are illustrated in Figs. 8 and 9. Figure 8 shows the absorption and photoionization cross-section functions of H_2 from the work of Cook and Metzger.[32, 73] Both curves show much structure. The threshold for ionization is in the 808 Å region, the ionization cross-section rising

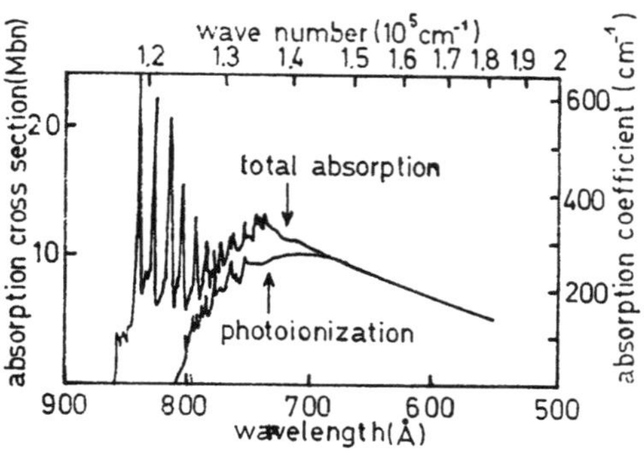

FIG. 8. Absorption cross-section and photoionization cross-section functions for H_2 (after Cook and Metzger[32, 73]).

to about 5.6 Mb at 780 Å. The absorption cross-section is much higher in this wavelength interval so that the corresponding photoionization efficiency γ_I is less than unity. γ_I approaches unity in the 680 Å region.

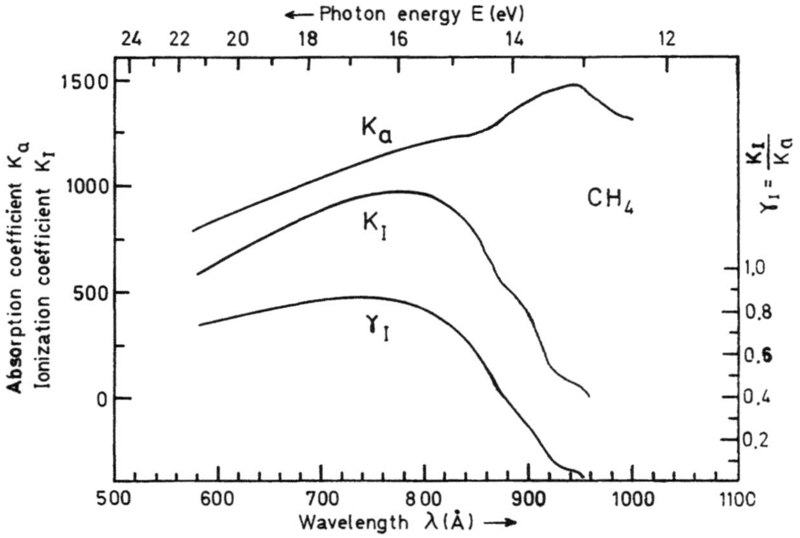

Fig. 9. Absorption and ionization coefficients and efficiencies for CH_4.[33]

Fig. 9 shows the absorption and ionization coefficients and the ionization efficiencies for methane measured by Metzger and Cook.[33]

Thresholds for the various ionization processes were determined with good precision by Chupka.[34] His results are as follows:

$$\lesssim 12.71 \text{ eV} \qquad CH_4 + h\nu \rightarrow CH_4^+ + e^-, \qquad (32a)$$

$$13.50 \pm 0.05 \text{ eV} \qquad CH_4 + h\nu \rightarrow CH_3^+ + H^-, \qquad (32b)$$

$$\sim 14.3 \text{ eV} \qquad CH_4 + h\nu \rightarrow CH_3^+ + H + e^-, \qquad (32c)$$

$$15.19 \pm 0.02 \text{ eV} \qquad CH_4 + h\nu \rightarrow CH_2^+ + H_2 + e^-, \qquad (32d)$$

The process:

$$CH_4 + h\nu \rightarrow CH_2^+ + 2H + e^- \qquad (32e)$$

has a threshold in the neighbourhood of 20 eV.[33]

The threshold appearance potential of process (32a) is probably greater than the adiabatic ionization potential; CH_4 undoubtedly undergoes important structural modifications on ionization so that the 0,0 transition may lie at energies lower than the observed threshold.

In Fig. 9 it can be seen that γ_I is lower than unity over the 600-1000 Å range studied by Metzger and Cook, attaining a maximum of ~ 0.9 at about 16.6 eV.

These results indicate that decay channels of superexcited states other than by autoionization, i.e. by processes (23-26), are important in H_2 and CH_4. The corresponding photoelectron yields are therefore expected to be less than unity at wavelengths above 680 Å for H_2, and within the whole wavelength region studied by Metzger and Cook for CH_4. In addition, ion pair production processes, e.g. (32b) would lower the γ_e values below those for γ_I at appropriate wavelengths.

Complementary data concerning methane are given in Fig. 10.

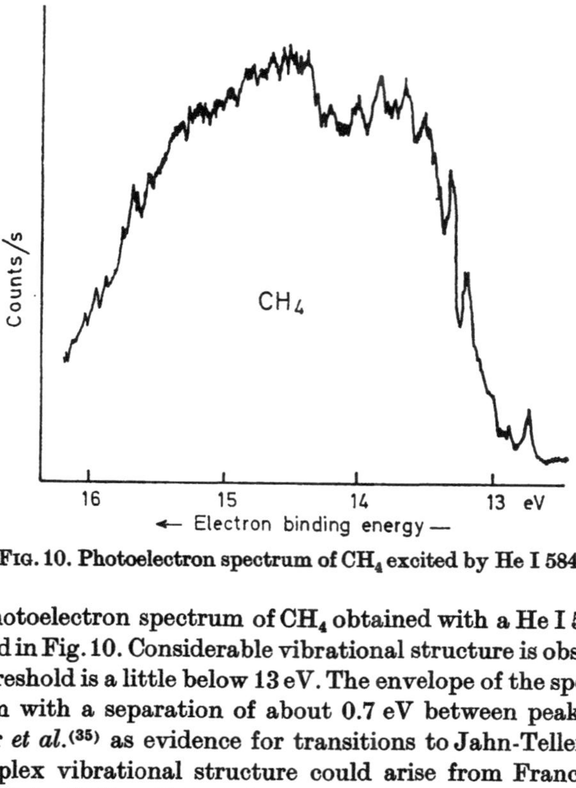

FIG. 10. Photoelectron spectrum of CH_4 excited by He I 584 Å (21.21 eV) photons.[35]

The photoelectron spectrum of CH_4 obtained with a He I 584 Å light source[35] is illustrated in Fig. 10. Considerable vibrational structure is observed in the first band whose threshold is a little below 13 eV. The envelope of the spectrum shows a double maximum with a separation of about 0.7 eV between peaks. This is interpreted by Baker *et al.*[35] as evidence for transitions to Jahn-Teller split states of CH_4^+. The complex vibrational structure could arise from Franck-Condon transitions to these Jahn-Teller distorted states and also from reduction of vibrational symmetries with respect to the T_d symmetry of the neutral CH_4 ground state giving rise to a greater number of independent vibrational modes and frequencies. It is probable that the higher Jahn-Teller state of CH_4^+ is predissociated, decomposing to $CH_3^+ + H$ (process (32c)).

The process (32b) with a threshold photon energy at 13.5 eV is an ion pair production process and should not produce photoelectrons. No evidence is seen in Fig. 10 for modifications in the photoelectron spectrum at the threshold energy of 15.19 eV corresponding to the formation of CH_2^+ by process (32d). However, this is not surprising since the electronic states of CH_4 and CH_4^+ are such that no new channel should exist for photoelectron production at this energy. CH_2^+

formation therefore probably occurs as a result of Herzberg case II predissociation of the ground electronic state of CH_4^+ containing high vibrational energy.

It should be noted that the electron energy spectrum of CH_4, in coincidence with mass analysed photoions (see § 2.2) obtained by Brehm and von Puttkamer[54] with He I 584 Å excitation, does indicate the production of photoelectrons from the CH_2^+ formation process at an onset energy of about 15.2 eV, but with a very low relative yield. These authors observe the process (32c) at an onset energy of about 14.3 eV.

The threshold electron energy measurements of Villarejo *et al.*[36] as a function of photon energy over the 1000-600 Å region (12-20 eV) show no discontinuities corresponding to processes (32c) and (32d) but it should be noted that the electron energy resolution was quite low in these experiments ($\Delta E \sim 130$ meV).

2. Experimental methods

Photoelectron spectroscopy with u.v. sources was introduced in the early 1960's by Vilesov *et al.*[37] and by Turner and Al-Joboury.[38] There has been a rapid development in technique during recent years. This section will be devoted to a review of some of the techniques used and a discussion of the problem of energy resolution in photoelectron spectroscopy. The principal instrumental components are the photon source and the electron energy analyser. These will now be discussed in turn.

2.1. PHOTON SOURCES

Helium atom resonance radiation at 584.334 Å (21.21 eV) has been the principal light source used in photoelectron spectroscopy. It is usually excited by direct current discharge[1, 39] or by microwave powered discharge[40] in flowing helium gas.

It is of interest to examine briefly the spectral emission in such discharges. Figures 11 and 12 show the emission spectra obtained by Samson[41] for helium excited in a Hunter-type discharge tube (4 mm bore quartz capillary) at 400 mA.

The first resonance line of He I $1s^2\,{}^1S \leftarrow 1s2p\,{}^1P$ at 584.33 Å is sixty times more intense than the second resonance line $1s^2\,{}^1S \leftarrow 1s3p\,{}^1P$ at 537.03 Å. The latter will therefore not be an appreciable source of error in photoelectron spectroscopy with a He I source. Other He I resonance lines (Fig. 12) are even weaker. Some weak impurity lines due to excitation of oxygen, hydrogen, and nitrogen are also observed. The most intense impurity line, the Lyman-α line of H I at 1215.67 Å, has a peak intensity about 10% that of He I 584.33 Å. The H I 1215.67 Å line corresponds to an energy of 10.21 eV and could be a source of error, under these conditions, in photoelectron spectroscopy if ion levels exist below this energy.

The spectra shown in Figs. 11 and 12 were obtained at a helium pressure of 0.16 torr. Samson has observed that the intensity of the He I 584 Å line continually increases with pressure over the range 0.06 torr to 0.6 torr, with a con-

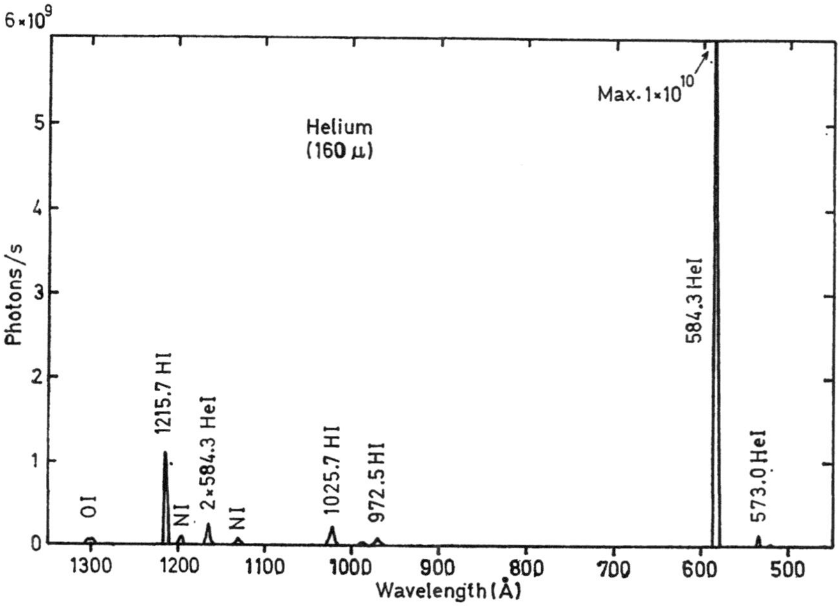

Fig. 11. Helium emission spectrum excited in a Hunter-type discharge tube[41]: wavelength region 450-1350 Å; Helium gas pressure = 0.16 torr; discharge current = 400 mA.

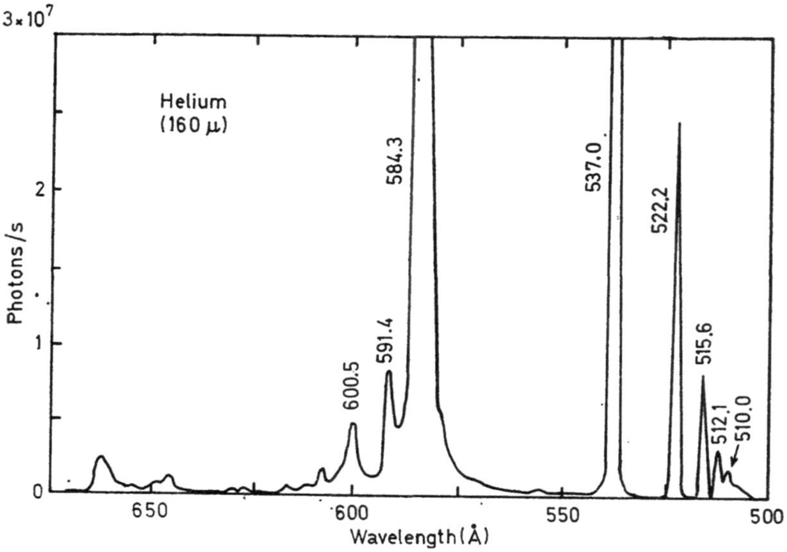

Fig. 12. Helium emission spectrum excited in a Hunter-type dicharge tube:[41] wavelength region 475-675 Å; helium gas pressure = 0.16 torr; discharge current = 400 mÅ.

comitant decrease in the 537 Å and other He I resonance lines. The latter lines are due to emission from He I levels of longer radiative lifetime than the $1s2p^1S$ upper state of the 584 Å transition. Their decrease in intensity with increasing pressure is suggested by Samson to be due to collisional processes intervening during the lifetime of these states, leading to ionization via processes of the type

$$He^* + He \rightarrow He_2^+ + e^-.$$

Other rare gases, pure or mixed with helium, have been excited by direct current or microwave discharges, giving resonance emission lines which may be used as photoionization sources in photoelectron spectroscopy (Table 1).

Discharges through hydrogen produce strong emission of the Lyman-α H I line at 1215.67 Å (10.21 eV) and the much less intense Lyman-β line at 1025.72 Å (12.09 eV).

Price[16] has been successful in using the He II $1s^2S \leftarrow 2s^2P$ resonance line at 303.78 Å (40.8 eV) as a light source in photoelectron spectroscopy. Appreciable emission from the helium ion was obtained by operation at lower helium pressures and greater excitation power than for optimum production of the He I 584 Å line. However, the peak intensity of He II 303.78 Å is rather smaller than for He I 584 Å under the previous conditions.

TABLE 1. RESONANCE SOURCES FOR PHOTOELECTRON SPECTROSCOPY

Resonance source	Wave-length (Å)	Photon energy (eV)
He I	584.334	21.21
Ne I	735.895	16.85
	743.718	16.67
Ar I	1048.219	11.84
	1066.659	11.62
Kr I	1164.867	10.64
	1235.838	10.03
Xe I	1295.586	9.57
	1469.610	8.44

The wide doublet structure of rare gas resonance emission, in the case of Ne, Ar, Kr, and Xe, is a disadvantage from the viewpoint of energy discrimination in photoelectron spectroscopy. This can be overcome by using v.u.v. monochromators to isolate particular lines but with a loss in useful photon intensity.

The discussion in § 1 of this chapter has demonstrated the interest of carrying out photoionization and photoelectron spectroscopy studies as a function of photon energy. This may be accomplished by using multiline or continuum sources and filtering the exciting radiation with a monochromator, or by simply using a set of resonance line sources. In the latter case the photon intensity will be relatively high, but the number of different wavelengths available (Table 1) is severely restricted.

Vilesov *et al.*,[37] who measured the photoelectron spectra of aromatic molecules, used a hydrogen continuum source and a vacuum monochromator. In their instrument a lithium fluoride window separated the monochromator from the ionization chamber so that work could be done only with photons of less than about 11 eV in energy.

Monochromator studies with photons of higher energies require windowless systems in which differential pumping is used to isolate the photon source, monochromator, and ionization chamber. As examples, can be mentioned Comes and Sälzer[42] and Doolittle and Schoen[43] who used pulsed sources emitting multiline spectra; Blake and Carver[44] who worked with a capillary discharge through helium gas at 50 torr at which pressure the 615-900 Å Hopfield continuum is emitted; Berkowitz and Chupka[29] and Villarejo *et al.*[25, 45] who used pulsed high-voltage helium sources emitting the Hopfield continuum.

Schoen[46] has done photoelectron spectroscopy, using a monochromator, at the wavelengths given in Table 2. These correspond to lines emitted in repetitive high-voltage sparks through air at a few hundred microns pressure in a ceramic capillary.

TABLE 2. WAVELENGTHS AT WHICH PHOTOELECTRON DATA WERE OBTAINED BY SCHOEN[46]

λ (Å)	Energy (eV)	λ (Å)	Energy (eV)
990	12.5	703	17.6
980	12.7	686	18.1
955	13.0	630	19.7
924	13.4	625	19.8
834	14.9	617	20.1
790	15.7	609	20.4
780	15.9	598	20.7
771	16.1	583	21.3
760	16.3	554	22.4
746	16.6	526	23.6
719	17.2	508	24.4

The radiation intensities obtained in monochromator studies are much lower than in resonance lamp work so that the smaller number of photoelectrons produced require sophisticated detection techniques in order to overcome signal-to-noise ratio problems.

An example of such sophisticated techniques is provided by the work of Berkowitz and Chupka.[29]

2.2. ELECTRON ENERGY ANALYSERS

There are two general classes of electron energy analyser used in photoelectron spectroscopy:

(a) the retarding field type, with which one measures the total electron current as a function of a retarding potential;

(b) the electron deflection type, in which the dispersing element is a magnetic or an electric field and which gives the electron current as a function of electron kinetic energy so that the spectrum appears as a series of peaks.

The resolution P_R of an electron energy analyser is defined as $E/\Delta E$, where E is a "monochromatic" electron energy and ΔE the half-width of the experimental energy distribution recorded by the analyser.

Retarding field analysers

In the retarding field analyser the electrons admitted are subjected to a retarding field and their kinetic energy is determined by the height of the maximum potential barrier that they overcome. This measures the momentum of the electrons in a perpendicular direction to the equipotential lines of the field. The *total* kinetic energy will therefore be determined only if the trajectories of the electrons are normal to the equipotential lines. If this is not so there is a loss of energy resolution. Obtaining maximum electron current and maximum energy resolution involves additional general design problems.[48] These include minimization of space charge effects, which may be particularly severe for low electron energies, and electron reflection at metal surfaces.

A *cylindrical grid electron energy analyser* was used by Al-Joboury and Turner to measure the ionization potentials of a large number of molecules.[39, 49] This consisted of two cylindrical grids and a cylindrical collector electrode which were each coaxial with the light beam. A suitable detector, in this case a vibrating reed electrometer, measured the electron current.

With zero retarding potential V_R between the grids, and an electron collection potential of about 20 V between the outer grid and the collector electrode, all the photoelectrons produced by interaction between the light beam and the sample gas reach the collector electrode. When V_R is regularly increased from zero, the photoelectron current decreases sharply as each group of electrons of different kinetic energy just fails to surmount the potential barrier and is so prevented from reaching the collector electrode. One therefore measures a photoelectron stopping curve expressed as electron current i as a function of $(h\nu - V_R)$.

A plot of the differential di/dV_R as a function of $(h\nu - V_R)$ gives a series of peaks whose positions on the energy scale correspond to ionization potentials and whose amplitudes are proportional to photoelectron cross-sections.

The energy half-width resolved in this instrument was of the order of 100 meV. This was insufficient for accurate determination of the vibrational levels of the ions produced by photon collision.

One of the principal sources of loss of resolution in the cylindrical grid device is that the electrons are not all ejected in a perpendicular direction to the retarding grid. Some improvement in resolution was made by the use of slotted cylindrical grids which confine the distribution of photoelectrons accepted by the collector electrode to those emitted normal to the photon beam and along the radius of the cylindrical retarding field.[16] However, the photoelectron currents are then smaller than in the mesh type grid used by Al-Joboury and Turner.

Increased resolution can be obtained by dispensing with the outer grid and applying the retarding potential between the remaining grid and the collector electrode.(50) This decreases electron scattering between the central ionization region and the collector electrode. In this case the positive ions reach the collector, in contrast to the two-grid system in which they are repelled by a suitable positive potential between the grids. This is claimed to be only a minor disadvantage

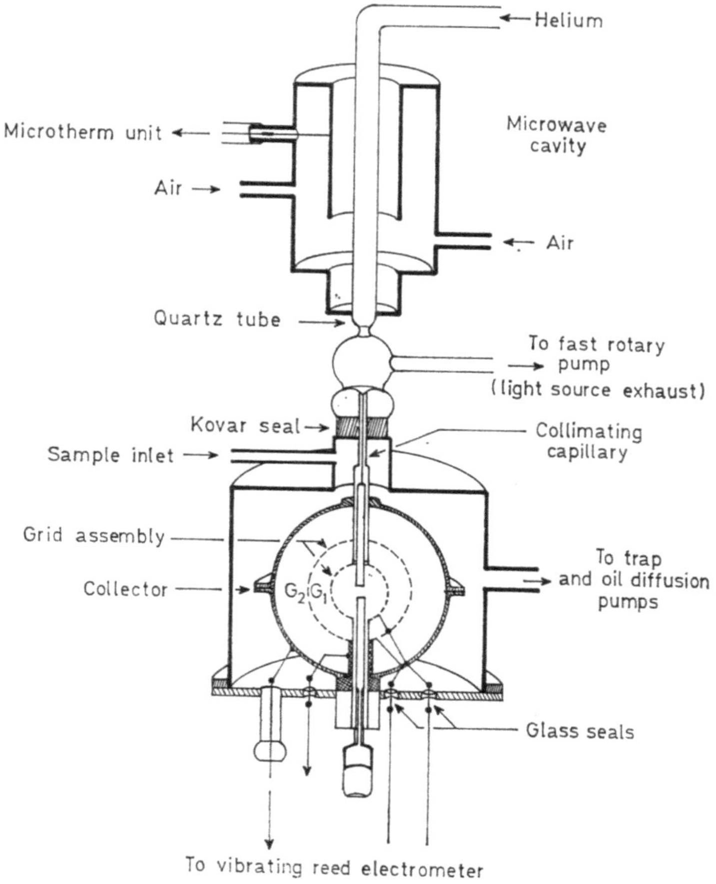

FIG. 13. Microwave He 584 excitation source and spherical grid retarding potential electron energy analyser.(40)

since the positive ions create a constant background which can be allowed for. Single-grid operation improves electron energy resolution by a factor of two.(50, 51)

In the single-grid device of Marmet *et al.*,(52) the signal was differentiated directly in the ionization chamber and a lock-in amplifier detector was used.

Use of ***spherical retarding grids*** and collector further improves energy resolution. Frost *et al.*[40, 53] have developed the instrument shown schematically in Fig. 13.

The photon source was a microwave powered discharge through flowing helium gas at about 0.02 torr pressure. The light beam enters the photoionization chamber through a gold-plated brass tube, 5 mm in diameter, which is equipotential to the inner grid G_1. A similar tube provides an exit for the light beam and is so

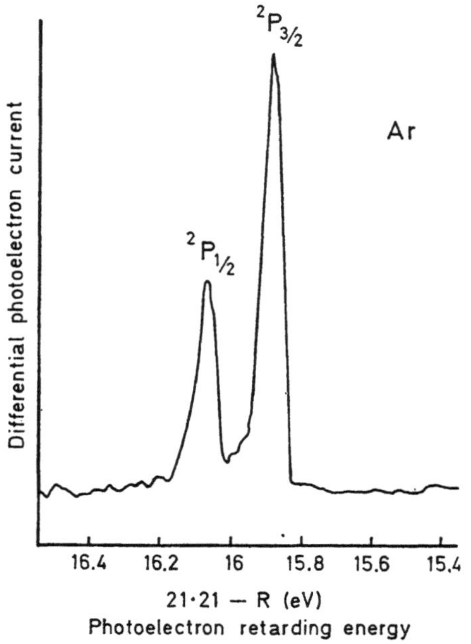

FIG. 14. First differential of the photoelectron stopping power curve for argon excited by He I 584 Å photons: measurements by Frost *et. al.*[40]

arranged as to delimit a small interaction zone at the centre of the spherical grid system. In this way the electrons collected are those photoejected normal to the retarding field.

Each of the electrodes of the photoelectron energy analyser was constructed as two separate hemispheres. The grids G_1 and G_2 were made of 30×30 mesh 0.127 mm gauge brass and were gold-plated after being electrolytically reduced in gauge to an optical transparency of 75%. The spherically symmetric electron collector was also made of brass. The internal diameter of grid G_1 was 38 mm, of G_2 51 mm, and of the collector electrode 76 mm.

Positive ions were stopped from entering the electron retarding region by applying a constant potential difference such that G_2 was 3 V positive with respect to G_1. The photoelectron retarding potential V_R was applied between G_1 and the collector electrode. Electron currents were of the order of 10^{-11} to 10^{-10}A as measured with a Cary model 31 vibrating reed electrometer.

After each scan as a function of V_R the run is repeated with no sample gas present. The measured background is then subtracted from the original signal to obtain the true photoelectron spectrum.

For highest resolution it was found important to minimize the magnetic field within the apparatus by shielding the analyser with several layers of Mumetal. The resolution obtained with this instrument is illustrated by the differential retarding curve for argon[40] given in Fig. 14.

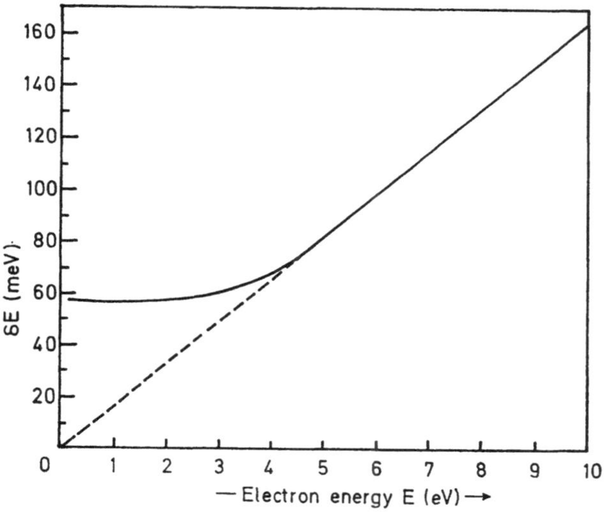

FIG. 15. Spherical grid retarding potential electron energy analyser of Samson and Cairns[15]. Total energy spread δE of the electron energy analyser as a function of energy E.

The half width of the $^2P_{1/2}$ peak is 0.062 ± 0.002 eV at an electron energy of 5.1 eV. This is equivalent to a resolving power $P_R \sim 82$.

With a similar spherical grid analyser, but using a monochromatized light source and an electron multiplier as detector, Samson and Cairns[15] obtained a resolving power $P_R \sim 125$ for electron energies greater than 4 eV. For lower energies the total energy spread δE tended to a constant value of 56 meV (Fig. 15) which corresponds to a full-width half-height resolution of $\Delta E = 28$ meV.

Still higher resolution can be obtained with the retarding field method by using an electrostatic einzel lens to collimate the photoelectrons into a parallel beam and analysing their energies by a plane-retarding field. In the instrument described by Brehm and von Puttkamer,[54] a half-width resolution of 13 meV was obtained. Using a molecular beam of argon as a target, and counting the electrons by means of a 400-channel analyser Rose *et al.* were able to improve the full-width half-height resolution to 10 meV.[55]

It should be noted that the instrument of Brehm and von Puttkamer[54] can be used not only to measure photoelectron energies *per se* but also to measure them in coincidence with mass-analysed photoions. Under these conditions the low signal level, about 3 coincidences per second, requires an elaborate detection system and limits considerably the sensitivity of the instrument. Optimum operating conditions for coincidence work require the use of a draw-out field of about 1 V cm^{-1} in the ion source. This results in a limiting resolution of about 100 meV in the electron spectrometer due to the finite depth of the helium radiation beam in the direction of the draw-out field.

A similar instrument has been used for photoelectron and photoion coincidence measurements by Berkowitz and Chupka.[29] The photon source is a helium continuum, and the wavelengths (0.12-0.4 Å band-width) are selected with a 1 meter normal-incidence monochromator with a 1200 line per mm diffraction grating. A Bendix channel multiplier was used as a photoelectron detector. The full-width half-height resolution of electron energy was about 50 meV.

Electron deflection energy analysers

In deflection analysers the ionization chamber and the electron energy analyser are separated by a slit or a circular aperture. This arrangement permits the pressure in the analyser to be kept much lower ($\sim 10^{-5}$ torr) than in the ionization chamber (10^{-2}-10^{-1} torr). Chemical deterioration of the analyser by the vapour under study is therefore reduced. Electron multipliers can be safely used as detectors in these circumstances.

Turner and May[56] have developed a 180° *magnetic field deflection analyser*, schematically illustrated in Fig. 16.

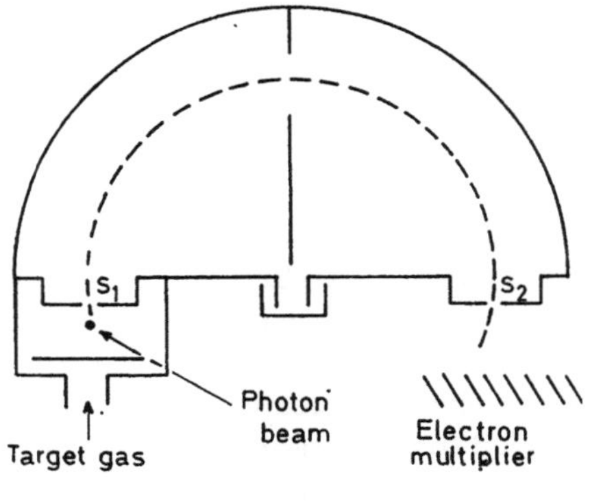

FIG. 16. Schematic representation of a 180° magnetic field deflection electron energy analyser.[56]

A beam of 21.21 eV photons from a helium resonance lamp is directed into the ionization chamber in a direction parallel to the electron velocity analyser entrance slit S_1 and 1.5 mm away from it. The geometry of the system is such that only those electrons photoejected radially within an arc of $\pm 7°$ perpendicular to the light beam can pass through the exit slit S_2 after entering S_1 and selection by suitable deflection by the magnetic field. The latter is directed parallel to the photon beam. The slits are each 0.2 mm wide and 8 mm high. The radius of the electron orbit is 20 mm. A 17-stage electron multiplier is used to detect the photoelectrons emerging from S_2. On increasing the magnetic field strength, electrons of increasing kinetic energy pass successively through S_2. Defocusing effects due to the earth's magnetic field are reduced by appropriate orientation of the axis of the analyser and by compensating magnetic fields.

The resolution obtained with this instrument is $P_R \sim 40$ for electrons of 5.1 eV energy. The energy peaks are much more symmetrical than those obtained by differentiation of the signal in the grid retarding field analyser used by Al-Joboury and Turner.[39] However, this instrument has the disadvantage of introducing large discrimination effects since the collection efficiency varies considerably with electron kinetic energy.

Double focusing magnetic field electron spectrometers have been designed and built, principally for ESCA work, by Siegbahn *et al.*[1] Such instruments have high luminosity, defined as the product of source area and transmission.

Electrostatic deflection analysers are now becoming extensively used in photoelectron spectroscopy. Turner[24] has developed an analyser in which 127° deflection of electrons takes place in a cylindrical electrostatic field. The principal orbit radius of the electrons is 10 cm. Limitations in energy resolution in earlier forms of the cylindrical condenser analyser were overcome by using split sectors and a set of correcting electrodes which are used to correct the electron trajectories. Cuspiform slits were also used to reduce field distortion resulting from accumulation of static charges on the slit system.

The half-width energy resolution obtained with this instrument was 15 meV for the $^2P_{3/2}$ argon ion peak at an electron energy of 5.455 eV. At 9 eV a resolving power of $P_R = 450$ was obtained for the $^2P_{3/2}$ xenon ion peak.[57]

A commercial instrument is available (Perkin-Elmer) which uses a He I 584 Å line source and analyses the photoelectrons by means of a cylindrical electrostatic field which deflects the electrons through 127°. The electron flux is detected by an electron multiplier which is coupled to an amplifier and a rate-meter.

Siegbahn *et al.* have built a double focusing electrostatic field deflection analyser, again mainly for ESCA work.[1] A sector angle of 157.5° was chosen for a practical optimum ratio of transmission to resolution. The electrostatic field is produced between sectors of two concentric spherical aluminium electrodes whose radii are 32 and 40 cm respectively. The theory and problems encountered in the design of spherical electrostatic field deflection analysers have been discussed by Siegbahn *et al.*[1] and by Simpson.[58] It is probable that spherical and hemispherical electrostatic field deflection analysers will be considerably used in photoelectron spectroscopy in the near future since they are capable of high transmission and high resolution.[58–60]

A *parallel plate electrostatic field* electron energy analyser has been built for photoelectron spectroscopy work by Eland and Danby[61] and is schematically shown in Fig. 17. In this instrument the photoelectrons are collimated by slits S_2 and S_3 so as to enter the electrostatic analyser at an incidence angle of 45° via slit S_4. The angular divergence, which must be kept limited because of the weak focusing power of the parallel plate analyser, is essentially controlled by the width of slit S_4. Under the influence of the constant opposing field, the photoelectrons follow a parabolic trajectory. The velocity selected electrons pass through slits S_5 and S_6 after which they are accelerated by a potential of about 200 V before striking the first dynode of a 19-stage particle multiplier (EMI 9641/2B). Magnetic screening is achieved through the use of Mumetal.

The photoelectron spectrum is obtained either by varying the potential applied to the analyser or by adjusting the analyser for a fixed electron energy and then varying the draw-out potential on the plate carrying slit S_2. Optimum conditions for routine operation were found to be a fixed draw-out potential of 2.5 V and scanning the spectrum by varying the analyser potential.

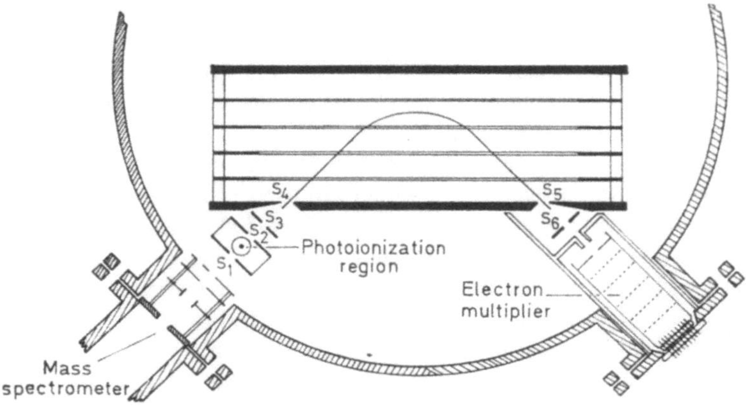

FIG. 17. Schematic representation of a parallel-plate electrostatic field electron energy analyser. The instrument built by Eland and Danby[61] is designed for obtaining photoelectron and photoion spectra in coincidence. S_1 is the photoion source slit, S_2 the photoelectron source slit, S_3 and S_6 are the 45° angle defining slits, S_4 and S_5 are adjustable analyser slits.

The highest resolution obtained was a half-width ΔE of 27 meV at an electron energy of 10 eV. The resolving power $P_R \sim 370$ thus corresponds to $\Delta E \sim 15$ meV at an electron energy of 5.1 eV.

It was found that the apparent electron energy for a particular ionization process decreased notably with decreasing gas pressure in the ionization chamber, due to positive ion space charge. The photoionization chamber has since been modified so as to reduce considerably these effects without loss of resolution. An energy resolution of $\Delta E = 20$ meV was then obtained over the whole electron energy range.[62]

2.3. ENERGY RESOLUTION IN PHOTOELECTRON SPECTROSCOPY

The experimental parameters determining energy resolution in photoelectron spectroscopy will now be examined. These parameters are:

(a) line width of the exciting radiation;
(b) translational motion of the absorbing species;
(c) resolving power of the electron energy analyser.

Line width of exciting radiation

In the case of a *continuous source* the effective energy spread of the radiation which ionizes the target atoms or molecules will depend on instrumental factors in the monochromator used for isolating a particular band of radiation.

For *line sources*, the principal factors responsible for the photon energy distribution are Doppler effects due to thermal motion of the emitting atoms, the non-thermal recoil of the atom as it emits, and pressure broadening effects. The latter include van der Waals broadening due to collisions with foreign gases, resonance broadening resulting from collisions with atoms of the same kind, and Stark effect broadening which occurs through interaction of the emitting atom with electrostatic fields of charged particles in the electric discharge. An additional broadening effect, particularly marked for resonance lines, can occur in the event of self-absorption. These effects are cumulative and they cause the observed line-width to be greater than the natural line-width for resonance radiation of atoms, the latter being of the order of 10^{-2} cm^{-1} ($\sim 10^{-6}$ eV) for spontaneous emission in the far u.v.

The recoil energy W_r taken up by the emitting atom is given by

$$W_r = \frac{(h\nu)^2}{2Mc^2}, \tag{33}$$

where M is the mass of an isolated atom, c is the velocity of light, and ν the emitted frequency. In losing the recoil energy W_r, the emitted photon will undergo a red shift $\Delta\nu_r$. The relative recoil shift $\Delta\nu_r/\nu$ for an atomic weight of unity (the most unfavourable case) and a photon energy of 10 eV will be only 10^{-8}, and so the red shift will be negligible in the u.v. region.

The thermal Doppler half-width $\Delta\nu_D$ of the emitting atom radiation is given by

$$\Delta\nu_D = 2\nu\left[2\ln\frac{2kT}{Mc^2}\right]^{1/2} = 7.16\times 10^{-7}\,\tilde{\nu}\,\sqrt{\left(\frac{T}{M^*}\right)}, \tag{34}$$

where $\tilde{\nu}$ is the peak radiation frequency in cm^{-1} and M^* the atomic weight. T is the temperature for a thermal equilibrium corresponding to a statistical distribution of velocities of the radiating atoms given by the Maxwell-Boltzmann relation. For helium 584 Å radiation at a gas temperature of 300 K, $\Delta\nu_D = 1.04$ cm^{-1} (6.54×10^{-5} eV).

In the resonance emission sources used so extensively in photoelectron spectroscopy, resonance broadening will generally be much more important than van der Waals broadening. For He I 584 Å radiation emitted by a source at a pressure of 1 torr, the half-width due to resonance broadening would be of the order of 3×10^{-7} eV and would thus be several orders of magnitude smaller than the Doppler broadening half-width $\Delta\nu_D$.

Interatomic Stark broadening is due to effects on the radiating atoms from the electric fields of ions or electrons in the discharge source. These effects are most marked for emission lines which are subject to first order Stark splittings and displacements in an electric field than for lines exhibiting only second-order Stark displacements. Since the electric field strength is statistically distributed over all emitting atoms, Stark line broadening will reflect the probability distribution of the various frequency displacements. Estimations of Stark broadening would be extremely difficult to achieve under the experimental conditions prevailing in line sources used for photoelectron spectroscopy. However, the relative importance of Stark broadening can be determined by measuring emission line widths as a function of charged particle density in experiments where the discharge current is varied at a constant voltage.

Broadening due to self-absorption can become very important even at pressures at which pressure broadening is negligible. The amount of self-absorption is proportional to the square root of N, the number of like atoms in the beam path. N is directly proportional to the product of the length of the beam path and the vapour pressure of the absorbing gas. Self-absorption will be maximum at the resonance frequency peak and fall off rapidly at higher and lower frequencies. The effect is to reduce the total emission intensity observed and to increase the apparent half-width of the emission line. As N increases, self-reversal can occur manifesting itself as a dip in the centre of the emission line profile.

Samson[63] has examined the profile of the He I 584 Å resonance line emitted by four different line sources. The measurements were made with a 1 m vacuum monochromator operating in the fifth order with a 1200 line per mm grating. Instrumental resolution was about 0.04 Å ($\sim$ 12 cm^{-1}; $\sim$ 1.5 $\times$ 10^{-3} eV).

A detailed description of the four line sources has been given by Samson.[41] Their principal parameters as used in the line profile experiments are given in Table 3.

TABLE 3. CHARACTERISTICS OF HELIUM SOURCES STUDIED BY SAMSON[63]

He source	Length of capillary of discharge l (cm)	Diameter of capillary (mm)	Length of neutral gas volume[a] L (cm)	Discharge conditions
A: dc glow with uncooled capillary	7.6	4	1.9	250 W, 500 mA
B: microwave discharge, aircooled	4.6	10	4.4	100 W, 2450 MHz
C: dc glow with water-cooled capillary	10.2	6	2.5	250 W, 500 mA
D: condensed spark with watercooled capillary	15.2	10	4.8	15 kV (ac), 50 mA, 0.005 μF

[a] L is the length of the column of unexcited helium lying between the entrance slit of the monochromator and the capillary discharge.

The line profiles shown in Fig. 18 were observed with lamp *C* at helium gas pressures of 0.2, 0.4 and 0.8 torr respectively. They indicate the increasing pressure dependence of the self-absorption due to unexcited helium in the region between the capillary discharge and the monochromator entrance slit.

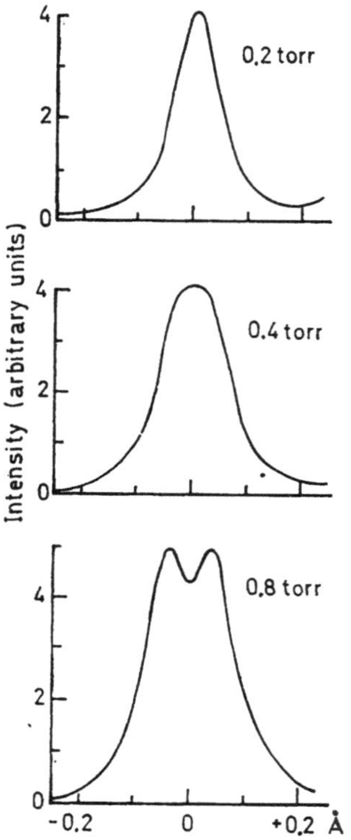

FIG. 18. Profiles of the He I 584 Å resonance emission observed for lamp *C* at helium gas pressures of 0.2, 0.4 and 0.8 torr.[63]

Data on the observed half-widths of the 584 Å resonance line as a function of pressure are given in Fig. 19 for the four lamps. The microwave discharge lamp *B* gave the narrowest line over much of the pressure range. Comparison of the data for the four lamps shows that the rate of increase of line half-width with pressure is greater with a greater length of neutral gas between the capillary and the monochromator slit, indicating the importance of self-absorption in this region. The line half-width is also roughly inversely proportional to the length of the discharge capillary. The diameter of the capillary and whether it is cooled or uncooled do not seem to be important variables. Stark broadening appeared to be negligible in determining observed line-widths since with the glow discharge source there was no effect on measured line-widths when the current was varied from 100 to 500 mA at constant voltage across the discharge.

From these results Samson concluded that the line-width is determined by a combination of natural, Doppler, and self-absorption broadening, the latter occurring both within the capillary and in the non-excited region between the capillary and the monochromator entrance slit.

The smallest line half-width of about 1 meV was obtained with the microwave powered lamp B at a pressure of 0.2 torr but the radiant energy was not

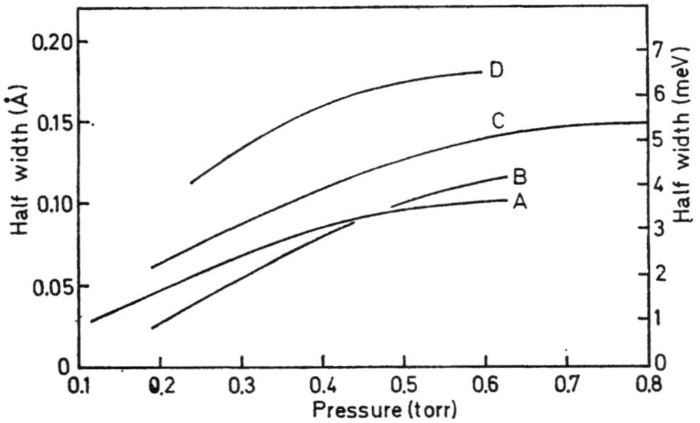

Fig. 19. Observed half-widths of the He I 584 Å resonance emission line as a function of pressure for lamps A, B, C and D. The characteristics of these lamps are given in Table 8.3.

high. The best compromise between the various parameters would be a d.c. glow discharge with a capillary length of about 30 mm and so designed as to have a minimum length of unexcited gas between the capillary and the entrance slit of the monochromator or photoelectron spectrometer.

Translational motion of absorbing species

In the non-thermal recoil of species which eject photoelectrons, the recoil energy carried off by the ion is determined by the principle of conservation of momentum. The maximum recoil energy will occur when the recoil is in the direction of the incident photon. For conditions occurring in low-energy photoelectron spectroscopy, the situation is a non-relativistic one and the maximum fraction of photon energy carried off as recoil energy by the newly formed ion is

$$\frac{m}{M} \times \frac{\frac{1}{2}mv^2}{h\nu},$$

where m and M are the mass of the photoelectron and the ion respectively, v the velocity of the ejected photoelectron, and $h\nu$ the energy of the incident photon. This fraction will be less than 5×10^{-8}; in the case of a hydrogen molecule ionized by 21.21 eV photons, the ion recoil energy will have a maximum value of 1.5×10^{-4} eV. It is clear that the ion recoil energy can be neglected in present-day photoelectron spectroscopy experiments where instrumental energy resolution is no better than 10^{-2} eV.

The thermal translational motion of the absorbing species gives rise to a spread of the energy of ejected photoelectrons. For a Maxwell-Boltzmann distribution of molecular velocities the maximum of the distribution curve corresponds to a velocity

$$v_M = \sqrt{\left(\frac{2kT}{M}\right)}.$$

If the gas molecules are assumed to absorb monochromatic radiation, then the emitted photoelectrons will have an energy spread whose half-width ΔE_t is given by

$$\Delta E_t = \left(\frac{2m^2v^2T}{M}\right)^{1/2} \tag{35}$$

where m, v, T and M are as previously defined.

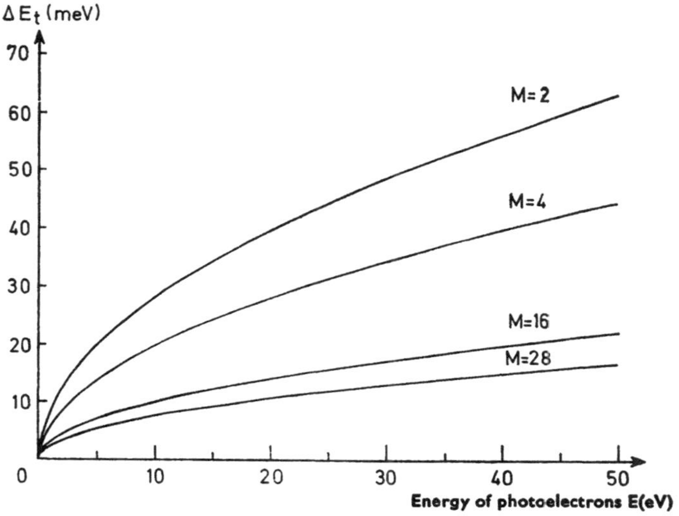

FIG. 20. Effect on electron energy resolution due to thermal translational motion of the photoionized species: half-width ΔE_t of the photoelectron energy spread as a function of photoelectron energy E. The curves are for mass units = 2 (H_2), 4 (He), 16 (CH_4) and 28 (N_2; CO).

Values of ΔE_t as a function of the energy of the ejected photoelectrons for a gas temperature of 300 K are given in Figs. 20 and 21 for various gases.[63] In practice, for aromatic molecules ($M^* \gtrsim 80$) with a minimum ionization potential of 7 eV, ΔE_t will be less than 10 meV for room temperature experiments with a He I 584 Å source. For H_2 however, ΔE_t will have a maximum value of 21 meV ($\sim$ 170 cm^{-1}) for electrons produced by He I 584 Å photons, and 44 meV ($\sim$ 350 cm^{-1}) with He II 304 Å photons. These values are greater than present-day resolution in photoelectron spectrometers.

If one is interested in resolution or partial resolution of rotational structure, then it will be necessary to reduce ΔE_t as much as possible. This could be achieved by working at low temperatures or through the use of molecular beams. The practical problems involved will certainly be studied in the near future. Some work using multichannel molecular jets has been carried out by Rose *et al.*[55]

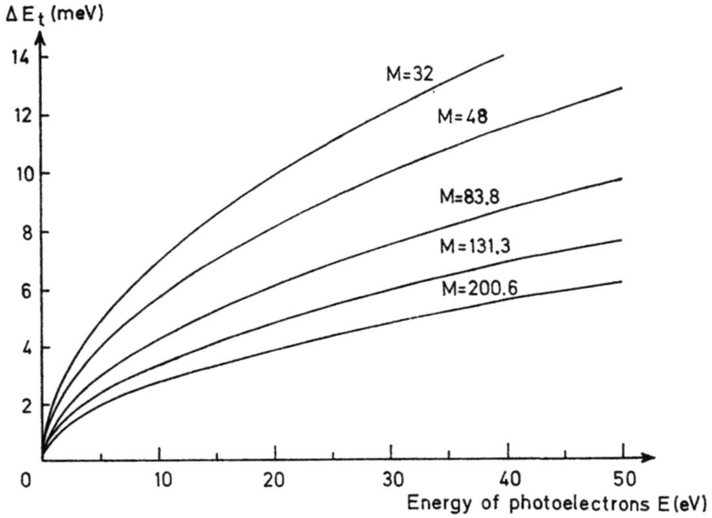

Fig. 21. Effect on electron energy resolution due to thermal translational motion of the photoionized species: half-width ΔE_t of the photoelectron energy spread as a function of photoelectron energy E. The curves are for mass units = 32 (O_2), 48 (O_3), 83.8 (Kr), 131.3 (Xe) and 200.6 (Hg).

Electron energy analyser resolving power

(i) *Theoretical resolving power: analyser geometry parameters.* The maximum theoretical resolving power of the different types of electron energy analyser used in photoelectron spectroscopy can be defined in terms of geometrical parameters of the analyser.

(a) For cylindrical and spherical retarding potential analysers, the theoretical resolving power is

$$P_R = \frac{E}{\Delta E} = \frac{R^2}{r^2}, \tag{36}$$

where r is the radius of the ionizing volume and R the radius of the first grid. Equation (36) was first obtained by Lukirsky.[64] The theoretical analysis was extended by Soboleva[65] for the case of non-spherical electron sources in a spherical analyser. Her results showed that for disc- or cube-shaped sources, ΔE can increase by as much as 20%. Simpson[48] has discussed the further theoretical limitations imposed by a hemispherical geometry.

(b) For electrostatic and magnetic deflection analysers, the resolving power depends not only on slit widths and electron orbit but also on the angular distribution of the electrons at the entrance slit.

For an electrostatic or magnetic deflection analyser of the 180° or 127° type, the ideal resolving power is given by[1]

$$P_R = \frac{4R_e}{s_1 + s_2}, \tag{37}$$

where R_e is the mean central orbit radius of the electron, s_1 and s_2 the entrance and detector slit widths respectively.

For a parallel plate electrostatic deflection analyser, the resolving power in the ideal case where all the electrons enter at an incidence angle of 45° is given by[66]

$$P_R = \frac{x}{S_1 + S_2},$$

where x is the distance between entrance and exit slits whose widths are S_1 and S_2 respectively.

(ii) *Experimental limitations to electron analyser resolving power*. The theoretical resolving powers given in the above discussion are not achieved in practice. The experimentally determined resolving power has been given for various instruments in previous sections, and some of the factors responsible for deviation from the ideal values have been mentioned. These factors are recapitulated below.

In the case of cylindrical and spherical retarding potential analysers an important factor in loss of theoretical resolving power is the existence of transverse components of the electron kinetic energy.

The energy resolution of deflection analysers depends not only on the geometrical factors given above but also on the angular distribution of electrons at the entrance slit. These instruments also suffer from aberrations in the electron optics which may be reduced by suitable design.

Reflection of electrons at metal surfaces and production of secondary electrons can be troublesome, especially since the reflection coefficient increases with incident electron energy. These effects can be reduced by suitable coating of the metal surfaces, e.g. by gold blacking[67] and by special design of the electron collector.

Space charge effects lead to electron-electron repulsion and consequent spreading of the electron beam. The defocusing effects of space charge can be reduced by careful design of the analyser and limiting the electron current transmitted through the instrument.

Other important factors in limiting the theoretical performance of electron energy analysers are: field penetration effects, especially when draw-out fields are used in the ionization chamber; contact potential effects; perturbing magnetic field and field gradients which must be compensated for.

Detailed discussions of the problems of energy resolution in electron energy analysers have been given by Siegbahn *et al.*,[1] by Simpson *et al.*,[58–60] and by Harrower.[66] Specific design problems in instruments built for photoelectron spectroscopy are considered in the references relevant to the instruments discussed in previous sections, in particular § 2.2.

The question of transmission of electron energy analysers has been mentioned only incidentally in this review. In practice a compromise between maximum transmission and maximum energy resolution must be made. The type of detector is therefore also a critical factor. Modern counting techniques and statistical analysis of data play a considerable role in recent instruments, especially for threshold experiments.[29, 44]

3. Applications of photoelectron spectroscopy

Although this review is mainly concerned with the principles and instrumentation of photoelectron spectroscopy with far u.v. light sources, it is appropriate to conclude with a brief discussion of applications of this technique. Some of the physical and spectroscopic applications have already been referred to in detail in earlier parts of this chapter.

There are three main fields of application of photoelectron spectroscopy. The first concerns *physical* processes such as photoionization, autoionization, and the angular distribution of photoelectrons. The second involves *spectroscopic* processes and information: ionization potentials, electronic energy levels and electron configurations of ions, spin-orbit coupling, vibrational frequencies of molecular ions, molecular ion geometries as determined from the Franck-Condon principle, orbital energies. It should be mentioned that particular care is necessary in the determination of orbital energies from the ionization potentials measured by photoelectron spectroscopy since Koopmans' theorem may not always be valid.[68]

A third domain of application is to *chemistry*. Investigations have been made of systematic trends in orbital energies and symmetries as related to molecular structure, in particular in relation to conjugation and substitution effects.[16, 47, 70, 71] Information obtained from photoelectron spectroscopy can be useful in radiation chemistry and in mass spectrometry, in particular for elucidating fragmentation processes of molecular ions. Photoelectron spectra may be used as "fingerprints" in analytical chemistry.[70] Possible applications of photoelectron spectroscopy to the study of species bonded in surface layers have been discussed and preliminary experiments carried out by Bordass and Linnett.[72] Further developments along this line could lead to useful information concerning intermediates in surface catalytic processes.

Acknowledgement

This review was begun at the National Research Council in Ottawa, Canada, where the author spent the summer of 1969. Thanks are due to Dr. Gerhard Herzberg, Dr. Alec Douglas, and Dr. D. A. Ramsay for having made this visit possible and for their hospitality.

References

1. Siegbahn, K., Nordling, C., Fahlman, A., Nordberg, R., Hamrin, K., Hedman, J., Johansson, G., Bergmark, T., Karlsson, S. E., Lindgren, I., and Lindberg, B., ESCA, Atomic, molecular and solid state structure studied by means of electron spectroscopy, *Nova Acta R. Soc. Scient. upsal.* ser. IV, **20**, (1967).
2. Branscomb, L. M., in *Atomic and Molecular Processes* (ed. D. R. Bates) Academic Press, New York, 1962, chap. IV, pp. 100-40.
3. Berry, R. S., *Proc. VIème Conf. Int. Phénomènes d'ionization dans les gaz,* CIPIG, Paris, 1963, vol. 1, p. 13.
4. Cooper, J., and Zare, R. N., *J. Chem. Phys.* **48**, 942 (1968).
5. McGowan, J. W., Vroom, D. A., and Comeaux, A. R., *J. Chem. Phys.* **51**, 5626 (1969).
6. Lowry, J. F., Tomboulian, D. H., and Ederer, D. L., *Phys. Rev.* **137**, A 1054 (1965).
7. Ederer, D. L., and Tomboulian, D. H., *Phys. Rev.* **133**, A 1525 (1964).
8. Hasted, J. B., *Physics of Atomic Collisions*, Butterworths, London, 1964, p. 330.
9. Samson, J. A. R., *Advances in Atomic and Molecular Physics* (ed. D. R. Bates and I. Estermann), Academic Press, New York, 1966, vol. 2, p. 192.
10. Fano, U., and Cooper, J. W., *Phys. Rev.* **137**, A 1364 (1965).
11. Herzberg, G., *Molecular Spectra and Molecular Structure*, vol. II, *Diatomic Molecules*, van Nostrand, New York, 1950, p. 419; vol. III, *Electronic Spectra of Polyatomic Molecules*, van Nostrand, New York, 1966, p. 455.
12. Berry, R. S., *J. Chem. Phys.* **45**, 1228 (1966).
13. Fano, U., *Phys. Rev.* **124**, 1866 (1961).
14. Comes, F. J., and Elzer, A., *Phys. Letters* **25** A, 334 (1967).
15. Samson, J. A. R., and Cairns, R. B., *Phys. Rev.* **173**, 80 (1968).
16. Price, W. C., in *Molecular Spectroscopy* (ed. P. Hepple), Institute of Petroleum, London 1968, p. 221.
17. Asundi, R. K., *Current Sci.* **37**, 160 (1968).
18. Albritton, D. L., Schmeltekopf, A. L., and Zare, R. N., *J. Chem. Phys.* **51**, 1667 (1969).
19. Gilmore, F. R., *J. Quant. Spectrosc. Rad. Transf.* **5**, 369 (1965).
20. Yoshino, K., and Tanaka, Y., *J. Chem. Phys.* **48**, 4859 (1968).
21. Huffmann, R. E., Larrabee, J. C., and Tanaka, Y., *J. Chem. Phys.* **40**, 356 (1964).
22. Collin, J. E., and Natalis, P., *Int. J. Mass Spectrosc. Ion Physics* **2**, 231 (1969).
23. Herzberg, G., *Phys. Rev. Letters* **23**, 1081 (1969).
24. Turner, D. W., *Proc. Roy. Soc. London* A, **307**, 15 (1968).
25. Villarejo, D., *J. Chem. Phys.* **48**, 4014 (1968).
26. Bardsley, J. N., *Chem. Phys. Letters* **1**, 229 (1967).
27. Russek, A., Patterson, M. R., and Becker, R. L., *Phys. Rev.* **167**, 17 (1968).
28. Nielsen, S. E., and Berry, R. S., *Chem. Phys. Letters* **2**, 503 (1968).
29. Berkowitz, J., and Chupka, W. A., *J. Chem. Phys.* **51**, 2341 (1969).
30. Tully, J. C., Berry, R. S., and Dalton, B. J., *Phys. Rev.* **176**, 95 (1968).
31. Schneider, B., and Berry, R. S., *Phys. Rev.* **182**, 141 (1969).

32. Cook, G. R., and Metzger, P. H., *J. Opt. Soc. Am.* **54**, 968 (1964).
33. Metzger, P. H., and Cook, G. R., *J. Chem. Phys.* **41**, 642 (1964).
34. Chupka, W. A., *J. Chem. Phys.* **48**, 2337 (1968).
35. Baker, A. D., Baker, C., Brundle, C. R., and Turner, D. W., *Int. J. Mass Spectrosc. Ion Physics* **1**, 285 (1968).
36. Villarejo, D., Stockbauer, R., and Inghram, M. G., *J. Chem. Phys.* **50**, 4599 (1969).
37. Vilesov, F. I., Kurbatov, B. L., and Terenin, A. N., *Dokl. Akad. Nauk SSSR* **138**, 1329 (1961); Kurbatov, B. L., Vilesov, F. I., and Terenin, A. N., *Dokl. Akad. Nauk SSSR* **140**, 797 (1961).
38. Turner, D. W., and Al-Joboury, M. I., *J. Chem. Phys.* **37**, 3007 (1962).
39. Al-Joboury, M. I., and Turner, D. W., *J. Chem. Soc.* 5141 (1963).
40. Frost, D. C., McDowell, C. A., and Vroom, D. A., *Proc. Roy. Soc. London* A, **296**, 566 (1967).
41. Samson, J. A. R., *Techniques of Vacuum Ultraviolet Spectroscopy*, Wiley, New York, 1967, pp. 140 and 141.
42. Comes, F. J., and Sälzer, H. G., *Z. Naturforsch.* **19a**, 1230 (1964).
43. Doolittle, P. H., and Schoen, R. I., *Phys. Rev. Letters* **14**, 348 (1965).
44. Blake, A. J., and Carver, J. H., *J. Chem. Phys.* **47**, 1038 (1967).
45. Villarejo, D., Herm, R. R., and Inghram, M. G., *J. Opt. Soc. Am.* **56**, 1574 (1966).
46. Schoen, R. I., *J. Chem. Phys.* **40**, 1830 (1964).
47. Turner, D. W., in *Molecular Spectroscopy* (ed. P. Hepple), Institute of Petroleum, London, 1968, p. 209.
48. Simpson, J. A., *Rev. Sci. Inst.* **32**, 1283 (1961).
49. Al-Joboury, M. I., and Turner, D. W., *J. Chem. Soc.* 4434 (1964).
50. Frost, D. C., Sandhu, J. S., and Vroom, D. A., *Nature* **212**, 604 (1966).
51. Collin, J. E., and Natalis, P., *Bull. Acad. r. Belg. Cl. Sci.* 5e sér. **55**, 352 (1969).
52. Marmet, P., Natalis, P., and Dumont, A., *Rev. Sci. Inst.* **39**, 686 (1968).
53. Frost, D. C., McDowell, C. A., Sandhu, J. S., and Vroom, D. A., *Advances in Mass Spectrometry*, Institute of Petroleum, London, 1968, vol. 4, p. 781.
54. Brehm, B., and von Puttkamer, E., *Advances in Mass Spectrometry*, Institute of Petroleum, London, 1968, vol. 4, p. 591.
55. Rose, T., Frey, R., and Brehm, B., *Chem. Commun.* 1518 (1969).
56. Turner, D. W., and May, D. P., *J. Chem. Phys.* **45**, 471 (1966).
57. Turner, D. W., *Nature* **213**, 795 (1967).
58. Simpson, J. A., *Rev. Sci. Inst.* **35**, 1698 (1964).
59. Kuyatt, C. E., and Simpson, J. A., *Rev. Sci. Inst.* **38**, 103 (1967).
60. Hafner, H., Simpson, J. A., and Kuyatt, C. E., *Rev. Sci. Inst.* **39**, 33 (1968).
61. Eland, J. D. H., and Danby, C. J., *J. Sci. Inst. (J. of Phys. E.)*, ser. 2, **1**, 406 (1968).
62. Eland, J. D. H., and Danby, C. J., *Int. J. Mass Spectrosc. Ion Physics* **1**, 111 (1968).
63. Samson, J. A. R., *Rev. Sci. Inst.* **40**, 1174 (1969).
64. Lukirsky, P., *Z. Phys.* **22**, 351 (1924).
65. Soboleva, N. A., *Radiotekh. Élektron.* **3**, 339 (1958).
66. Harrower, G. A., *Rev. Sci. Inst.* **26**, 850 (1955).
67. McGowan, J. W., *Rev. Sci. Inst.* **38**, 285 (1967).
68. Richards, W. G., *Int. J. Mass Spectrosc. Ion Physics* **2**, 419 (1969).
69. Turner, D. W., *Advances in Physical Organic Chemistry* (ed. V. Gold), Academic Press, London, 1966, vol. 4, p. 31.
70. Turner, D. W., *Advances in Mass Spectrometry*, Institute of Petroleum, London, 1968, vol. 4, p. 755.
71. Dewar, M. J. S., and Worley, S. D., *J. Chem. Phys.* **51**, 263 (1969).
72. Bordass, W. T., and Linnett, J. W., *Nature* **222**, 660 (1969).
73. Marr, G. V., *Photoionization Processes in Gases*, Academic Press, New York, 1967, p. 166.

Additional bibliography

In a field of research as rapidly developing as photoelectron spectroscopy, it is particularly frustrating to both author and reader to be confronted with the extremely long delay in publication since the completion of this review in February 1970. The following supplementary bibliography, though non-exhaustive, should bring the reader up to date to mid-1972. The references relate to the subjects treated in the present review and new trends in the field.

Rapid developments are taking place in theoretical and experimental work on autoionization processes and on intensities of photoelectron spectra. Studies of the angular distribution of photoejected electrons are being actively and fruitfully pursued. Some effort is being given to the observation of the photoelectron spectra of both stable and transient free radicals. Attempts to understand the photoelectron spectra of small molecules and the electronic and vibrational structure of molecular ions have induced a considerable amount of theoretical calculations using methods which range from highly approximate to highly refined *ab initio*.

Finally, it should be mentioned that in the future it is very it probable that some of the more fundamental and physical aspects of photoelectron spectroscopy will be studied using as a light source the continuous spectrum of polarized light emitted as synchrotron radiation in electron synchrotrons and storage rings. With a suitable monochromator this source should be particularly useful in the study of autoionization and ion fragmentation processes.

The supplementary bibliography has been divided into a number of sections. A, Books. B, Conference reports (the communications given are not listed separately; these reports contain many valuable papers). C, Review articles (reviews in some overlapping but relevant fields are also cited). D, Articles mainly on experimental and instrumental aspects (cf. section 2 of this chapter). E, Articles illustrating fundamental principles (cf. section 1 of this chapter). F, Studies of molecular structure of small molecules and free radicals.

A, BOOKS

Sevier, K. D., *Low Energy Electron Spectrometry*, Wiley-Interscience, New York, 1972.

Siegbahn, K., Nordling, C., Johansson, G., Hedman, J., Hedan, P., Hamrin, K., Gelius, U., Bergmark, T., Werme, L., Manne, R., and Baer, Y., *ESCA Applied to Free Molecules*, North-Holland, American Elsevier, Amsterdam, New York, 1969.

Turner, D. W., Baker, C., Baker, A. D., and Brundle, C. R., *Molecular Photoelectron Spectroscopy. A Handbook of the* 584 Å *Spectra*, Wiley-Interscience, London, 1970.

B, CONFERENCE REPORTS

A Discussion on Photoelectron Spectroscopy, a Royal Society discussion organized by W. C. Price and D. W. Turner, *Phil. Trans. Roy. Soc. London* A, **268** (1184) 1-175 (1970).

Electron Spectroscopy, Proceedings of an International Conference held at Asilomar, Pacific Grove, California, 7-10 September 1971 (ed. D. A. Shirley), North-Holland, Amsterdam, 1972.

The Photoelectron Spectroscopy of Molecules, discussion of the Faraday Division of the Chemical Society, London, held at the University of Sussex, 12-14 September 1972 (in press).

C, REVIEW ARTICLES

BAKER, A. D., Photoelectron spectroscopy, *Accounts Chem. Res.* **3**, 17-25 (1970).
BERKOWITZ, J., Photoionization mass spectrometry and photoelectron spectroscopy in high temperature vapors, *Adv. High Temp. Chem.* (in press).
BERRY, R. S., Electronic spectroscopy by electron spectroscopy, *Ann. Rev. Phys. Chem.* **20**, 357-406 (1969).
BRUNDLE, C. R., The application of electron spectroscopy to surface studies, chapter 6, *Surface and Defect Properties of Solids*, vol. 1, Specialist Periodical Reports, the Chemical Society, London, 1972.
DELGASS, W. N., HUGHES, T. R., and FADLEY, C. S., X-Ray photoelectron spectroscopy: a tool for research in catalysis, *Catalysis Rev.* **4**, 179-220 (1970).
HERCULES, S. H., and HERCULES, D. M., Electron spectroscopy (ESCA): a novel technique for studying molecular structure, *Record Chem. Prog.* **32**, 183-213 (1971).
HOLLANDER, J. M., and JOLLY, W. L., X-Ray photoelectron spectroscopy, *Accounts Chem. Res.* **3**, 193-200 (1970).
HOLLANDER, J. M., and SHIRLEY, D. A., Chemical information from photoelectron and conversion-electron spectroscopy, *Ann. Rev. Nucl. Sci.* **20**, 435-466 (1970).
REID, N. W., Photoionisation in mass Spectrometry, *Int. J. Mass Spectrom. Ion Phys.* **6**, 1-31 (1971).
WORLEY, S. D., Photoelectron spectroscopy in chemistry, *Chem. Rev.* **71**, 295-314 (1971).

D, ARTICLES MAINLY ON EXPERIMENTAL AND TECHNICAL ASPECTS

AKSELA, S., Analysis of the energy distribution in cylindrical electron spectrometers, *Rev. Scient. Instrum.* **42**, 810-12 (1971).
AKSELA, S., KARRAS, M., PESSA, M., and SUONINEN, E., Study of the electron optical properties of an electron spectrograph with coaxial cylindrical electrodes, *Rev. Scient. Instrum.* **41**, 351-5 (1970).
ALLARD, P., and CARETTE, J. D., Spectromètres électrostatiques: II, *Can. J. Phys.* **49**, 2132-7 (1971).
ALLEN, J. D., JR, WOLFE, J. P., and SCHWEITZER, G. K., A new electron spectrometer design, *Int. J. Mass Spectrom. Ion Phys.* **8**, 81-83 (1972).
ARNOW, M., and JONES, D. R., Reanalysis of the trajectories of electrons in 127° cylindrical spectrometers, *Rev. Scient. Instrum.* **43**, 72-75 (1972).
BAER, T., PEATMAN, W. B., and SCHLAG, E. W., Photoionization resonance studies with steradiancy analyser: II, The photoionization of CH_3I, *Chem. Phys. Letters* **4**, 243-7 (1969).
BALLARD, R. E., and GRIFFITHS, G. A., Photoelectron spectrum of potassium hexacyanoferrate(II) measured by a Millikan condenser method, *Chem. Commun.* 1472-3 (1971).
BERKOWITZ, J., Photoelectron spectroscopy of high-temperature vapors: I, TlCl, TlBr, and TlI, *J. Chem. Phys.* **56**, 2766-74 (1972).
BERKOWITZ, J., and GUYON, P. M., On the interpretation of peak intensities from energy analyzers, *Int. J. Mass Spectrom. Ion Phys.* **6**, 302-4 (1971).
BOSI, G., Fringing field of electrostatic analyzers, *Rev. Scient. Instrum.* **43**, 475-8 (1972).
CAIRNS, R. B., HARRISON, H., and SCHOEN, R. I., Photoelectron spectroscopy with undispersed ultraviolet radiation, *Appl. Opt.* **9**, 605-8 (1970).
DANBY, C. J., and ELAND, J. H. D., Photoelectron-photoion coincidence spectroscopy: II, Design and performance of a practical Instrument, *Int. J. Mass Spectrom. Ion Phys.* **8**, 153-61 (1972).

DELAGE, Y., and CARETTE, J. D., Spectromètres électrostatiques: I, *Can. J. Phys.* **49**, 2118-31 (1971).

ELAND, J. H. D., Photoelectron-photoion coincidence spectroscopy: I, Basic principles and theory, *Int. J. Mass Spectrom. Ion Phys.* 8, 143-51 (1972).

EVANS, S., ORCHARD, A. F., and TURNER, D. W., A simple, medium resolution Helium(I) photoelectron spectrometer, *Int. J. Mass Spectrom. Ion Phys.* 7, 261-7 (1971).

FADLEY, C. S., HEALEY, R. N., HOLLANDER, J. M., and MINER, C. E., Design of a high-resolution high-efficiency magnetic spectrometer for electron spectroscopy, *J. Appl. Phys.* **43**, 1085-1102 (1972).

GREEN, T. S., and PROCA, G. A., A parallel plate electrostatic spectrograph, *Rev. Scient. Instrum.* **41**, 1409-14 (1970).

HEDDLE, D. W. O., A comparison of the *étendue* of electron spectrometers, *J. Phys. E. Scient. Instrum.* **4**, 589-92 (1971).

HUCHITAL, D. A., and RIDGEN, J. D., High sensitivity electron spectrometer, *Appl. Phys. Letters* **16**, 348-51 (1970).

KINSINGER, J. A., STEBBINGS, W. L., VALENZI, R. A., and TAYLOR, J. W., Spectral evaluation of a sealed helium discharge lamp for studies in photoelectron spectroscopy, *Anal. Chem.* **44**, 773-7 (1972).

KRAUSE, M. O., The $M\zeta$ X-Rays of Y to Rh in photoelectron spectrometry, *Chem. Phys. Letters* **10**, 65-69 (1971).

LLOYD, D. R., Calibration of a He(I) photoelectron spectrometer, *J. Phys. E. Scient. Instrum.* **3**, 629-32 (1970).

PEATMANN, W. B., BORNE, T. B., and SCHLAG, E. W., Photoionization resonance spectra: I, Nitric oxide and benzene, *Chem. Phys. Letters* **3**, 492-7 (1969).

PULLEN, B. P., CARLSON, T. A., MODDEMAN, W. E., SCHWEITZER, G. K., BULL, W. E., and GRIMM, F. A., Photoelectron spectra of methane, silane, methyl fluoride, difluoromethane and trifluoromethane [Reviewer's note: this article contains a description of an electron spectrometer which employs double focusing electrostatic plates] *J. Chem. Phys.* **53**, 768-82 (1970).

ROY, D., and CARETTE, J. D., Optimum deflection angle for cylindrical and spherical electrostatic spectrometer, *Appl. Phys. Letters* **16**, 413-16 (1970).

ROY, D., and CARETTE, J. D., Spectromètres électrostatiques: III, *Can. J. Phys.* **49**, 2138-59 (1971).

ROY, D., and CARETTE, J. D., Methods of measuring the performance of an electrostatic spectrometer, *Rev. Scient. Instrum.* **42**, 1122-7 (1971).

ROY, D., and CARETTE, J. D., Improvement of the resolving power and transmission of electrostatic spectrometers, *J. Appl. Phys.* **42**, 3601-15 (1971).

ROY, D., DE CELLES, M., and CARETTE, J. D., Influence de la longueur des fentes sur la résolution des spectromètres électrostatiques cylindriques et plan, *Rev. Phys. Appliquée*, **6**, 51-54 (1971).

SIEGBAHN, K., HAMMOND, D., FELLNER-FELDEGG, H., and BARNETT, E. F., Electron spectroscopy with monochromatized X-rays, *Science* **176**, 245-52 (1972).

SPOHR, R., GUYON, P. M., CHUPKA, W. A., and BERKOWITZ, J., Threshold photoelectron detector for use in the vacuum ultraviolet, *Rev. Scient. Instrum.* **42**, 1872-9 (1971).

VON PUTTKAMER, E., Koinzidenzmessung von Photoionen und Photoelektronen, *Z. Naturf.* **25a**, 1062-71 (1970).

YAVOR, S. YA., PETROV, I. A., and DENISOV, E. P., High resolution electrostatic spectrometer for the Siegbahn ESCA method, *Sov. Phys. Tech. Phys.* **16**, 1455-60 (1972).

E, ARTICLES ILLUSTRATING FUNDAMENTAL PRINCIPLES

AMUSIA, M. YA., CHEREPKOV, N. A., and CHERNYSHEVA, L. V., Angular distribution of photoelectrons with many-electron correlations, *Phys. Letters* **40** A, 15-16 (1972).

ASBRINK, L., The photoelectron spectrum of H_2, *Chem. Phys. Letters* **7**, 549-52 (1970).

BAHR, J. L., BLAKE, A. J., CARVER, J. H., and VIJAY KUMAR, Photoelectron spectra and partial cross-sections for carbon dioxide, *J. Quant. Spectrosc. Radiative Transfer* **9**, 1359-64 (1969).

BAHR, J. L., BLAKE, A. J., CARVER, J. H., GARDNER, J. L., and VIJAY KUMAR, Autoionization in diatomic molecules studies by photoelectron spectroscopy, *J. Quant. Spectrosc. Radiative Transfer* **11**, 1839-52 (1971) [see also *ibid.* **11**, 1853-61 (1971)].

BAHR, J. L., BLAKE, A. J., CARVER, J. H., GARDNER, J. L., and VIJAY KUMAR, Photoelectron spectra and partial photoionization cross sections for NO, N_2O, CO, CO_2 and NH_3, *J. Quant. Spectrosc. Radiative Transfer* **12**, 59-73 (1972).

BERRY, R. S., and NIELSEN, S. E., Dynamic coupling phenomena in molecular excited states: I, General formulation and vibronic coupling in H_2, *Phys. Rev.* A, **1**, 383-94 (1970).

BERRY, R. S., and NIELSEN, S. E., Dynamic coupling phenomena in molecular excited states: II, Autoionization and predissociation in H_2, HD and D_2, *Phys. Rev.* A, **1**, 395-411 (1970).

BREHM, B., and FREY, R., Rotationsstruktur im 21,22 eV photoelektronenspektrum von H_2, *Z. Naturforsch.* **26 a**, 523-526 (1971).

CARLSON, T. A., Angular dependence of vibrational structure in the photoelectron spectra of N_2 and O_2, *Chem. Phys. Letters* **9**, 23-26 (1971).

CARLSON, T. A., and ANDERSON, C. P., Angular distribution of the photoelectron spectrum for benzene, *Chem. Phys. Letters* **10**, 561-4 (1971).

CARLSON, T. A., and JONAS, A. E., Angular distribution of the photoelectron spectra for Ar, Kr, Xe, H_2, N_2 and CO, *J. Chem. Phys.* **55**, 4913-24 (1971).

CARVER, J. H., and GARDNER, J. L., Photoelectron spectroscopy and partial photoionization cross sections for iodine and bromine, *J. Quant. Spectrosc. Radiative Transfer* **12**, 207-18 (1972).

COLLIN, J. E., DELWICHE, J., and NATALIS, P., Energy levels of NO^+ ion by He and Ar resonance lines photoelectron spectrometry, *Int. J. Mass Spectrom. Ion Phys.* **7**, 19-26 (1971).

CORNFORD, A. B., FROST, D. C., MCDOWELL, C. A., RAGLE, J. L., and STENHOUSE, I. A., The vibrational constants of the ground states of H_2^+, HD^+ and D_2^+ from photoelectron spectroscopic measurements, *Chem. Phys. Letters* **5**, 486-8 (1970).

COX, P. A., and ORCHARD, F. A., On band intensities in the photoelectron spectra of open-shell molecules, *Chem. Phys. Letters* **7**, 273-5 (1970).

ELAND, J. H. D., Predissociation of triatomic ions studied by photoelectron-photoion coincidence spectroscopy and photoion kinetic energy analysis I. CO_2^+, *Int. J. Mass. Spectrom. Ion. Phys.* **9**, 397-406 (1972).

FAISAL, F. H. M., Autoionization of H_2 near threshold, *Phys. Rev.* A, **4**, 1396-1400 (1971).

HARRISON, H., Angular distribution of photoelectrons: I, Cadmium and zinc atoms at 584 and 1048 Å, *J. Chem. Phys.* **52**, 901-5 (1970).

HEINZMANN, U., KESSLER, J., and LORENZ, J., Wavelength dependence of the Fano effect, *Phys. Rev. Letters* **25**, 1325 (1970).

JONATHAN, N., MORRIS, A., ROSS, K. J., and SMITH, D. J., High resolution vacuum ultraviolet photoelectron spectra of transient species: O_2 ($^1\Delta_g$) and previously unobserved states of O_2^+, *J. Chem. Phys.* **54**, 4954-5 (1971).

JONATHAN, N., SMITH, D. J., and ROSS, K. J., High resolution vacuum ultraviolet photoelectron spectroscopy of transient species: O_2 ($^1\Delta_g$), *J. Chem. Phys.* **53**, 3758-9 (1970).

LEFEBVRE-BRION, H., Intensity anomaly in the photoelectron spectrum of NO, *Chem. Phys. Letters* **9**, 463-4 (1971).

LORQUET, J. C., and CADET, C., Configuration interaction intensity borrowing in photoelectron spectroscopy, *Chem. Phys. Letters* **6**, 198-9 (1970).

LORQUET, J. C., and CADET, C., Excited states of gaseous ions: I, Selection rules in photoelectron spectroscopy and photoionization: the case of N_2O^+, *Int. J. Mass Spectrom. Ion Phys.* **7**, 245-54 (1971).

MILLER, W. H., Computation of autoionization lifetimes via a "Golden rule"-like formula, *Chem. Phys. Letters* **4**, 627-31 (1970).

MITCHELL, P., and CODLING, K., Variation of the angular distribution of photoelectrons from argon in the photon energy range 0-23 eV above threshold, *Phys. Letters* **38** A, 31-32 (1972). [Reviewer's note: the light source was synchrotron radiation from a 300 MeV electron synchrotron.]

MORGENSTERN, R., NIEHAUS, A., and RUF, M. W., Angular distribution of photoelectrons, *Chem. Phys. Letters* **4**, 635-8 (1970).

NATALIS, P., DELWICHE, J., and COLLIN, J. E., Enhancement of vibrational level population of N_2^+ and CO^+ by photoelectron spectrometry, *Chem. Phys. Letters* **13**, 491-5 (1972).

NAZAROFF, G. V., General theory of autoionizing states of two-electron systems, *J. Chem. Phys.* **52**, 5873-83 (1970).

NIEHAUS, A., and RUF, M. W., Angular distribution of photoelectrons from different rotational $H_2 \rightarrow H_2^+$ transitions, *Chem. Phys. Letters* **11**, 55-57 (1971).

RITCHIE, B., Semiclassical theory for molecular auto-ionization, *Phys. Rev.* A, **3**, 95-100 (1971).

SAMSON, J. A. R., Observation of double electron excitation in H_2 by photoelectron spectroscopy, *Chem. Phys. Letters* **12**, 625-7 (1972).

SHAW, G. B., and BERRY, R. S., Photoionization and autoionization of H_2, *J. Chem. Phys.* **56**, 5808-17 (1972).

SICHEL, J. M., Angular distribution and intensity in molecular photoelectron spectroscopy: II: Information contained in rotational fine structure, *Mol. Phys.* **18**, 95-100 (1970).

THOMAS, I. L., Angular dependence of the vibrational and rotational excitations seen in photoelectron spectroscopy, *Phys. Rev.* A **4**, 457-9 (1971).

YARIS, R., and BOEHM, R., Many-body theory of autoionization, *J. Chem. Phys.* **55**, 1802-10 (1971).

F, STUDIES OF MOLECULAR STRUCTURE OF SMALL MOLECULES AND FREE RADICALS

ÅSBRINK, L., EDQVIST, O., LINDHOLM, E., and SELIN, L. E., The electronic structure of benzene, *Chem. Phys. Letters* **5**, 192-4 (1970).

ÅSBRINK, L., LINDHOLM, E., and EDQVIST, O., Jahn-Teller effect in the vibrational structure of the photoelectron spectrum of benzene, *Chem. Phys. Letters* **5**, 609-12 (1970).

BRANTON, G. R., FROST, D. C., MCDOWELL, C. A., and STENHOUSE, I. A., The photoelectron spectra of phosphine and arsine, *Chem. Phys. Letters* **5**, 1-2 (1970).

BRANTON, G. R., FROST, D. C., MAKITA, T., MCDOWELL, C. A., and STENHOUSE, I. A., Photoelectron spectra of ethylene and ethylene-d_4, *J. Chem. Phys.* **52**, 802-6 (1970).

BREHM, B., MENZINGER, M., and ZORN, C., The photoelectron spectrum of XeF_2, *Can. J. Chem.* **48**, 3193-6 (1970).

BROGLI, F., and HEILBRONNER, E., The competition between spin orbit coupling and conjugation in alkyl halides and its repercussion on their photoelectron spectra, *Helv. chim. Acta* **54**, 1423-34 (1971).

BRUNDLE, C. R., On the first ionization potential of nitrogen dioxide and nitric oxide, *Chem. Phys. Letters* **5**, 410-15 (1970).

BRUNDLE, C. R., Ionization and dissociation energies of HF and DF and their bearing on $D_0^0(F_2)$, *Chem. Phys. Letters* **7**, 317-22 (1970).

BRUNDLE, C. R., The vibrational structure in the photoelectron spectra of ethylene and ethylene-d_4, and its relationship to the vibrational spectrum of Zeise's salt $K[PtCl_3(C_2H_4)] \cdot H_2O$, *Spectrochim. Acta* **27** A, 2491-8 (1971).

BRUNDLE, C. R., JONES, G. R., and BASCH, H., He I and He II photoelectron spectra and the electronic structures of XeF_2, XeF_4, and XeF_6, *J. Chem. Phys.* **55**, 1098-1104 (1971).

BRUNDLE, C. R., NEUMAN, D., PRICE, W. C., EVANS, D., POTTS, A. W., and STREETS, D. G., Electronic structure of NO_2 studied by photoelectron and vacuum UV spectroscopy and gaussian orbital calculations, *J. Chem. Phys.* **53**, 705-15 (1970).

BRUNDLE, C. R., ROBIN, M. B., and BASCH, H., Electronic energies and electronic structures of the fluoromethanes, *J. Chem. Phys.* **53**, 2196-2213 (1970).

BRUNDLE, C. R., ROBIN, M. B., and JONES, G. R., High-resolution He I and He II photoelectron spectra of xenon difluoride, *J. Chem. Phys.* **52**, 3383-6 (1970).

BRUNDLE, C. R., ROBIN, M. B., and KUEBLER, N. A., Perfluoro effect in photoelectron spectroscopy: II, Aromatic molecules, *J. Am. Chem. Soc.* **94**, 1466-75 (1972).

BRUNDLE, C. R., ROBIN, M. B., KUEBLER, N. A., and BASCH, H., Perfluoro effect in photoelectron spectroscopy: I, Nonaromatic molecules, *J. Am. Chem. Soc.* **94**, 1451-65 (1972).

CORNFORD, A. B., FROST, D. C., HERRING, F. G., and McDOWELL, C. A., Ionization potentials of the difluoroamino radical by photoelectron spectroscopy and INDO calculations, *J. Chem. Phys.* **54**, 1872-3 (1971).

CORNFORD, A. B., FROST, D. C., HERRING, F. G., and McDOWELL, C. A., Photoelectron spectra of F_2O and Cl_2O, *J. Chem. Phys.* **55**, 2820-2 (1971).

CORNFORD, A. B., FROST, D. C., HERRING, F. G., and McDOWELL, C. A., The photoelectron spectrum of the free radical chlorine dioxide, *Chem. Phys. Letters* **10**, 345-6 (1971).

CORNFORD, A. B., FROST, D. C., McDOWELL, C. A., RAGLE, J. L., and STENHOUSE, I. A., Photoelectron spectra of the halogens, *J. Chem. Phys.* **54**, 2651-7 (1971).

COWAN, D. O., GLEITER, R., GLEMSER, O., HEILBRONNER, E., and SCHÄUBLIN, J., The photoelectron spectrum of thiazyl fluoride (NSF), *Helv. chim. Acta* **54**, 1559-64 (1971).

DIXON, R. N., On the Jahn-Teller distortion of CH_4^+, *Molec. Phys.* **20**, 113-26 (1971).

DIXON, R. N., DUXBURY, G., FLEMING, G. R., and HUGO, J. M. V., The photoelectron spectrum of thiazyl fluoride, *Chem. Phys. Letters* **14**, 60-63 (1972).

DIXON, R. N., DUXBURY, G., HORANI, M., and ROSTAS, J., The H_2S^+ radical ion: a comparison of photoelectron and optical spectroscopy, *Molec. Phys.* **22**, 977-92 (1971).

DIXON, R. N., MURRELL, J. N., and NARAYAN, B., The photoelectron spectra of the halomethanes, *Molec. Phys.* **20**, 611-23 (1971).

DURMAZ, S., KING, G. H., and SUFFOLK, R. J., Calculation of the vibrational envelope and bond angle in the first excited state of H_2S^+ from the photoelectron spectrum of H_2S, *Chem. Phys. Letters* **13**, 304-6 (1972).

EDQVIST, O., ÅSBRINK, L., and LINDHOLM, E., On the photoelectron spectrum of NO, *Z. Naturf.* **26a**, 1407-10 (1971).

EDQVIST, O., LINDHOLM, E., SELIN, L. E., and ÅSBRINK, L., On the photoelectron spectrum of O_2, *Physica Scripta* **1**, 25-30 (1970),

EVANS, S., JOACHIM, P. J., ORCHARD, A. F., and TURNER, D. W., A study of the orbital electronic structure of the P_4 molecule by photoelectron spectroscopy, *Int. J. Mass Spectrom. Ion Phys.* **9**, 41-49 (1972).

FROST, D. C., KATRIB, A., McDOWELL, C. A., and McLEAN, R. A. N., The $^2A_1$ band in the photoelectron spectrum of hydrogen sulphide, *Int. J. Mass Spectrom. Ion Phys.* **7**, 485-8 (1971).

GOFFART, C., MOMIGNY, J., and NATALIS, P., Photoionization studies by total ionization measurements and photoelectron spectra: II, Pyridine, *Int. J. Mass Spectrom. Ion Phys.* **3**, 371-84 (1969).

HILLIER, I. H., and SAUNDERS, V. R., A theoretical interpretation of the bonding, and the photoelectron and ultra-violet spectra of sulphur dioxide, *Mol. Phys.* **22**, 193-201 (1971).

HOLLAS, J. M., and SUTHERLEY, T. A., The geometry of the ground state of $C_2H_2^+$ from photoelectron spectroscopy compared with that of C_2H_2 in some Rydberg states, *Molec. Phys.* **21**, 183-5 (1971).

JONATHAN, N., MORRIS, A., OKUDA, M., SMITH, D. J., and ROSS, K. J., Photoelectron spectroscopy of transient species: the CS molecule, *Chem. Phys. Letters* **13**, 334-6 (1972).

JONATHAN, N. MORRIS, A., SMITH, D. J., and ROSS, K. J., Photoelectron spectra of ground state atomic hydrogen, nitrogen and oxygen, *Chem. Phys. Letters* **7**, 497-9 (1970).

JONATHAN, N., SMITH, D. J., and ROSS, K. J., The high resolution photoelectron spectra of transient species: sulphur monoxide, *Chem. Phys. Letters* **9**, 217-18 (1971).

KING, G. H., KROTO, H. W., and SUFFOLK, R. J., The photoelectron spectrum of a short-lived species in the decomposition products of CS_2, *Chem. Phys. Letters* **13**, 457-8 (1972).

LINDHOLM, E., and SAHLSTROM, G., The mass spectrum of NO_2, *Int. J. Mass Spectrom. Ion Phys.* **4**, 465-73 (1970).

MAIER, J. P., and TURNER, D. W., Photoelectron spectroscopy and the geometry of the phosphine and phosphorous trifluoride molecular ions, *J. Chem. Soc. Faraday Trans.* II **68**, 711-19 (1972).

MUSZKAT, A., and SCHÄUBLIN, J., Correlation between photoelectron ionization bands and electronic levels in pyrazine, *Chem. Phys. Letters* **13**, 301-3 (1972).

POTTS, A. W., and PRICE, W. C., Photoelectron spectra of the halogens and mixed halides ICl and IBr, *Trans. Faraday Soc.* **67**, 1242-52 (1971).

POTTS, A. W., and PRICE, W. C., The photoelectron spectra of methane, silane, germane and stannane, *Proc. Roy. Soc. London* A, 326, 165-79 (1972).

POTTS, A. W., and PRICE, W. C., Photoelectron spectra and valence shell orbital structures of groups V and VI hydrides, *Proc. Roy. Soc. London* A, **326**, 181-97 (1972).

RABALAIS, J. W., BERGMARK, T., WERME, L. O., KARLSSON, L., and SIEGBAHN, K., The Jahn-Teller effect in the electron spectrum of methane, *Physica Scripta* **3**, 13-18 (1971).

RAFTERY, J., and RICHARDS, W. G., A theoretical calculation of the states of O_2^+ accessible in the photoelectron spectrum of $O_2(^1\Delta_g)$, *Int. J. Mass Spectrom. Ion Physics* **6**, 269-77 (1971).

RAGLE, J. L., STENHOUSE, I. A., FROST, D. C., and McDOWELL, C. A., Valence-shell ionization potentials of halomethanes by photoelectron spectroscopy I. CH_3Cl, CH_3Br, CH_3I. Vibrational frequences and vibronic interaction in CH_3Br^+ and CH_3Cl^+, *J. Chem. Phys* **53**, 178-84 (1970).

WALKER, T. E. H., and HORSLEY, J. A., Non-empirical calculation of the electronic energy levels of the BF_3^+ ion. Interpretation of the photoelectron spectrum of BF_3, *Molec. Phys.* **21**, 939-42 (1971).

WEISS, M. J., and LAWRENCE, G. M., Photoelectron spectroscopy of NH_3 and ND_3 using molecular beams, *J. Chem. Phys.* **53**, 214-18 (1970).

WEISS, M. J., LAWRENCE, G. M., and YOUNG, R. A., Photoelectron spectroscopy of HCl and DCl using molecular beams, *J. Chem. Phys.* **52**, 2867-70 (1970).

PART III

VACUUM ULTRAVIOLET EMISSION FROM HOT PLASMAS

CHAPTER 8

Vacuum ultraviolet emission from hot plasmas

C. Breton

Section de Physique des Plasmas, Centre d'Études Nucléaires, 92 260, Fontenay aux Roses, France
and

J. L. Schwob*

Laboratoire des Interactions Moléculaires et des Hautes Pressions du CNRS, Bellevue, 92 190 Meudon, France

Contents

* Present address: The Racah Institute of Physics, The Hebrew University, Jerusalem, Israel.

Introduction

During the last two decades considerable effort has been devoted to the creation and study of very high-temperature plasmas produced in the laboratory. This work has been, to a large extent, the result of research on controlled thermonuclear fusion, marked by the construction of a large number of magnetically confined assemblies. Particular attention has been paid to improve the conditions of stability and increase the temperature of hot plasmas at relatively low densities ($10^{12} < n_e < 10^{15}$ cm^{-3}, $10 < kT_e < 500$ eV), produced by discharges in closed magnetic configurations (n_e and T_e are the electronic density and temperature respectively; the latter is expressed in electron volts, the unit corresponding to 11,600 K). More recently, strongly ionized plasmas of higher density have been produced by focusing a high-intensity laser beam on a solid target.

In this chapter we are particularly interested in these two types of very high-temperature plasmas. For spectroscopists they are excellent sources in the far u.v. down to the soft X-ray region. The utility of such plasmas in the field of atomic physics is towards the determination of energy levels in the spectra of highly ionized atoms in the first place. Useful applications are also towards determination of fundamental atomic data—still rather scarce—for example, transition probabilities and the cross-sections for inelastic collisions. In such experiments advantage is taken of the fact that the plasma is optically thin for low enough densities. The results thus obtained could complete, particularly in the region of high energies, the data already obtained from more traditional sources, like vacuum sparks, which have been extensively used for the classification of lines in the far u.v. Apart from fundamental atomic physics which was, until recently, the principal preoccupation of the spectroscopist, there is another aspect involved in the study of emission by plasmas. In the former case the source represents merely a method for exciting the particular atom or ion under study, and plasma effects are usually disturbing—causing, for example, line-broadening and thus limiting high resolution measurements (Doppler and Stark effects). However, the study of these causes of broadening—closely related to collision phenomena in the plasma—is the subject of extensive work. For the plasma spectroscopist knowledge of physical processes within the source itself becomes one of the principal aims. In most cases densities are too low for local thermodynamic equilibrium (LTE) to be valid, and ionization and excitation processes are not described by equilibrium equations. Accordingly, we attempt to construct models, based on atomic data, which fit the observations—as has been done by the astrophysicist for a long time. In these models, electron temperature and density are parameters determining the state of ionization and excitation of the plasma in terms of populations of the various possible species and line intensities. Knowing the relevant atomic data, it is possible, by suitable measurements, to obtain these two fundamental parameters. Spectroscopy thus provides valuable diagnostic techniques which have the advantage of not disturbing the plasma. Concerning such aims it is clear that the division of the spectral range into different regions—designated by soft X-rays, far u.v., visible, infrared, etc.—appears arbitrary except for the diversity of the techniques employed. Thus the free-free transition spectrum extends over all these ranges. Moreover, the method of electron temperature measurement based on line intensity ratios may use one line located in the Schumann region and the other in the visible. However, the vacuum ultraviolet (v.u.v.) remains the most important area because it contains the resonance transitions of hydrogen (the gas generally used to support the discharge) and of the common impurities (oxygen, nitrogen, carbon) or of added test elements (helium, rare gases). As a consequence, the principal radiation losses through these resonance lines, as well as through free-free transitions, occur in this wavelength region.

In order to facilitate comprehension of this chapter we shall start with a general description of the various models used, followed by the possible diagnostic methods based on these models and employing far u.v. techniques and, finally, we shall give indications on measurements related to atomic physics obtained with high-temperature plasma devices.

THEORETICAL MODELS

In this section we shall only give a brief review of the atomic processes taking place in hot plasmas and of theoretical models worked out for interpreting emission phenomena in such plasmas. More detailed reviews on this subject have been provided in the past by Wilson,[1] McWhirter,[2] and Cooper;[3] Griem[4] has treated mainly the case of LTE plasmas.

Thermodynamic equilibrium in plasmas

There is a *thermodynamic equilibrium* in a plasma when every process is in equilibrium with its converse. Since this statement includes radiation processes, optical depths must be large at all wavelengths. In such case, well-known laws for radiation and particles are valid.

The radiation field is isotropic, and its intensity $J_\nu(T)$, is given by Planck's law:

$$J_\nu(T) = B_\nu(T) = \frac{2h\nu^3}{c^2}\frac{1}{\exp(h\nu/kT) - 1}, \tag{1}$$

where ν is the frequency, h Planck constant, k Boltzmann constant, and c the velocity of light. J_ν is a spectral radiance that is expressed in power unit per cm^2 per unit solid angle and per unit frequency range. B_ν is the Planck function.

The ratio of the emissivity j_ν to the absorption coefficient k_ν is given by the Kirchhoff law:

$$\frac{j_\nu}{k_\nu} = B_\nu(T). \tag{2}$$

Particle velocity distribution follows Maxwell's law:

$$f(v)\,dv = \left(\frac{M}{2\pi kT}\right)^{3/2} 4\pi v^2 \exp\left(-Mv^2/2kT\right) dv, \tag{3}$$

where $f(v)\,dv$ is the probability of a particle having its velocity between v and $v + dv$, M being its mass.

Saha's equation governs the population distribution of every species in each degree of ionization:

$$\frac{n_{r+1}}{n_r} = \frac{1}{n_e}\,\frac{2U_{r+1}(T)}{U_r(T)}\,\frac{(2\pi m_e kT)^{3/2}}{h^3} \exp\left(-\chi_r/kT\right), \tag{4}$$

where n_r, n_{r+1}, n_e are the densities (number of particles per cm^3) for ions of charge r and $r+1$, and for electrons. U_r and U_{r+1} are partition functions for each stage of ionization. χ_r is the ionization energy of the ions of charge r in the ground state and m_e the electron mass. Data on partition functions and Saha equilibrium for a number of elements can be found in ref. 5.

Boltzmann's law gives the particle density in each excited level:

$$\frac{n_{r,j}}{n_{r,f}} = \frac{g_{r,j}}{g_{r,f}} \exp\left(-\chi_{fj}/kT\right); \quad \frac{n_{r,j}}{n_r} = \frac{g_{r,j}}{U_r(T)} \exp\left(-\chi_{fj}/kT\right), \tag{5}$$

where $n_{r,f}$ and $n_{r,j}$ are the ion densities in the ground state f and in the level j, and the g's are the corresponding statistical weights; χ_{fj} is the excitation energy from the ground state to the excited level j.

A laboratory plasma is never in complete thermodynamic equilibrium. Radiation escapes and temperature gradients are present; the Planck law is no longer valid. Nevertheless, the Kirchhoff law, which is a local relationship, may still be applicable. In such a case, the plasma is said to be in LTE. In other words, if any collision process is balanced by its converse and if, in comparison, radiation processes are of small importance, the collisional equilibrium is not appreciably modified, and the various population densities are governed by the Saha and Boltzmann laws. Radiation intensity, at a given wavelength, can be easily calculated if the optical depth is small.

Non-local thermodynamic equilibrium plasmas

When the total plasma density is small enough, radiative processes will again compete with collision processes. For example, excitation will take place through electron impacts and de-excitation through radiative decay, the rate of the latter being larger than the rate of collisional de-excitation. There is no longer equilibrium between any process and its converse and, as a consequence, equilibrium laws are no longer valid. To describe processes in these *non-LTE plasmas*, balance equations must be written for each species, the rate of change in its density being equal to the difference between the rate of the processes populating and depopulating the species. Nevertheless, the collision frequency among the free electrons will be large enough for a Maxwellian velocity distribution to apply in practically all cases of interest, and only the populations of excited levels of atoms and ions have to be studied in detail through rate equations.

I. ELEMENTARY PROCESSES

A. The possible atomic processes

An atom or ion in a state of ionization r has bound energy levels $\ldots i, j, k, \ldots$ $(i < j < k)$ and virtual levels in the continuum $\varkappa, \varkappa'$ (Fig. 1). A bound level j can be populated through:

1. Electron impacts:
   - (a) Excitation $i \to j$.
   - (b) De-excitation $k \to j$.
   - (c) Three-body recombination $\varkappa \to j$.
2. Radiative processes:
   - (d) Photoexcitation (by a photon of energy $h\nu_{ji}$) $i \to j$.
   - (e) Spontaneous decay (emission of a photon $h\nu_{kj}$) $k \to j$.
   - (f) Radiative recombination $\varkappa \to j$.

This level j can be depopulated through:

3. Electron impacts:
   - (a′) De-excitation $j \to i$.
   - (b′) Excitation $j \to k$.
   - (c′) Ionization $j \to \varkappa$.
4. Radiative processes:
   - (d′) Spontaneous decay (photon $h\nu_{ji}$) $j \to i$.
   - (e′) Photoexcitation (by a photon $h\nu_{kj}$) $j \to k$.
   - (f′) Photoionization $j \to \varkappa$.

There are two additional free-free transitions

- (g) Bremsstrahlung $\varkappa \to \varkappa'$.
- (g′) Continuous absorption (inverse bremsstrahlung) $\varkappa' \to \varkappa$.

If, in these processes, each pair is balanced (a ↔ a′, b ↔ b′, etc.) we have a plasma in thermodynamic equilibrium. In a collision dominated plasma only the three pairs a ↔ a′, b ↔ b′, and c ↔ c′ are balanced. This is LTE.

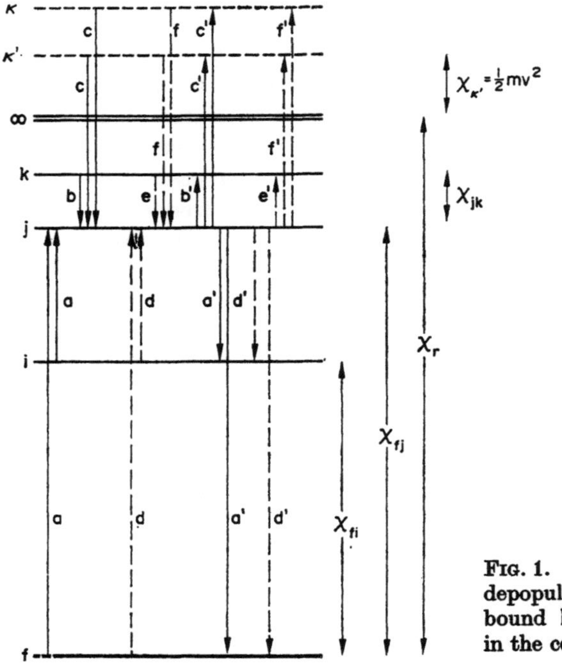

FIG. 1. Processes populating and depopulating a level j; i, j, k are bound levels, $\varkappa$, $\varkappa'$, virtual levels in the continuum.

Problems are more complicated if there is more than one outer electron. For example, in a double excitation the energy of an atom or ion can be higher than its ionization energy and there is a subsequent autoionization (the reverse process, "dielectronic recombination", may be important in plasmas).

B. Inelastic collisions: cross-sections and rate coefficients

Each collisional process has a cross-section $\sigma(v)$ in cm^2 depending on the electron velocity. In order to compute the rate of a given process we need to know the average of the product $v\sigma(v)$ over the velocity distribution of the electrons. This is the so-called rate coefficient written as follows:

$$X = \int_{v_0}^{\infty} \sigma(v)\, v\, f(v)\, dv\,, \qquad (6)$$

where v_0 is the threshold velocity such that $\frac{1}{2} m_e v_0^2 = \chi$, χ being the energy involved in the process (ionization, excitation, etc.); we assume that $f(v)\, dv$ is given by eqn. (3). The rate coefficient only depends, in this case, on the electron temperature T_e. The validity of the assumption of a Maxwellian distribution has been discussed in recent papers[6, 7] as well as the role of two electron populations of different temperature.[8] In this chapter a Maxwellian distribution is assumed throughout.

The symbols $Q_{r,ij}$ and $Q_{r,ji}$ will be used for excitation and de-excitation rate coefficients by electron impacts of an ion of charge r. These two are related by the equation

$$Q_{r,ji}(T_e) = (g_{r,i}/g_{r,j}) \exp (\chi_{ij}/kT_e) Q_{r,ij}(T_e) \quad (i < j). \tag{7}$$

$S_{r,j\infty}$ will be used for the rate coefficient for ionization by electron impact of an ion of charge r in the excited level j, the ion of charge $r + 1$ being in the ground state.

The symbol $\alpha_{r+1 \to r,j}$ will represent the rate coefficient of radiative recombination for an ion of charge $r + 1$ in its ground state to form the ion of charge r in level j. The total rate coefficient is

$$\alpha_{r+1} = \sum_{j=f}^{\infty} \alpha_{r+1 \to r,j}. \tag{8}$$

$\beta_{r+1 \to r,j}$ will denote the coefficient for three-body recombination of the ion of charge $r + 1$ to level j of the ion of charge r; this has to be multiplied by the electron density n_e for comparison with the other rate coefficients. $\beta_{r+1 \to r,j}$ and $S_{r,j\infty}$, relative to converse processes, are related as follows:

$$\beta_{r+1 \to r,j} = (g_{r,j}/2g_{r+1,f}) h^3 (2\pi m_e k T_e)^{-3/2} \exp (\chi_{j\infty}/kT_e) S_{r,j\infty}. \tag{9}$$

Cross-sections for these inelastic processes are of fundamental importance for solving the rate equations, computing the intensity of spectra, or for deriving plasma parameters from observed spectra. Some 15 years ago very few data were available on the subject. With growing interest in plasma physics, due to the intensive work on fusion programmes, the number of measured cross-sections has steadily increased, especially for light elements—hydrogen (H_2, H_2^+, H, H*), helium, nitrogen, oxygen, and the rare gases. A large number of excitation and ionization cross-sections have been measured for the ground states of neutral atoms. A few ionization cross-sections for singly ionized elements have been recently obtained. Reviews on this subject have been provided by Kieffer and Dunn,[9] Laborie *et al.*,[10] Moiseiwitsch and Smith,[11] and Massey *et al.*[12] During the same period, extensive theoretical work on excitation and ionization cross-sections has been performed.[11–14] Calculations are for simple atomic models: for practical cases one has to use semi-empirical formulae in order to compute rate coefficients for more complicated atoms or ions.[15–19]

The problem of recombination has been reviewed by Bates and Dalgarno.[20] Calculations are possible for radiative recombination cross-sections in the case of hydrogen-like ions. For non-hydrogenic ions one generally takes recombination coefficients of the hydrogen-like ion having the same charge. Recent theoretical papers on dielectronic recombination[20–22] show that this process is more important than radiative recombination at high temperatures.

C. Spontaneous decay: spectral line intensities

In an optically thin plasma, the power per cm^3 W_{ji} emitted in a spectral line for the transition from level j to level i is given by

$$W_{ji} = h\nu_{ji} n_{r,j} A_{ji}, \tag{10}$$

where A_{ji} is Einstein's coefficient for the transition $j \to i$ and $n_{r,j}$ the density of ions of charge r in level j. This can be given in terms of radiance in power unit per cm^2 and per unit solid angle:

$$J_{ji} = \frac{1}{4\pi} h\nu_{ji} n_{r,j} A_{ji} l, \tag{11}$$

l being the thickness of the source. To obtain the actual intensity at the detector, the apparatus function has to be taken into account.

In order to compute line intensities through theoretical models, one needs the transition probabilities. Information on these is scarce in the v.u.v. Theoretical calculations have been made by Bates and Damgaard.[23] Griem[4] gives numerical values obtained by this Coulomb approximation method for the visible and u.v. ranges for atoms or ions of low degree of ionization. A list of experimental or calculated values of oscillator strengths can be found in ref. 24. Complete tables[25, 26] give oscillator strengths and Einstein coefficients for atoms and ions (hydrogen through calcium).

D. Bremsstrahlung or free-free continuum

In the field of an ion of charge r, a free electron radiates a continuum. If the electrons have a Maxwellian velocity distribution, the power in erg s^{-1} emitted per cm^3 at frequency ν in the interval $d\nu$ is given by

$$W_\nu \, d\nu = 6.8 \times 10^{-38} \times n_e n_r \frac{r^2}{T_e^{1/2}} g_{ff} \times \exp(-h\nu/kT_e) \, d\nu, \tag{12}$$

where g_{ff} is the Gaunt factor for free-free transitions, of the order unity, and T_e is in K. This spectrum, as a function of wavelength, has an intensity maximum for wavelength λ_M given by the equation

$$\lambda_M \, (\text{Å}) = \frac{6200}{kT_e \, (\text{eV})}. \tag{13}$$

For plasmas of interest here, kT_e is of the order of 100 eV and λ_M ($\simeq$ 62 Å) is in the extreme u.v. range. Figure 2 gives an example of the spectrum for a hydrogen plasma. The temperature has been taken small enough to show radiative recombination spectra.

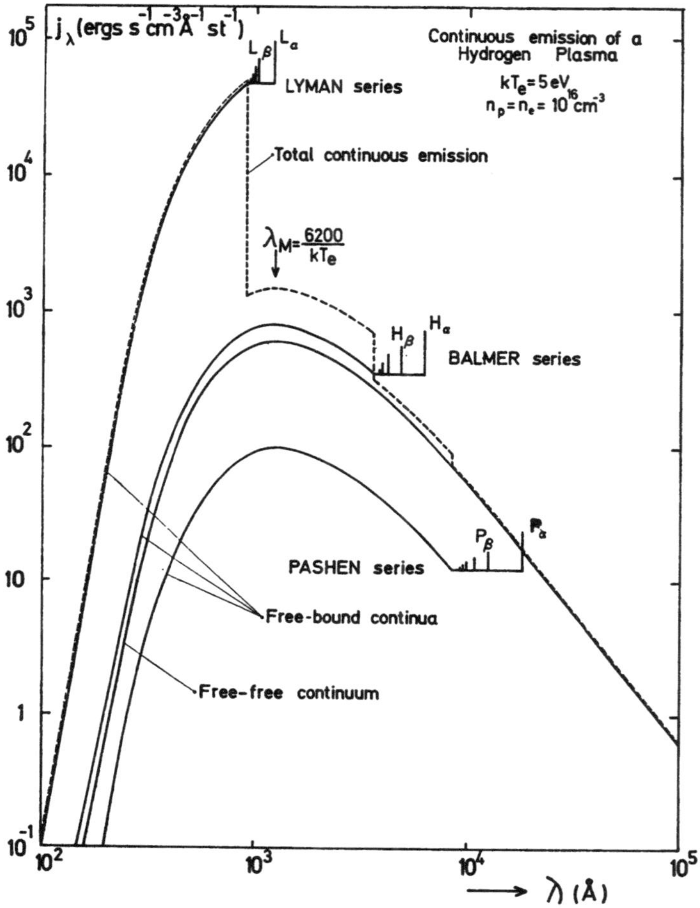

FIG. 2. Free-free and free-bound continua for a hydrogen plasma ($j_\lambda = \frac{1}{4\pi} W_\lambda$). Line series intensities are only schematically represented.

E. Recombination spectra or free-bound continua

A free electron of kinetic energy E_e can be radiatively captured by an ion of charge r into any excited level j of the ion of charge $r - 1$. The emitted photon has an energy equal to the sum $E_e + \chi_{j\infty}$, $\chi_{j\infty}$ being the ionization energy of the ion of charge $r - 1$ from level j. The recombination spectrum can be calculated for hydrogen-like ions. In this case $r = Z$, the atomic number of the element and the hydrogenic ion of charge $Z - 1$, has an ionization energy from level of principal quantum number j given by

$$\chi_{j\infty} = \chi_H Z^2/j^2. \tag{14}$$

Assuming a Maxwellian velocity distribution for free electrons, the power $W_{\nu,j}\, d\nu$, radiated between ν and $\nu + d\nu$, accompanying capture into level j

is given by

$$W_{\nu,j}\,d\nu = 2.15 \times 10^{-32}\, n_e n_Z\, \frac{Z^4}{T_e^{3/2}}\, \frac{g_{fb}}{j^3} \exp\left(\frac{\chi_{j\infty} - h\nu}{kT_e}\right) d\nu, \tag{15}$$

where T_e is in K, $W_{\nu,j}\,d\nu$ in erg s^{-1} cm^{-3}, and g_{fb} the Gaunt factor for free-bound transitions is of the order unity. The total intensity $W_\nu\,d\nu$ is obtained by summing over the j's larger than $j_m = (\chi_H Z^2/h\nu)^{1/2}$:

$$W_\nu\,d\nu = 2.15 \times 10^{-32}\, \frac{Z^4}{T_e^{3/2}}\, n_e n_Z \left\{\sum_{j>j_m}^{\infty} \frac{g_{fb}}{j^3} \exp\left(\chi_H Z^2/j^2 kT_e\right)\right\} \exp\left(h\nu/kT_e\right) d\nu. \tag{16}$$

An example of recombination continua is given in Fig. 2 for hydrogen.

Between two discontinuities, free-free and free-bound continua have the same shape. The free-free spectrum dominates for high temperatures ($kT_e \gg \chi_H Z^2$).

F. Radiation trapping

Depending on plasma temperature, particle density and thickness, radiation can be self-absorbed to a degree which also depends on the process involved. Details on radiative transfer problems can be found in ref. 27. The relationship between the intensity J_ν from the plasma and the source function j_ν/k_ν can be written as follows:

$$J_\nu = J_{0\nu} \exp(-\tau_{\nu l}) + \exp(-\tau_{\nu l}) \int_0^{\tau_{\nu l}} \frac{j_\nu}{k_\nu} \exp(\tau_{\nu l})\, d\tau_{\nu l}, \tag{17}$$

were $\tau_{\nu l} = k_\nu l$ is the optical thickness for the frequency ν ($d\tau_{\nu l} = k_\nu dl$), l is the geometric thickness. The first term of eqn. (17) is included only if there is an external source. The source function j_ν/k_ν is generally not known, and hypotheses have to be made in order to solve the transfer equation.

A mean escape factor $\mathscr{E}(\tau_{\nu l})$ along a line of sight of the source of geometric thickness l and for a frequency ν can be defined by the ratio of the emergent intensity J_ν to the intensity without self-absorption. This coefficient, for a plasma of homogeneous density and temperature, is

$$\mathscr{E}(\tau_{\nu l}) = \frac{1 - \exp(-\tau_{\nu l})}{\tau_{\nu l}}. \tag{18}$$

Absorption processes can take place through free-free transitions (inverse bremsstrahlung) or free-bound transitions (photoionization). These two processes will be negligible in laboratory plasmas of low density. Even at low number densities of the absorbing species, optical depths can be large in resonance line radiation (photoexcitation) and self-absorption can become appreciable.

Assuming the emission profile to be identical with the absorption profile[28] and to have a Gaussian shape, the mean escape factor for a transition $j \rightarrow i$, after integrating $\mathscr{E}(\tau_{\nu l})$ over the whole line profile, is as follows:

$$\left.\begin{aligned} \mathscr{E}_{ij}(\tau_l) &= \pi^{-1/2}\tau_l^{-1}\int\{1-\exp[-\tau_l\exp(-y^2)]\}\,dy,\\ \tau_l &= 2.5\times10^{-2}\,\frac{n_{r,i}f_{ij}}{\Delta\nu_{ji}}\,l,^* \end{aligned}\right\} \tag{19}$$

where $\Delta\nu_{ji}$ is the width of the line (FWHM), $n_{r,i}$ the particle density of the absorber in level i, f_{ij} the oscillator strength, and $\tau_l = k_0 l$ the optical thickness at the centre of the line. y is the reduced frequency such as $y = 2\,\frac{\nu-\nu_0}{\Delta\nu}\sqrt{(\ln 2)}$. The line intensity from eqn. (11) has to be multiplied by the mean escape factor in order to get the actual intensity of the transition $j \rightarrow i$:

$$J_{ji} = \frac{1}{4\pi}\,h\nu_{ji}n_{r,j}A_{ji}\times\mathscr{E}_{ij}(\tau_l)\times l. \tag{20}$$

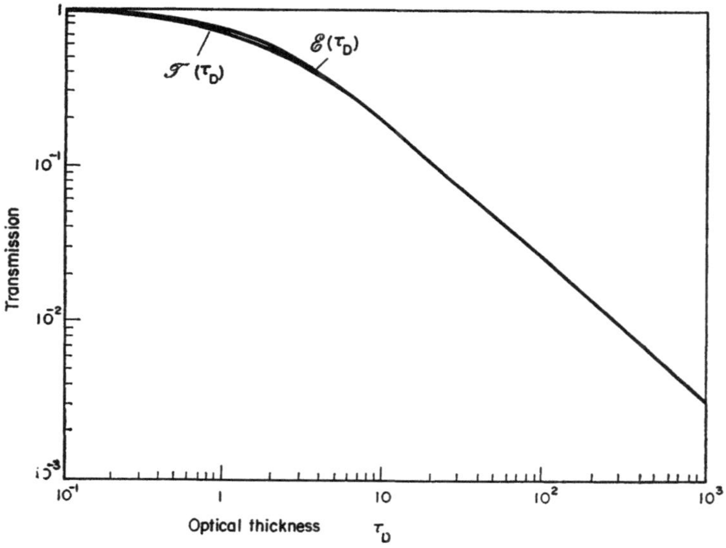

FIG. 3. Mean escape factors for a cylindrical plasma of diameter D and infinite length. $\mathscr{E}(\tau_D)$ is for a diametrical line of sight and $\mathscr{T}(\tau_D)$ is for the whole volume by integrating over all directions (see ref. 42).

The mean escape factor $\mathscr{E}(\tau_l)$ is given in Fig. 3 as a function of τ_l; on this figure, l is equal to the diameter D of a cylindrical plasma. As an example, for the Lyman-α line ($\lambda = 1216$ Å, $f_{12} = 0.416$), for a neutral density of 10^{14} cm^{-3} and an atomic temperature of 2 eV ($\Delta\nu = 2.7\times10^{11}$ s^{-1}), the optical thickness is $\tau_l = 10$, and $\mathscr{E}(\tau_l) = 0.2$ for $l = 10$ cm.

* See Mitchell and Zemansky,[29] or Breton and Schwob[42].

II. CORONAL MODEL

In the coronal model the following assumptions are made:

(1) negligible optical thickness at all wavelengths;

(2) rate of spontaneous decay larger than collisional rate (excitation or de-excitation) for any level;

(3) Maxwellian velocity distribution for free electrons.

Among the processes listed in § I, the only remaining ones are (Fig. 4):

ionization from the ground state by electron impact (c′: $f \to \varkappa$);

excitation from the ground state (a: $f \to j$) by electron impact with subsequent spontaneous decay to the ground state or by cascade (d′: $j \to i$);

free-free (g: $\varkappa \to \varkappa'$) and free-bound (f: $\varkappa \to j$) transitions.

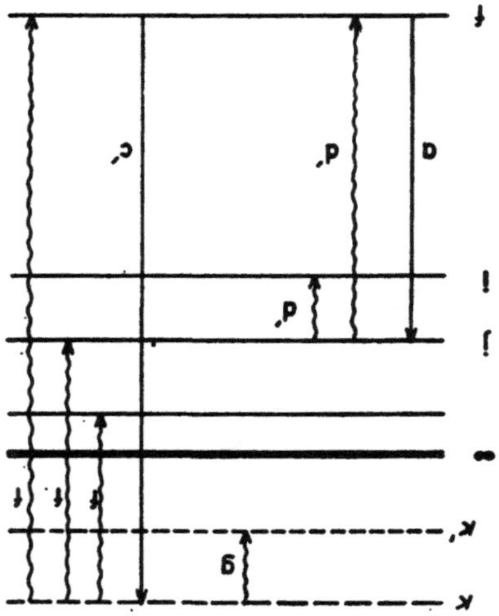

Fig. 4. Processes involved in the coronal model.

Every stage of ionization r will be populated and depopulated at a rate given by the following equation:

$$\frac{dn_r}{dt} = n_e n_{r-1} S_{r-1} + n_e n_{r+1} \alpha_{r+1} - n_e n_r S_r - n_e n_r \alpha_r \tag{21}$$

with $\Sigma r n_r = n_e$.

Rate coefficients for ionization from the ground state and for radiative recombination to all levels are functions of the electron temperature.

At any time the rate of spontaneous decay of a level j is equal to the collisional excitation rate from the ground state. The instantaneous population density of an excited level j is thus given by the following equation:

$$\frac{n_{r,j}}{n_{r,f}} = \frac{n_e Q_{fj}(T_e)}{\sum\limits_{x=f}^{j-1} A_{jx}}. \tag{22}$$

In this model $n_{r,f}$ is much greater than the population density of any excited level; thus it is approximately equal to the total number density $n_r = \sum_{j=f}^{\infty} n_{r,j}$ of ions of charge r.

The line intensity for the transition $j \rightarrow i$, in terms of radiance [eqn. (11)] is

$$J_{ji} = \frac{1}{4\pi} n_e n_r h\nu_{ji} \frac{A_{ji}}{\sum_{x=f}^{j-1} A_{jx}} Q_{fj}(T_e) l. \tag{23}$$

After a sufficient time, the transient coronal model will end in a stationary state given by eqn. (21) in which $dn_r/dt = 0$. It follows that the ratio of the densities of two consecutive degrees of ionization in a stationary state is given by

$$\frac{n_{r+1}}{n_r} = \frac{S_r(T_e)}{\alpha_{r+1}(T_e)}. \tag{24}$$

The transient coronal model has been used for interpreting spectra of high-temperature plasmas at low density for hydrogen and impurities[30–32] and the stationary coronal model for the solar corona.[33–36]

III. COLLISIONAL RADIATIVE MODEL

The coronal model is only valid for rather low density plasmas. In fact, collisional processes involving excited levels are not negligible for the majority of non-LTE laboratory plasmas. This becomes more important for high quantum numbers. The collisional-radiative (CR) model takes every collisional process into account. Only two radiative processes are considered, viz. spontaneous decay and radiative recombination. A detailed study of this problem has been made by Bates *et al.*[37, 38] and by Hearn and McWhirter.[39]

A. Equations

Every level population is described by a differential equation, expressing the algebric sum of the rates of processes populating and depopulating the level:

$$\begin{aligned} \frac{dn_{r,j}}{dt} = & -n_{r,j}\left[n_e\left(S_{j\infty} + \sum_{k=j+1}^{\infty} Q_{jk} + \sum_{i=f}^{j-1} Q_{ji}\right) + \sum_{i=f}^{j-1} A_{ji}\right] \\ & + n_e\left[\sum_{k=j+1}^{\infty} n_{r,k} Q_{kj} + \sum_{i=f}^{j-1} n_{r,i} Q_{ij}\right] + \sum_{k=j+1}^{\infty} n_{r,k} A_{kj} \\ & + n_e^2 n_{r+1} \beta_{\kappa j} + n_e n_{r+1} \alpha_{\kappa j} \qquad (i < j < k). \end{aligned} \tag{25}$$

There are as many equations as levels. In transient regimes the rate coefficients S, Q, α, β—functions of temperature—will in general depend on time. It is clearly impossible to solve such a system of equations. Fortunately, their number can be limited. For quantum numbers larger than a certain value q, a critical level, Saha equilibrium will apply.

B. Critical level, partial local thermodynamic equilibrium

With increasing quantum number, the probability of collisional processes becomes larger and the probability of radiative processes becomes smaller. Thus to any desired accuracy there is always some critical level q for which the effect of the radiative processes may be neglected. For j equal or greater than q collisional processes dominate, and the Saha equation may be used to calculate population densities of these levels. Hinnov and Hirschberg[40, 41] have shown the existence of such a critical level in a helium plasma of low density.

The number of equations of type 25 reduces to $q - 1$. A level such as $j \geqq q$ is populated according to

$$\frac{n_{r+1,f}}{n_{r,j}} n_e = 2 \frac{g_{r+1,f}}{g_{r,j}} \frac{(2\pi m_e k T_e)^{3/2}}{h^3} \exp\left(-\chi_{j\infty}/kT_e\right). \tag{26}$$

In practice increasing values of q are tried until comparison of the results of the calculations shows that the change is smaller than a given figure.

C. Quasi-stationary state

If one compares the relaxation time of energy exchange of bound electrons in excited levels with each other or with the free electrons, one finds that this time is far smaller than the relaxation time for the ground level. As a result, any change in the population density in the ground level is followed by a practically instantaneous rearrangement in excited level populations. The set of eqns. (25) reduces to one differential equation governing the ground-level population, $q - 2$ equations such as

$$\frac{dn_{r,j}}{dt} = 0 \qquad (f < j < q),$$

and some Saha equations of type (26) for levels j equal or greater than q.

The collisional radiative model reduces to the LTE model for high densities and to the coronal model for low densities. In between, choice of the critical level q has to be made.

IV. EFFECT OF RADIATION TRAPPING

Resonance lines may be strongly self-absorbed even for plasmas of quite low density. In transient regimes non-resonance lines may also be trapped to a significant degree due to the fact that the populations of excited levels are substantially larger than the populations in stationary conditions. Radiation trapping acts in two ways: (1) modification of the excited level populations, and (2) reduction of the amount of the emergent light intensity [eqn. (20)].

To take into account the effect of radiation trapping on the level populations, the simplest way is to introduce a "trapping coefficient" $\mathcal{T}_{ij}$ $(o < \mathcal{T}_{ij} < 1)$ for a given transition $j \rightarrow i$, which reduces the corresponding Einstein coefficient A_{ji} so that the terms $\sum A_{ji}$ in eqn. (25) become $\sum A_{ji}\mathcal{T}_{ij}$. The coefficients $\mathcal{T}_{ij}$ are of the same nature as the $\mathscr{E}_{ij}$'s [eqn. (19)]. $\mathscr{E}_{ij}(\tau_l)$ is the mean escape factor for a thickness l of the plasma along a line of sight. To obtain $\mathscr{T}_{ij}$, one has to calculate mean values of $\mathscr{E}_{ij}(\tau_l)$ over all possible dimensions of the plasma in all directions. Such a calculation has been carried out by Breton and Schwob[42] for a cylindrical plasma of diameter D and of infinite length.

$\mathscr{T}_{ij}(\tau_D)$ is given in Fig. 3 together with the function $\mathscr{E}_{ij}(\tau_D)$ relative to a diametrical line of sight ($l = D$). These two coefficients are equivalent for large optical thicknesses and of course for $\tau_D = 0$. The difference is never very large, and in the following rate equations $\mathscr{T}_{ij}(\tau_D)$ may be replaced by $\mathscr{E}_{ij}(\tau_D)$ making the calculation much easier, i.e.

$$\frac{dn_{r,j}}{dt} = -n_{r,j}\left[n_e\left(S_{j\infty} + \sum_{k=j+1}^{\infty} Q_{jk} + \sum_{i=f}^{j-1} Q_{ji}\right) + \sum_{i=f}^{j-1} A_{ji}\mathscr{E}_{ij}\right] + n_e\left[\sum_{k=j+1}^{\infty} n_{r,k}Q_{kj} + \sum_{i=f}^{j-1} n_{r,i}Q_{ij}\right] + \sum_{k=j+1}^{\infty} n_{r,k}A_{kj}\mathscr{E}_{jk} + n_e^2 n_{r+1}\beta_{\varkappa j} + n_e n_{r+1}\alpha_{\varkappa j}. \quad (27)$$

$\mathscr{E}_{ij}(\tau_D)$ depends on time [eqn. (19)] through (1) the density of the self-absorbing species, (2) the line width (slowly because $\Delta\nu_{ji}$ varies as $\sqrt{(T_r)}$, (3) the plasma diameter (in a pinch-discharge for example).

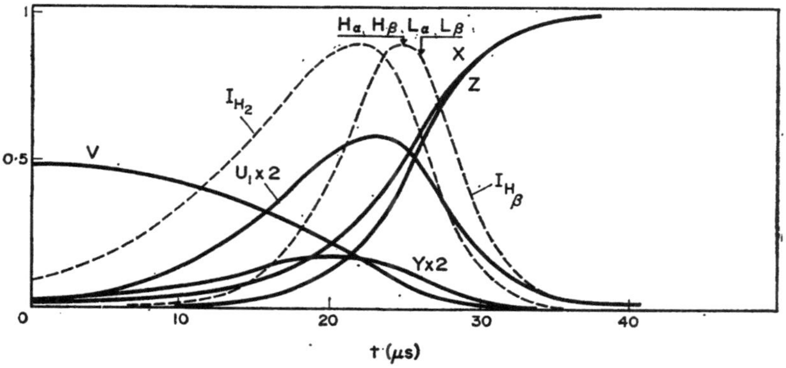

FIG. 5. Build-up phase of a proton-electron plasma from a molecular hydrogen gas. For each species, relative number densities are given in terms of hydrogen nucleus number density. V is for H_2, U_1 for H in the ground level (only represented), X is for electrons, Y for H_2^+ ions, and Z for protons. The initial density is $N = 10^{14}$ cm^{-3} and initial temperatures are $kT_e = 8$ eV and $kT_H = 1.5$ eV. Arrows indicate the intensity maxima of Balmer and Lyman lines.[42]

Models taking account of radiation trapping have been used for hydrogen. Bates and Kingston,[38] in studying plasma recombination with the CR model, introduce a total trapping for resonance lines by simply making $\mathscr{E}_{fj} = 0$.

Breton and Schwob[18, 42, 43] have used an absorption coefficient calculated for each line emitted during its build-up phase in a plasma starting from molecular hydrogen, so that the population density of a level j varies from zero to zero through a maximum value. Time histories of the various species are shown in Fig. 5 together with line intensities. Self-absorption experiments, interpreted in terms of this model, have been performed[18, 44] for a toroidal pinch (TA 2000).

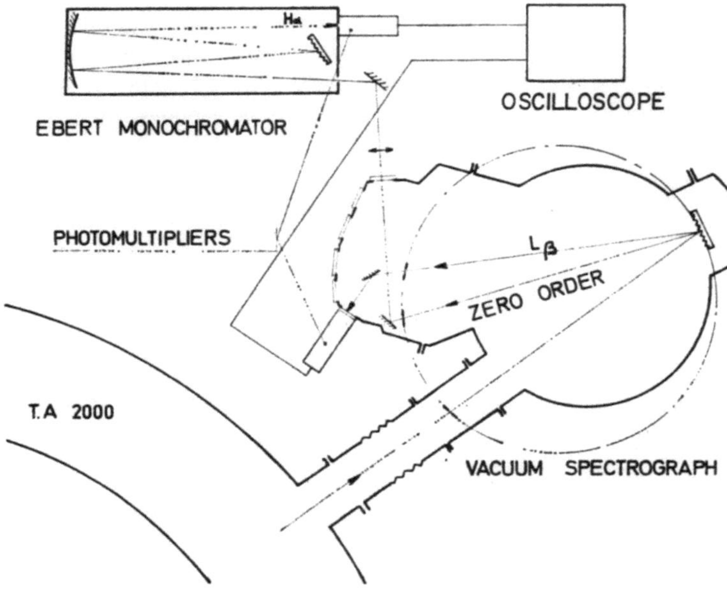

Fig. 6. Experimental set-up for studying the trapping effect on lines.

Measurements were made for Lyman and Balmer lines (L_α, L_β, H_α, H_β). The experimental set-up is shown in Fig. 6. The zero order of the vacuum spectrograph was used for visible measurements to make sure that the same plasma zone was observed in both spectral ranges.

Range of validity of the models

The range of validity of a model depends on the electron density and temperature. LTE will be valid for a quantum number equal or greater than q if collisional processes are preponderant for this level. This can be expressed by comparing the rate coefficient for collisional de-excitation with the rate of spontaneous decay towards any level j below q:

$$n_e Q_{qj} \gg A_{qj} \quad (j < q). \tag{28}$$

In order to evaluate this inequality we shall use a simple empirical formula for the excitation cross-section (deduced from a general empirical formula[45]) to get an analytical expression for Q_{jq} and to obtain the rate coefficient for collisional de-excitation Q_{qj} from eqn. (7). The resulting criterion for LTE to be valid is

$$n_e \gg 10^{14} \chi_{jq}^3 \, (kT_e)^{1/2}. \tag{29}$$

χ_{jq} and kT_e are in electron volts. Complete LTE is achieved if q is the first excited level. For hydrogen-like ions,

$$\chi_{12} = \frac{3}{4}\,\chi_H Z^2 = 10.2\, Z^2\,\mathrm{eV},$$

so that the complete LTE criterion is

$$n_e \gg 10^{17} Z^6 \,(kT_e)^{1/2}. \tag{30}$$

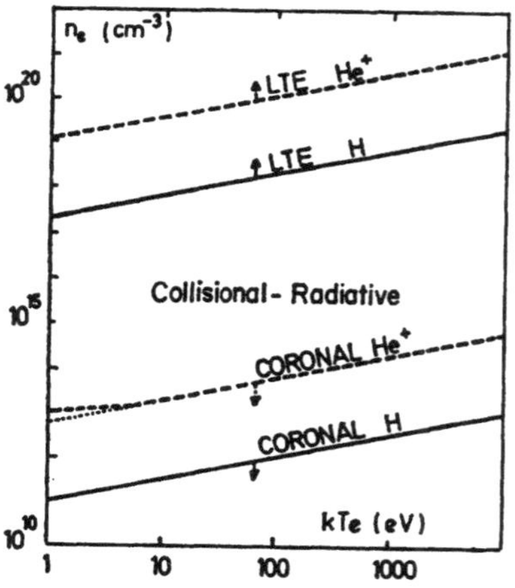

FIG. 7. Limits of validity of the models

Similar expressions have been found by Griem[4] and McWhirter.[2] The limit of the LTE model is shown in Fig. 7 for hydrogen and singly ionized helium.

Descending in the density scale we pass through the CR area, and then we reach the coronal domain, the limit of which is given by choosing a high enough value for the critical level q. Taking $q = 6$ after McWhirter[2] we get from inequality (29) and with $\chi_{56} = 0.17 Z^2$ eV, the following criterion for the coronal model:

$$n_e \ll 5 \times 10^{11} Z^6 \,(kT_e)^{1/2}. \tag{31}$$

The majority of the laboratory plasmas will lie in the CR area, with additional radiative transfer problems depending on the size of the plasma.

PLASMA DIAGNOSTICS BY VACUUM ULTRAVIOLET METHODS

Spectroscopic methods can be well adapted to derive plasma parameters such as electron and ion temperatures, electron and ion densities as function of time and space. Recycling of particles from the outside of the plasma, particle and power losses, and impurity levels provide other problems which can be studied spectroscopically. Reviews can be found in recent books.[2,46] As previously mentioned, the v.u.v. is the most important spectral range for hot plasmas containing highly stripped ions. In this section we shall deal with specific v.u.v. methods, mainly for electron temperature measurements. Other investigations have also been made in this spectral region such as:

(1) Determination of the amounts of impurities by use of resonance line intensities.

(2) Doppler broadening measurements. These are mainly performed in the visible range where high resolution is easily obtained (as an example, multichannel Fabry-Pérot measurements of ion temperature can be cited[47]). However, with increasing temperature only very weak lines of multiply-charged ions are available in the visible, and the stronger v.u.v. lines have to be used (despite the unfavourable $\Delta\lambda/\lambda$ ratio) for temperatures greater than 1 keV. Sawyer *et al.*[48] have made such measurements in the extreme u.v. using the Lyman-α line of O VIII (19 Å) or the helium-like O VII line at 18.6 Å; ion temperatures between 1 and 10 keV were so obtained in a theta-pinch experiment (Scylla I).

(3) Stark broadening measurements. The majority of these n_e measurements are performed in the visible. Nevertheless, some experiments have been done in the v.u.v., on the Lyman lines of hydrogen,[49–51] on O VI and K IX lines between 200 and 100 Å,[52] and on He II lines ($\lambda = 1215, 1085, 1025, 992$ Å).[53]

(4) Confinement time and recycling problems may be studied through the temporal behaviour of spectral line intensities.[54]

(5) Power losses through lines and continua, which maximize in the v.u.v. range, have been measured by Post,[55] Bishop *et al.*,[56] and Kolb and McWhirter.[57]

I. CALIBRATION OF VACUUM ULTRAVIOLET SPECTROGRAPHS

The spectrograph and detector have a response depending on the wavelength, and for many problems this response has to be known in either relative or absolute terms. For example, line intensity ratio measurements need a relative calibration whereas measurements of energy losses through radiation necessitate an absolute calibration. In this section we shall give a brief survey of the calibration problem, recently reviewed by Samson.[58]

The effective intensity at a wavelength λ per unit of wavelength $I_\lambda(\lambda, k)$, measured, for example, as the current from a photoelectric detector, is proportional to the product of three functions of the wavelength, i.e.:

(1) the spectral radiance of the source in photons, $J_\lambda/h\nu$, per unit of wavelength, per cm^2, per second, and per unit solid angle;

(2) the spectrograph transmission $T(\lambda, k)$, k being the order of diffraction of the grating ($0 < T < 1$);

(3) the quantum efficiency of the detector $R(\lambda)$,

$$I_\lambda(\lambda, k) = \frac{1}{h\nu} J_\lambda \, G \, T(\lambda, k) R(\lambda) g e_0; \tag{32}$$

the geometrical factor G is the spectrograph *étendue* (which is the product of the entrance slit area by the solid angle subtended by the grating at the slit), and g the detector gain. If e_0 is the electron charge in coulombs, $I_\lambda(\lambda, k)$ is the output current in amperes. The spectrograph-detector assembly is often calibrated as a whole and the apparatus function, $\mathscr{A}(\lambda, k)$, is $T(\lambda, k)R(\lambda)g$.

$$I_\lambda(\lambda, k) = \frac{1}{h\nu} J_\lambda G \mathscr{A}(\lambda, k) e_0. \tag{33}$$

Calibration can be achieved with a known source J_λ. Use of a continuum can be difficult because of order overlapping. It is often more convenient to make measurements with a line source at a number of points. In this case, the intensity can be measured for the whole line profile and then it is related to the radiance J_{fi} given by eqn. (20). The order sorter problem in the v.u.v. region has been studied by Jaeglé.[59]

A. *Transmission measurements*

$T(\lambda, k)$ can be measured with an unknown line source by using a v.u.v. monochromator, the exit slit of which is located at the entrance slit of the spectrograph to be calibrated. A movable detector, following the entrance slit, gives the input intensity. A second detector, calibrated against the first one, gives the output intensity. The ratio of the two intensities is $T(\lambda, k)$. With a suitable arrangement only one detector may be necessary.[60–62] By using an absolute detector at the entrance slit of the spectrograph one obtains directly the apparatus function $\mathscr{A}(\lambda, k)$.[60]

B. *Detectors*

The most commonly used are photomultipliers which have the advantage of fast response and high gain. For short wavelengths ($\lambda < 1500$ Å) open multipliers with metallic photocathodes have the added advantage of being blind to visible light. Quantum efficiencies are known within $\pm 20\%$. For wavelengths between 2000 and 1500 Å one has to use standard photomultipliers with phosphors which can be also used for the whole v.u.v. region. The efficiency of the phosphor-photocathode system is not accurately known.

C. *Sources*

Only a few sources have a calculable continuous spectrum suitable for calibration purposes in the v.u.v. range. The radiation from a synchrotron has been computed by Schwinger[63] and used by Codling and Madden,[64] and by Pitz.[65] Free-free or bremsstrahlung radiation which has a calculable intensity for a plasma of known parameters (n_e, T_e) has been proposed by Bogen *et al.*[66] as a standard source in this region. Other sources, if available, could then be used as secondary standards, for, in principle, the knowledge of $T(\lambda, k)$ and $R(\lambda)$ or $\mathscr{A}(\lambda, k)$ enables us to measure any source spectrum $J(\lambda)$.

D. Direct measurements of apparatus function $\mathscr{A}(\lambda, k)$

Methods of determination of the apparatus function $\mathscr{A}(\lambda, k)$, which do not necessitate an absolute intensity measurement in the v.u.v., have been proposed in the past few years.

(i) *The branching ratio method.* Consider a pair of lines with a common upper level, emitted by a plasma, observed by two different spectrometers. If trapping is negligible for both lines, their intensities do not depend on plasma conditions but only on the Einstein coefficients and some other known constants. The ratio of the actual intensities at the detectors will be [eqns. (11) and (33)]:

$$\frac{I(\lambda_{ji}, k)}{I(\lambda_{jm}, k')} = \frac{A_{ji} G \mathscr{A}(\lambda_{ji}, k)}{A_{jm} G' \mathscr{A}'(\lambda_{jm}, k')}. \tag{34}$$

One of the lines (λ_{jm}) is located in the visible and its intensity can be measured with a spectrometer of *étendue* G' calibrated again a tungsten ribbon lamp giving $\mathscr{A}'(\lambda, k')$. Thus from the measured intensity ratio [eqn. (34)], one deduces the apparatus function $\mathscr{A}(\lambda_{ji}, k)$ for the v.u.v. spectrometer. This method was first described by Zaidel *et al.*(67) and has been used by many others.(45, 68–71)

(ii) *The coincidence method.* Cristoferi *et al.*(72) use a 2 keV electron beam to excite molecular hydrogen. Among the processes occurring, the two lines H_α (transition $3 \rightarrow 2$) and L_α (transition $2 \rightarrow 1$) are emitted and their intensities can be measured by a photon-counting method. Some of the L_α photons are in coincidence with the H_α photons. These are L_α photons following H_α photons due to cascade processes in the same atoms. The number of H_α photons detected, in terms of pulses per second, is

$$n_{H_\alpha} = \frac{1}{h\nu_{H_\alpha}} J_{H_\alpha} G' \mathscr{A}'(\lambda_{H_\alpha}, k') \frac{1}{g'}$$

and the number of coincidences detected per second is

$$n_c = \frac{1}{h\nu_{H_\alpha}} J_{H_\alpha} G' \mathscr{A}'(\lambda_{H_\alpha}, k') \frac{1}{g'} \times \frac{1}{4\pi} G \mathscr{A}(\lambda_{L_\alpha}, k) \frac{1}{g}, \tag{35}$$

with the same notation as above. So the apparatus function in the v.u.v. $\mathscr{A}(\lambda_{L_\alpha}, k)$ for the particular wavelength $\lambda = 1216$ Å is

$$\mathscr{A}(\lambda_{L_\alpha}, k) = \frac{4\pi g}{G} \frac{n_c}{n_{H_\alpha}}. \tag{36}$$

II. METHODS OF PLASMA DIAGNOSTIC BASED ON LINE INTENSITIES

A. Time behaviour of line intensities

This method for pulsed discharges is based on the time history of line intensities corresponding to the successive stages of ionization of an element added in trace amount to a hydrogen plasma. One computes, with the appropriate model, the relationship between the time of appearence of a line (actually the time of the maximum intensity) and the temperature. One only needs relative intensity measurements without the necessity of calibrating the spectrograph, which is an appreciable advantage.

(i) *LTE plasmas*. In this case, we are dealing with plasmas at equilibrium, and any change in the various population densities, given by Saha and Boltzmann equations, is due to increasing or decreasing temperature or atom densities. The intensity of a line [eqn. (11)] is proportional to the population of the upper level given by eqn. (5), so that

$$J_{ji} = \frac{1}{4\pi} h\nu_{ji} n_r A_{ji} \frac{g_{r,j}}{U_r(T)} \exp - (\chi_{fj}/kT)\, l. \tag{37}$$

n_r, the total number density of the ions of charge r, is given by solving a system of Saha equations [eqn. (4)] for the additional element and for hydrogen, which is assumed, to simplify, to be initially an atomic gas:

$$\left.\begin{aligned}
\frac{n_1}{n_0} &= \frac{1}{n_e} \times \frac{2U_1(T)}{U_0(T)} \times \frac{(2\pi m_e kT)^{3/2}}{h^3} \exp - (\chi_0/kT), \\
&\vdots \\
\frac{n_r}{n_{r-1}} &= \frac{1}{n_e} \times \frac{2U_r(T)}{U_{r-1}(T)} \times \frac{(2\pi m_e kT)^{3/2}}{h^3} \exp - (\chi_{r-1}/kT), \\
&\vdots \\
\frac{n_s}{n_{s-1}} &= \frac{1}{n_e} \times \frac{2U_s(T)}{U_{s-1}(T)} \times \frac{(2\pi m_e kT)^{3/2}}{h^3} \exp - (\chi_{s-1}/kT), \\
\frac{n_P}{n_H} &= \frac{1}{n_e} \times \frac{2U_P(T)}{U_H(T)} \times \frac{(2\pi m_e kT)^{3/2}}{h^3} \exp - (\chi_H/kT), \\
n_e &= n_P + \sum_{r=0}^{s} r n_r, \\
n_H + n_P &= N, \\
\sum_{r=0}^{s} n_r &= aN.
\end{aligned}\right\} \tag{38}$$

s is the highest stage of ionization observed in the plasma; n_H and n_P are the number densities of hydrogen atoms and protons. a is the amount of impurity, and N the instantaneous density of neutral and ionized hydrogen atoms.

Using eqns. (37) and (38) one deduces the temperature corresponding to appearance of the maxima of the line intensities, for the N and a experimental values. By comparison with the experimental time for occurrence of the maximum of the line intensity, one gets the plasma temperature at that particular time. For wellconfined plasmas, the total density of hydrogen N is a constant.

The problem may be complicated by radiation trapping. The effect of this can be accounted for by multiplying eqn. (37) by the appropriate escape factor given by eqn. (19).

(ii) *Transient coronal plasmas.* In coronal plasmas, excitation and ionization are due to electron impacts with the atoms or ions in their ground-states and, as a consequence, only the electron temperature can be obtained from measurements of line intensities. The ion temperature may be different.

Let us consider a pulsed hydrogen plasma with a density in the range 10^{12} to 10^{14} cm^{-3} and an initial temperature of the order of 10 eV. A small amount of an impurity is added (carbon, nitrogen, oxygen, ...). After a short time, the hydrogen is fully ionized but impurities need a much longer time for successive ionizations to a final state, which may be the completely stripped ion if the temperature reaches a high enough value. During these ionization phenomena, the lines emitted by the successive ions appear at successive times. A relationship can be found between temperature and the time of appearance of these lines. This method of measuring electron temperature from time history of line intensity has been developped by Breton[32] using lines of hydrogen, helium, and oxygen atoms and ions in toroidal discharges. Further work[45, 73, 74] was done for the cases of nitrogen and argon ions. Similar measurements were made by Goldman and Kilb[75] on a theta-pinch discharge.

The set of eqns. (21), valid for a confined plasma (without particle losses or recycling), can be written as follow:

$$\left.\begin{aligned}
\frac{dv_0}{dt} &= Nx\,[v_1\alpha_1 - v_0S_0],\\
&\vdots\\
\frac{dv_r}{dt} &= Nx\,[v_{r-1}S_{r-1} + v_{r+1}\alpha_{r+1} - v_rS_r - v_r\alpha_r],\\
&\vdots\\
\frac{dv_s}{dt} &= Nx\,[v_{s-1}S_{s-1} - v_s\alpha_s],\\
x &= v_r + a\sum_{r=1}^{s} rv_r,\\
\sum_{r=0}^{s} v_P &= 1,
\end{aligned}\right\} \tag{39}$$

where relative electron and proton densities are expressed in terms of the instantaneous total density of hydrogen $N = n_H + n_P$: $x = n_e/N$, $v_P = n_P/N$. Relative densities of impurity ions are expressed in terms of the total density of impurity: $v_r = n_r/\sum_{r=0}^{s} n_r$. Simplification can be achieved by assuming a negligible amount of added impurity so that $x = v_P$; the latter reaches a constant value equal to the unit after the short transient period of ionization of hydrogen. It is possible to deduce x as a function of time from the observed temporal behaviour of hydrogen line intensities.[18, 43] The system of eqns. (39) can be solved, either with constant coefficients, α and S, signifying a constant temperature, or with α and S function of time, corresponding to a variable temperature. Line intensities are given by eqn. (23) depending on both electron and ion densities and on the temperature through the rate coefficients $Q(T_e)$.

The process of obtaining the electron temperature as a function of time, by solving the equations with a specified time dependence of T_e and adjusting this by comparing the results with the actual position of the line intensity maxima, is rather tedious. One rather can get the actual function $T_e(t)$ by an approximate method, making it always possible, if desirable, to check the results by solving eqns. (39) for this particular temperature dependence. There are two possibilities for this second method:

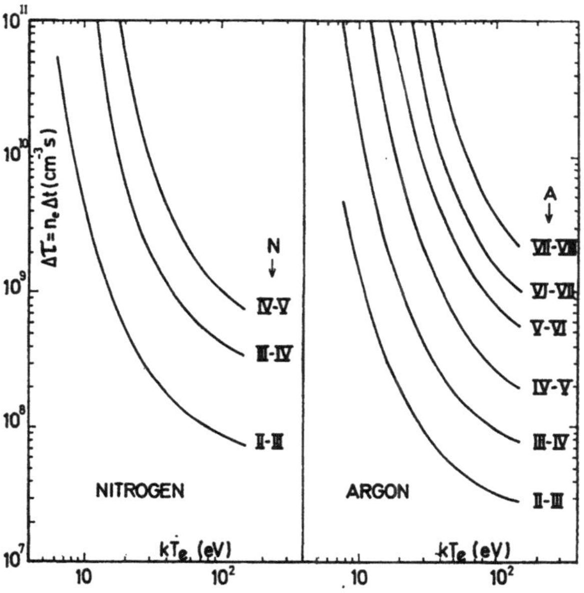

Fig. 8. $\Delta\tau_{r,r+1} = n_e \Delta t_{r,r+1}$ as a function of electron temperature for nitrogen and argon. Only ionizations of outer shell electrons are considered.[74]

(1) One computes the relative ion densities $v_r(t)$ from eqns. (39) for a set of values of the temperature (α and S constant) and for a constant value of the electron density Nx, assuming a negligible amount of impurity ($n_e = n_P$). Thus one obtains the time interval $\Delta t_{r,r+1} = t_{r+1} - t_r$ between two successive maxima of ion densities, and then plots the reduced time $\Delta\tau_{r,r+1} = Nx\, \Delta t_{r,r+1}$ as a function of temperature. Figure 8 shows such plots for nitrogen and argon.[74]

Because of the large jump in ionization energies after the last outer electron is stripped off, the number of differential equations was here limited to $s = 5$ for nitrogen and $s = 8$ for argon. Figure 9 gives the time history of nitrogen ion lines in the TA 2000 torus.[45, 74] On Fig. 10b are given the points obtained from Fig. 8 using the Δt's from Fig. 9. The full line represents the curve $T_e(t)$ (obtained after successive approximations), used in solving the set of eqns. (39); results are shown on Fig. 10a together with the measured position of the maxima of line intensities (arrowed).

(2) As indicated in ref. (54), v_r can be computed in an approximate form in the case of constant density and temperature and for a small amount of impurity. Using this formulation for $v_r(t, T_e)$ one gets the time t_r of appearance of the maximum of v_r by differentiating with respect to t and making $dv_r/dt = 0$. It follows

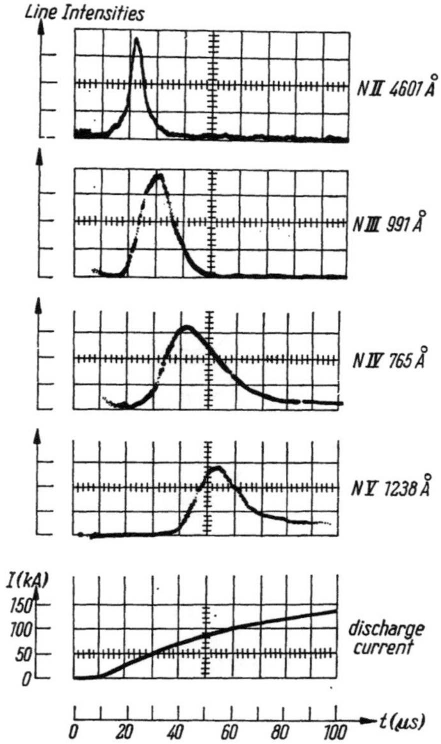

FIG. 9. Intensity versus time of nitrogen ion lines and discharge current in a toroidal pinch (TA 2000).[74]

the expression value for the reduced time $\Delta\tau_{r,r+1} = Nx\,(t_{r+1} - t_r)$, which is a function of T_e only[45, 74]

$$\Delta\tau_{r,r+1} = \frac{1}{X_{r+1} - X_r} \ln \frac{X_{r+1}Y_{r+1}}{X_r Y_r} - \frac{1}{X_r - X_{r-1}} \ln \frac{X_r Y_r}{X_{r-1}Y_{r-1}} \tag{40}$$

where

$$X_r = S_r + \alpha_r + \alpha_{r+1} \quad \text{and} \quad Y_r = \sum_{q=r+1}^{s} V_q .$$

V_q is the relative density of the ion of charge q, in terms of total density of impurity, for the steady state at the same temperature, as given by eqn. (24), i.e.

$$V_q = \frac{S_{q-1}}{\alpha_q} V_{q-1}, \quad \sum_{q=0}^{s} V_q = 1 . \tag{41}$$

This set of equations can be easily solved for various values of T_e knowing the temperature dependance of the rate coefficients. The accuracy of this method is limited by the approximate character of the relevant cross-sections, and the

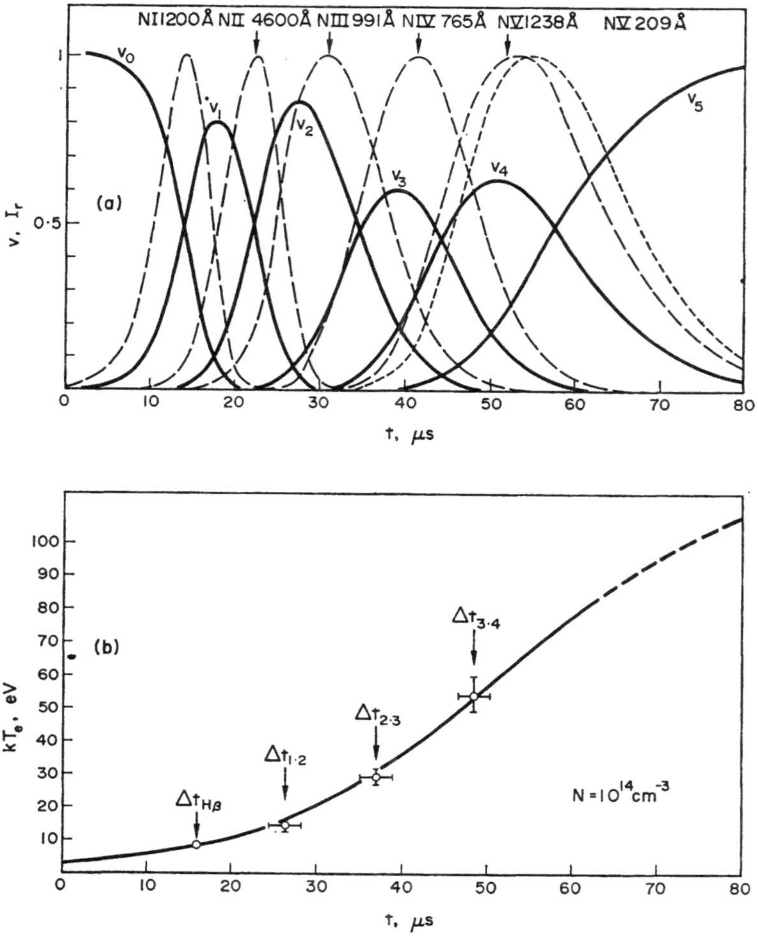

FIG. 10. Interpretation of line intensities of Fig. 9 in terms of the coronal model. In (b), the points are from the approximate method using the Δt's. The full $T_e(t)$ curve is used to obtain the results of (a). Line intensities are normalized to unity. The arrows indicate the measured position of the maxima (from Fig. 9).(74)

difference between results obtained by solving eqns. (40) and (41) rather than exact differential eqns. (39), is probably within this uncertainty. Other sources of error are of experimental nature, uncertainty of electron density as a function of time, spatial variations of the various parameters, and inaccuracies in the Δt's due to the limited amount of light in the impurity lines used.

As a conclusion, it should be noted that this method may be extended, in principle, to higher temperatures by use of elements of larger Z, for which strong lines lie at shorter wavelengths.

(iii) *Steady-state coronal plasmas.* If characteristic times for plasma parameter evolution are much larger than the time constant necessary to reach the steady state described by eqn. (24), any change in the various population densities is due to increasing or decreasing temperature. Relative line intensities may be used in the way indicated above for LTE plasmas but using eqn. (24) or (41) plus eqn. (23). Coronal steady-state regimes are rarely obtained in laboratory plasmas.

(iv) *Collisional-Radiative plasmas.* In this case we have to use the CR model. The set of eqns. (25), or possibly eqns. (27) if radiation trapping is important, has to be considered. As pointed out before, this model has only been treated for hydrogen or hydrogen-like ions, and seems to be too complicated to apply to a sequence of ions.

The complete model with transient self-absorption has been used[43] to measure the electron temperature in the ionization phase of a hydrogen plasma; comparison was made with the coronal model.[18] Suckewer[76, 77] gives an approximate method for calculating excited level populations in terms of ground-state population by using a recursion formula.

B. Line intensity ratio measurements

With the appropriate model we calculate the ratio of the intensity of two lines as a function of temperature and compare this with the instantaneous measured ratio. This method is valid for either transient or stationary conditions. It does not depend on confinement and is unaffected by particle losses or recycling. Furthermore, in some cases knowledge of the electron density is unnecessary. Relative calibration of the spectrograph is needed for the two wavelengths concerned.

(i) *LTE plasmas.* Details about the line intensity ratio method for temperature measurements of LTE plasmas can be found in ref. 4. The method has been mainly used in the visible range.

The intensity ratio of two lines of *the same ion* of charge r is [from eqns. (11) and (5)]

$$\frac{J_{ji}}{J_{nm}} = \frac{\nu_{ji}}{\nu_{nm}} \frac{A_{ji}}{A_{nm}} \frac{g_{r,j}}{g_{r,n}} \exp\left[(\chi_{fn} - \chi_{fj})/kT\right]. \tag{42}$$

This does not depend on electron or ion density, but the Einstein coefficients are in general not known experimentally. They are calculable for hydrogen-like ions and alkali-like ions. This method has been used in the v.u.v. with Al III lines[78, 79] and with Si IV and Cl VII lines,[79] for temperature measurements of vacuum sparks.

The two lines can be produced by *two consecutive ions*. From eqns. (4), (5), and (11),

$$\frac{J_{r+1,ji}}{J_{r,nm}} = \frac{\nu_{ji}A_{ji}g_{r+1,j}}{\nu_{nm}A_{nm}g_{r,n}} \times \frac{2(2\pi m_e kT)^{3/2}}{h^3} \times \frac{1}{n_e} \exp\left[(\chi_{f'n} - \chi_{fj} - \chi_r)/kT\right], \tag{43}$$

(f and f' refer to the ground levels of the ions of charge $r + 1$ and r). The intensity ratio of eqn. (43) is a sensitive function of temperature and the accuracy of the Einstein coefficients is of less importance than in the previous case, but the electron density has to be known from other measurements. Nevertheless, it is possible to eliminate the electron density by using two pairs of lines, thus yielding both temperature and density.[80]

For these two methods the plasma is assumed to be optically thin. This is often not the case, and a careful check is necessary when choosing the lines.

(ii) *Coronal plasmas.* Excitation processes from the ground state control the line intensities according to eqn. (23). Consequently, knowledge of impact excitation cross-sections is essential for any measurement based on line intensities.

As for LTE plasmas, one may compare two lines produced by *two consecutive ions*. The proposal has been made[4] to employ pair of lines O V 1371 Å and O VI 1032 Å—useful in the range of electron temperature 10-20 eV. Such a method depends on excitation rate coefficients and also on ionization and recombination rate coefficients governing the relative populations of the two ions; furthermore, the electron density is needed.

The method based on use of intensities of two lines of *the same ion* is much simpler. Their intensity ratio is [(eqn. (23)]

$$\frac{J_{ji}}{J_{nm}} = \frac{\nu_{ji}}{\nu_{nm}} \frac{A_{ji}}{A_{nm}} \frac{\sum_{y=1}^{n-1} A_{ny}}{\sum_{x=1}^{j-1} A_{jx}} \times \frac{Q_{fj}(T_e)}{Q_{fn}(T_e)}. \tag{44}$$

This is a function of temperature only, but a sensitive one if the excitation rate coefficients have different temperature dependences. Taking relative calibration of the spectrographs at the two wavelengths into account, the actual measured ratio, according to eqn. (33), is

$$\frac{I(\lambda_{ji}, k)}{I(\lambda_{nm}, k')} = \frac{G}{G'} \times \frac{\mathscr{A}(\lambda_{ji}, k)}{\mathscr{A}'(\lambda_{nm}, k')} \times \frac{K_{ji}}{K_{nm}} \times \frac{Q_{fj}(T_e)}{Q_{fn}(T_e)}, \tag{45}$$

where K_{ji} and K_{nm} are characteristic atomic constants for each line; G and G' are equal if the same apparatus is used.

Considering triplet and singlet levels, cross-sections for their excitation from the ground state are of very differing shape as functions of electron energy. Consequently, the triplet to singlet line intensity ratio should be a very sensitive function of temperature. This method, proposed for neutral helium by Cunningham[81] and refined by Sovie,[82] can be extended to ionized atoms of the He I sequence in the v.u.v. Excitations from metastable levels are neglected, and this seems to be an important limitation.[83–87]

Line intensity ratios in the v.u.v. have been used by Hinnov.[88] The measurements were performed on pairs of resonance lines of O III (703-508 Å) and of O IV (790-554 Å). Accuracy was limited by the small difference between the excitation energies of the upper levels of the lines and by uncertainties in the excitation cross-sections and Einstein coefficients for these ions, which have several outer-shell electrons.

An improvement consists in the use of alkali-like ions having one outer-shell electron.(16, 89) The Einstein coefficients are more reliably calculable in the Coulomb approximation,(90) and data on cross-sections are more secure. Furthermore, there

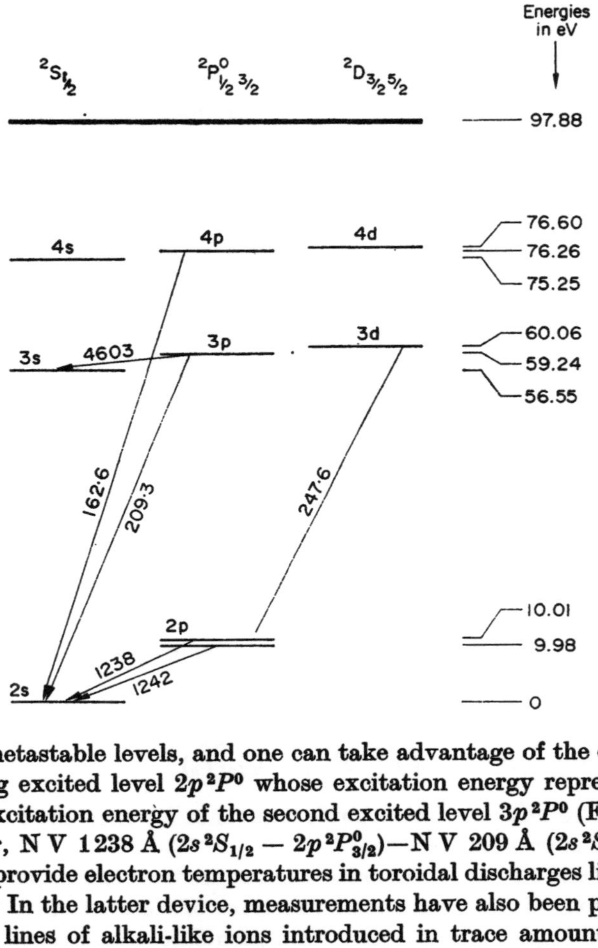

Fig. 11. Partial energy level scheme for alkali-like N^{4+} ions.

are no metastable levels, and one can take advantage of the existence of the first low-lying excited level $2p\,^2P^0$ whose excitation energy represents less than 20% of the excitation energy of the second excited level $3p\,^2P^0$ (Fig. 11). The nitrogen line pair, N V 1238 Å ($2s\,^2S_{1/2} - 2p\,^2P^0_{3/2}$)—N V 209 Å ($2s\,^2S - 3p\,^2P^0$) has been used to provide electron temperatures in toroidal discharges like ZETA(89) and TA 2000.(16) In the latter device, measurements have also been performed with other pairs of lines of alkali-like ions introduced in trace amounts in hydrogen. The following carbon and oxygen pairs have been choosen:(45, 91)

$$\text{C IV } 1548.2\,\text{Å}\ (2s\,^2S_{1/2} - 2p\,^2P^0_{3/2}) - \text{C IV } 312.4\,\text{Å}\ (2s\,^2S_{1/2} - 3p\,^2P^0_{1/2,3/2}),$$

$$\text{O VI } 1031.9\,\text{Å}\ (2s\,^2S_{1/2} - 2p\,^2P^0_{3/2}) - \text{O VI } 150.1\,\text{Å}\ (2s\,^2S_{1/2} - 3p\,^2P^0_{1/2,3/2}).$$

It should be noted that the $3S - 3P^0$ lines, located in the visible (C IV 5801 Å, N V 4603 Å, O VI 3819 Å), can be used for calibrating the vacuum spectrograph at the $2S - 3P^0$ wavelengths (C IV 312 Å, N V 209 Å, O VI 150 Å), the branching ratio method. These lines may also be used instead of the $2S - 3P^0$ lines in measuring temperatures,[45,91] but in this case two spectrographs have to be calibrated against each other.

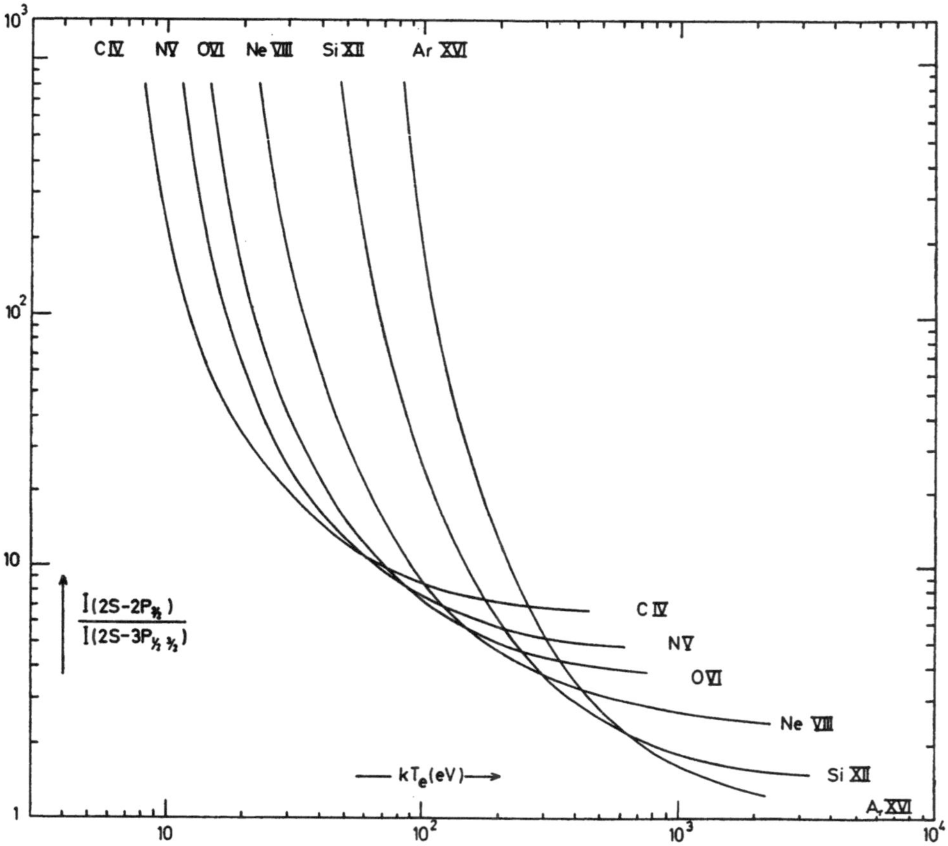

FIG. 12. The $2S - 2P^0/2S - 3P^0$ intensity ratio for C^{3+}, N^{4+}, O^{5+}, Ne^{7+}, Si^{11+}, Ar^{15+} alkali-like ions obtained by using calculated cross-sections.[93, 94]

As pointed out before, accuracy is limited by errors in the cross-sections for excitation of the two upper levels $2P^0$ and $3P^0$. No experimental measurements are available. General formulae by van Regemorter[92] and Seaton[13] were used by Heroux[89] and, with some modifications, by Schwob.[16] More recently calculations for these particular transitions have been worked out by Bely,[93] and by Burke *et al.*[94] The $2S - 2P^0$ to $2S - 3P^0$ line intensity ratios have been recalculated with these new cross-sections for C IV, N V, and O VI. The results are given in ref. 91 (Fig. 12). Use of the previous cross-sections led to underestimation of temperature by a factor of three.

Employing the curves for C IV, N V, and O VI, temperatures in the range 10-200 eV can be determined. Extension of the method to higher temperatures is, of course, feasible by using heavier alkali-like ions[89] such as Ne VIII, Si XII, and Ar XVI, with the possibility of reaching an electron temperature of 1 keV (Fig. 12).

(iii) *Collisional-Radiative plasmas.* Intensity ratios calculated through the CR model will depend on both electron temperature and density. Using a recursion formula, Suckewer[76, 77] was able to compute line intensity ratios for alkali-like ions C^{3+}, N^{4+}, O^{5+} for the $(2S - 2P^0)$-$(2S - 3P^0)$ pairs of lines used by Heroux[89] and Schwob.[16] The results fall between the curve calculated with Boltzmann populations for excited levels (LTE model) and the curve for the coronal model. Use of the intensity ratio of one line pair obviously cannot fix the values of both temperature and density. However, by choosing another pair of lines,[95] namely $(2P^0 - 3D)$-$(2S - 3P^0)$, it is possible to avoid this difficulty. The $3P^0$ and $3D$ levels have practically the same excitation energy but very different excitation processes (Fig. 11) and, as a result, the intensity ratio of the second line pair is weakly dependent on the electron temperature over a wide range of electron densities (typically between 10^{13} and 10^{16} cm^{-3}), and measurement of this ratio can be used for fixing the density, and the electron temperature is then obtained from the $2S - 2P^0/2S - 3P^0$ intensity ratio.

As regards the accuracy of measurements interpreted by the Suckewer model, it should be pointed out that the cross-sections used are of the type given by van Regemorter[92] and Seaton,[13] and large differences may be expected when more refined values[93, 94] are used. Furthermore, at large densities (10^{14}-10^{16} cm^{-3}) strong radiation trapping, depending on the size of the plasma, will occur for resonance transitions.

III. METHODS BASED ON CONTINUA

Formulae for free-free and free-bound continua have been given by eqns. (12) and (16). Their intensities depend on electron density and temperature; the spectral distributions are independent of any of the models described above.

The only conditions necessary are a Maxwellian electron velocity distribution and an optically thin plasma in the spectral region under consideration—two conditions easily fulfilled. Furthermore, the methods are based on relative intensity measurements. In practice, spectra are calculable only for simple cases (hydrogen and hydrogen-like ions), and for low density plasmas intensities are quite low and the results may be strongly affected by the presence of even weak lines.

The total power emitted per cm^3 at a frequency ν from free-free and free-bound transitions is given by summing eqns. (12) and (16):

$$W_\nu \, d\nu \simeq 6.8 \times 10^{-38} n_e n_z \frac{Z^2}{T_e^{3/2}} \exp(-h\nu/kT_e) \times \left[T_e + 3.16 \times 10^5 Z^2 \sum_{j>j_m}^{\infty} \frac{1}{j^3} \exp(\chi_H Z^2/j^2 kT_e)\right] d\nu. \tag{46}$$

A. Series discontinuity

One measures the continuum intensity at two frequencies, ν_1 and ν_2, one on either side of the series discontinuity ν_D. In the ratio $D_n = W_{\nu_1}/W_{\nu_2}$ the exponential factor is nearly equal to unity ($\nu_2 - \nu_1 \ll kT_e/h$), n being the lower quantum number of the series under consideration,

$$D_n = \frac{3.16 \times 10^{-6} T_e + Z^2 \sum\limits_{j=n+1}^{\infty} \frac{1}{j^3} \exp\,(\chi_{\rm H} Z^2/j^2 kT_e)}{3.16 \times 10^{-6} T_e + Z^2 \sum\limits_{j=n}^{\infty} \frac{1}{j^3} \exp\,(\chi_{\rm H} Z^2/j^2 kT_e)}. \tag{47}$$

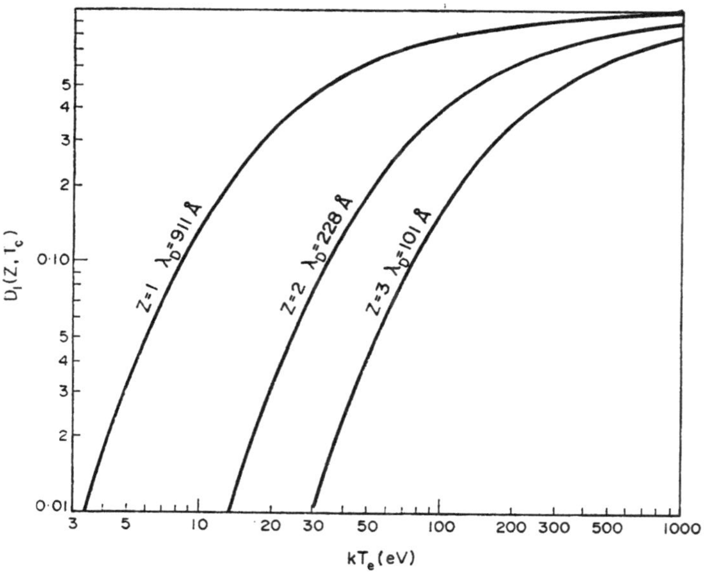

FIG. 13. Lyman discontinuity $D_1(Z, T_e) = W_{\nu_1}/W_{\nu_2}$, ($\nu_2 > \nu_D > \nu_1$). Gaunt factors have been taken equal to unity.

The Lyman discontinuity $D_1(Z, T_e)$ is shown on Fig. 13 for hydrogen and for hydrogen-like He^+ and Li^{++}. Gaunt factors have been taken equal to unity, and for more accurate calculations corrections can be found in ref. (96). Measurements by series discontinuities have been made by astrophysicists (see, for example, ref. 97) and on some laboratory plasmas.[98, 99]

B. Slope measurement

If there are no discontinuities in a portion of the continuum, its relative intensity can be written as follows:

$$W_\nu \, d\nu = C \exp\,(-h\nu/kT_e)\, d\nu.$$

The semi-logarithmic plot yields a straight line, with a slope depending only on electron temperature. Sawyer *et al.*[100, 101] have used this method in the extreme u.v. (20-5 Å) for temperature measurements in a theta-pinch (Fig. 14), and Boland *et al.*[102] have applied it to a laser-produced plasma, using recombination spectra of C^{6+} and C^{5+} (20-35 Å). The measurement may be limited to two wavelengths whose intensity ratio is

$$\exp\left[\frac{h(\nu_2-\nu_1)}{kT_e}\right].^{(103)}$$

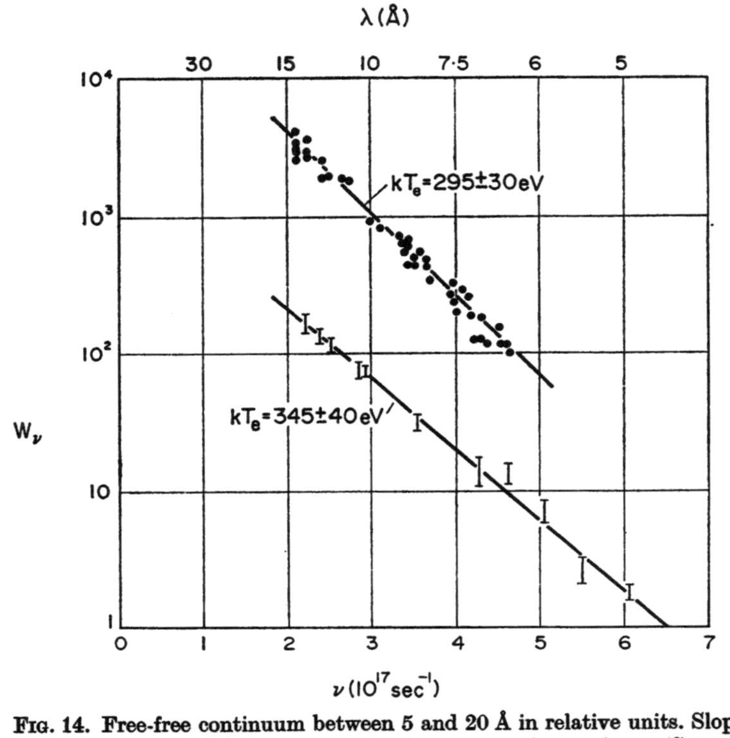

FIG. 14. Free-free continuum between 5 and 20 Å in relative units. Slope measurement of electron temperature in the Scylla theta-pinch experiment (Sawyer *et al.*[101])

C. Measurement of the maximum intensity wavelength

The free-free transition spectrum W_λ has a maximum at $\lambda_M = 6200/kT_e$ (λ_M in Å, kT_e in eV) [eqn. (13)]; a free-bound transition continuum has a maximum at the same wavelength if $kT_e > \frac{1}{2}(Z^2\chi_H/j^2)$. This maximum may be shifted by the apparatus function or by the presence of discontinuities in the neighbourhood.

Ehler[104] used this method for temperature estimation on a laser-produced tungsten plasma, assuming a pure free-free continuum.

Schwob *et al.*[105] have used the same method, with more refinements, on a laser-produced aluminium plasma (Fig. 15). The calculated spectrum (Fig. 16)[106] shows no discontinuity near λ_M. The measured spectrum, together with the contribution of overlapping orders and the net intensity are shown in Fig. 15. The net intensity, obtained after calibration of the v.u.v. spectrograph by the branching ratio method, exhibits an increase toward short wavelengths corresponding to recombination continua (Fig. 16).

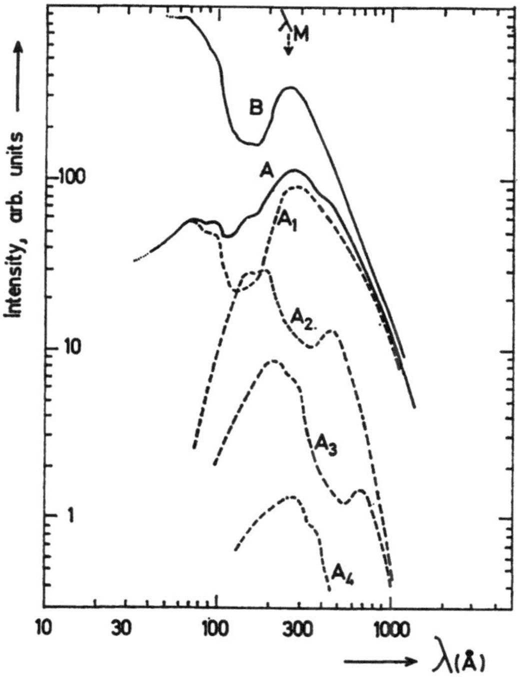

FIG. 15. Continua in relative units obtained from photoelectric measurements on a laser-produced aluminium plasma using a grazing-incidence vacuum spectrograph. A is the measured intensity; A_1, A_2, A_3, A_4 are contributions of successive orders in A, and B is the net intensity taking into account relative intensity calibration. Temperature is deduced from the maximum intensity wavelengths λ_M of the free-free continuum. The rise towards short wavelengths corresponds to recombination spectra.[105]

D. Use of absorbers in the soft X-ray range

Radiation over a wide frequency band can be selected by use of an absorber of known absorption coefficient $\mathscr{K}_\nu$.[107] Clearly this method is only applicable for wavelengths smaller than a limit of some tens of Å and, hence, for plasmas of density and temperature high enough to have measurable emission in this region. Usually, only the free-free continuum is considered. The transmitted intensity at a frequency ν is $W_\nu \exp(-\mathscr{K}_\nu l)\, d\nu$, where $W_\nu\, d\nu$ is the intensity given by eqn. (12) and l the thickness of the absorber. The intensity measured at the detector is the product of the emergent intensity and the response $R(\nu)$ of the

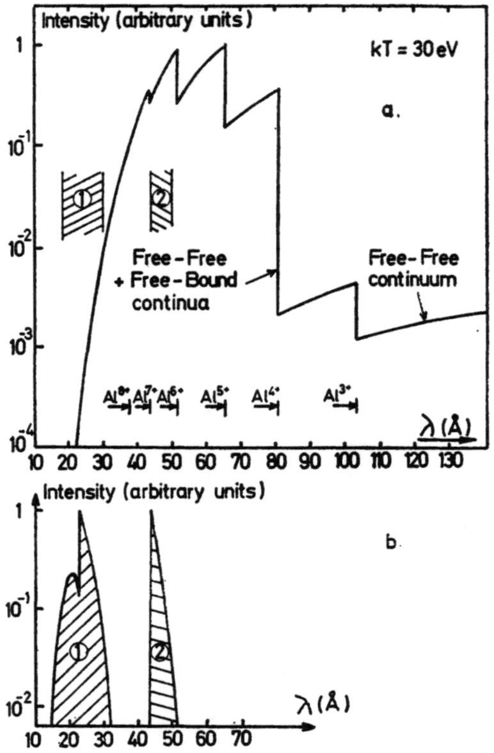

FIG. 16. (a) Calculated spectrum for an aluminium high density plasma. Recombination spectra of resonance series from Al^{8+} to Al^{3+} are included. (b) Normalized intensities transmitted through two filters of different type.(106)

detector. The ratio of measured intensities with and without the absorber is

$$\Psi(l, T_e) = \frac{\int_0^\infty R(\nu) \exp - (\mathcal{K}_\nu l + h\nu/kT_e)\, d\nu}{\int_0^\infty R(\nu) \exp(-h\nu/kT_e)\, d\nu}. \tag{48}$$

In practice one calculates isotherms $\Psi_{Te}(l)$(108, 109) which are compared with the experimental curve obtained in varying the absorber thicknesses. In an experiment two channels are used in the same geometry, one of them containing the absorbers. This method was first applied to theta-pinches,(108, 110) and has since been applied to many high-density high-temperature plasmas; namely other theta-pinches ($kT_e \sim 1$ keV, $n_e \sim 10^{16}$ cm^{-3}),(111–113) "plasma focus" discharges ($kT_e \sim 5$ keV, $n_e \sim 10^{20}$ cm^{-3}),(108, 114–116) and laser-produced plasmas both from solid targets(117) and from gas break-down.(118–120) In all these experiments fully ionized plasmas and pure free-free continuum are considered, and temperatures always exceed 100 eV.

Seka *et al.*[106] have extended the method to lower temperatures ($kT_e \sim$ 10-30 eV) for laser-produced aluminium plasmas. In such cases, free-bound continua make significant contributions to the observed spectra. Experimentally, two channels are used, each having a filter of different characteristics so that the temperature is obtained with only one shot. Figure 16 shows the calculated spectrum for $kT_e = 30$ eV and the transmitted intensity through the two filters.

HOT PLASMAS AS FAR ULTRAVIOLET SOURCES

I. EMISSION SPECTRA OF HOT PLASMAS

A number of sources producing line spectra as well as continua in the far u.v. are available. It is out of the scope of this section to give a complete review of the subject. Details can be found in Samson's book.[58]

High-temperature plasmas produced in magnetic configurations* are very interesting for obtaining far u.v. spectra. Some of them operate at quite low densities ($n_e \simeq 10^{12}$-10^{14} cm^{-3}), the temperature being in the range of a few hundreds of electron volts. Such sources yield very weak continua and are chiefly useful as line sources (toroidal pinches, stellarators, tokamaks etc.). Other configurations have higher working densities ($n_e \simeq 10^{16}$ cm^{-3}) and can reach temperatures of 1 keV (theta-pinches, etc.). Focusing pinches or "plasma focus" devices provide densities of the order of 10^{20} cm^{-3} and temperatures of a few kilo electron volts; very strong continua and line spectra of highly stripped ions are emitted down to the soft X-ray range. In the last few years the use of powerful lasers (from megawatts to gigawatts), focused on solid targets has made it possible to obtain very intense spectra.

A. New lines from highly stripped ions

The first work was done by Swedish physicists[122] by use of vacuum sparks. A review article was published some years ago by Edlen.[123] More recently results have been obtained with a three-electrode spark,[124a] and researches in a number of laboratories have been conducted with such sources. In particular, Schwob and Fraenkel,[124b] with an improved mounting, have obtained in the spark a hot point plasma giving spectra of highly stripped ions (Fe XXVI, Ni XXVI) down to X-ray region.

Identification of the various degrees of ionization has been achieved (1) by varying the electrical characteristics of the spark; (2) by using spatial resolution,[125] the higher degree of ionization being closer to the cathode; (3) by time-resolved spectra[126]: the higher degrees of ionization are only present during the first cycle.

Vacuum spark techniques did not include the capability of producing spectra of ions of the rare gases. Furthermore it was sometimes not easy to separate the spectrum of each degree of ionization. Some of these difficulties were overcome by using new types of sources.

* For details see ref. 121.

(i) *Spectra from magnetically confined plasmas.* Studies have been performed mainly with *Z*-pinches, θ-pinches, and plasma-focus devices. The advantages of such light sources are the following. The element under study is added in small amount in hydrogen. An easy separation of the various stages of ionization is achieved by time resolving the line intensities,[100, 127] since all lines of a given degree of ionization achieve their intensity maxima at about the same time (Fig. 9 and 10a). Spatial resolution has also been used,[128] the lower degrees of ionization being located near the outside of the plasma. In these discharges, each ion species has its own temperature and can be identified by measuring the Doppler widths of the lines.[128] In some experiments the separation by ion stage has been achieved by varying the voltage or the pressure[129, 130] or the amount of the added element.[131] Measurements are made easier by the fact that time constants are longer than those of vacuum sparks. Reproducibility and stability are also superior and, finally, the large size of the plasma is an appreciable advantage.

Z-pinches. Early measurements were performed in the near u.v. on the SPECTRE III and IV devices[128, 131] for rare gas spectra (Ne IV, Ne V, Ne VI) and N V.[132] In the far u.v. ionized spectra of B, C, N, O, F, and Ne were observed and lines of Ne IV, Ne VI, and Ne VIII were identified.[133]

The ZETA torus was extensively used for classification work in the range 1000 to 400 Å[127] for Ne VI-VII-VIII, Ar VI-VIII, Kr V-VI-VII-VIII, and Xe V-VI-VII-VIII-IX and in the range 400-16 Å[134] for C-V-VI, N VI-VII, O VII-VIII, and Ar IX-X; comparison was also made in the range 210-170 Å, between the solar spectrum and the ZETA spectrum. Further work was performed in the range 170-30 Å for rare gases.[135] Iron lines of interest for the solar spectrum (220-144 Å), from Fe VIII to Fe XII and Fe XIV were identified together with the same transitions of the isoelectronic sequences for argon through nickel using the ZETA torus, a 5 kJ theta-pinch and a 70 kV vacuum spark.[136]

θ-pinches. Early measurements in the range 16-6 Å were performed by Sawyer *et al.*[100, 101] on the SCYLLA device using a crystal spectrometer. The Lyman series of O VIII, lines of helium-like ions such as Na X, Mg XI, Al XII, and Si XIII, and, finally, lines of Ne VIII, Ne IX, and Ne X were identified. The first series of O VII and O VIII were observed in the range 25-15 Å in further investigations.[48, 137]

The 13.55-13.44 Å lines of Ne IX[135] and the O VI and O VII spectra[138] between 150 and 78 Å were identified in the "MAGGI-I" thetatron, and "MAGGI-II" was used to study the resonance lines of Ar XI-XII, Kr IX-X,[139] and helium-like series of N VI and Ne IX, $2p\,^1P^0 - nd\,^1D$, $2s\,^3S - np\,^3P^0$ and $2p\,^3P^0 - nd\,^3D$, in the spectral range 200-40 Å.[140]

Identification of new O III, O V, N V, and N VI lines[130] was first reported from experiments with the Uppsala θ-pinch, and complete spectra for N V[141] and N IV[142] between 8000 and 200 Å, were published.

Plasma focus. Using this type of discharge, Peacock *et al.*[143] made new identifications in the Ne IX-X and Ar XVII-XVIII spectra. Lines involving inner-shells for ions of moderate degrees of ionization have also been observed down to the soft X-ray region.

(ii) *Laser impacts on solid targets.* The beam of a powerful laser is focused on a solid target located in a vacuum chamber. A small amount of material is vaporized and ionized, and a rapid expansion takes place. Near the target surface, the density is in the range 10^{20}-10^{21} cm^{-3} and temperatures are between a few tens of electron volts and a few hundreds of electron volts. The emitted spectrum is a very strong continuum with lines of the various stages of ionization of the target material.

This technique is suitable for study of highly ionized spectra of any solid element which is rather difficult in experiments on magnetically confined plasmas.

Separation of the spectra of different stages of ionization is achieved (1) by varying the laser power; (2) by a spatial resolution,[52, 126, 144]—using a second slit crossed with the entrance slit of the grazing incidence spectrograph, or by locating the source at the Sirks focus of a normal incidence spectrograph. An improvement lies in use of the multilayer technique,[145] which permits elimination of the peripheral zone of the laser impact area, where the lower power density produces spectra of lower stages of ionization; (3) by time resolving the line intensities,[102, 105] the lower stages of ionization appearing at later times; (4) by analysing the shapes of the lines. This gives information on the self-absorption from which one can deduce the ionization stage and even the type of transition.[146] In some cases, the highly ionized species appear in emission, and the lower degrees of ionization in absorption.[147]

The first observations in the far u.v. were made by Ehler and Weissler.[148] Lines of Fe XV-XVI and Ni XVII-XVIII in the range 65-25 Å have been identified by Fawcett *et al.*[147] by use of a 400 MW ruby laser. Classification of new lines of the $2p\,^1P^0 - nd\,^1D$, $2s\,^3S - np\,^3P^0$, and $2p\,^3P^0 - nd\,^3D$ series of C V between 270 and 130 Å were made with a more powerful laser[140] and a polythene target. The area 1000-40 Å has been studied[144] with a 10 J glass neodymium laser and aluminium or calcium targets, producing Al VI to Al XI or Ca XII to Ca XIV spectra. New spectra of Fe XII to Fe XVIII and their isoelectronic sequences have been identified.[126] New lines of Fe XII and Fe XIII complete the classification of the intense features of the solar spectrum between 220 and 167 Å. Transitions involving inner-shell electrons have been identified for fluorine-, oxygen-, and nitrogen-like ions in S X, K XI, XII, XIII, Ca XII, XIII, XIV, Sc XIII, and Ti XIV together with lines in S XII to S XIV and in Cl XII to Cl XV spectra[149] using a 500 MW ruby laser. Finally, spectral lines of Ti VII to XII and isoelectronic transitions for V, Cr, Mn, and Fe in the range 600-250 Å have been classified;[150] identifications of Fe XIV solar lines have been confirmed.

B. Sources of continua

Hot plasma sources emit a continuum. In any case the wavelength of the intensity maximum is roughly given by [eqn. (13)]: λ_M (Å) $= 6200/kT_e$ (eV), a formula strictly valid in the case of pure free-free continuum; thus a plasma at $kT_e \simeq 100$ eV and $n_e \simeq 10^{16}$-10^{20} cm^{-3} may be expected to have a strong continuous emission in the v.u.v. This radiation has been used for diagnostic purpose in various ways to determine the temperature (see p. 270) in θ-pinches, plasma focus experiments, and laser-produced plasmas; continuous spectra have also been proposed as standard sources (see p. 259).

The intensity of the continuum from a laser-produced plasma increases with the atomic number of the target element. This spectrum is either a free-free continuum as observed by Ehler[104] with a tungsten target, or a recombination spectrum from targets of low atomic number (Li, C, etc.). The Lyman continuum of the hydrogen-like carbon ion C VI has been observed,[102] or, more generally, a superposition of both free-free and free-bound continua.[52, 105, 106, 151] Esteva and Romand[152] made a comparison of this type of spectrum with spectra obtained with vacuum sparks, and Carillon *et al.*[153] have studied the optical absorption of a laser-produced plasma using a second similar plasma as a source of continuum.

II. DETERMINATION OF ATOMIC CONSTANTS

We have seen that rate coefficients for ionization and excitation are the main parameters needed in order to compute intensities with one of the non-LTE models. These are derived from cross-sections which are either measured or calculated. A number of reliable measurements exist, but only for neutral or, in a few cases, for singly ionized atoms. Therefore, even a crude estimation of these rate coefficients or cross-sections is of value. Such estimation is possible by using the appropriate model in which known plasma parameters are introduced (such as electron temperature and density, confinement time, etc.) in a reverse way to the procedure adopted in the previous section to get electron temperature. The cross-sections cannot be obtained from such measurements but only rate coefficients, which may be compared with rate coefficients calculated from adjusted cross-sections. In this way oscillator strengths can be derived from excitation rate coefficients.

Very few results have been reported so far from this technique of measurement. Hinnov[154] reports rate coefficients for ionization and oscillator strengths for neon ions. Other measurements have been made for He,[155] O VII,[156] C V,[157] and N V.[158]

We shall briefly give the principle of the method based on the coronal model. Let us consider a homogeneous hydrogen plasma with electron temperature and density known as function of time. A small amount of the element under study (He, O, N, Ne, ...) is added, and this impurity suffers successive ionizations according to coronal rate eqn. (21). In the ionization phase, recombination can be neglected so that

$$\frac{dn_r}{dt} = n_e n_{r-1} S_{r-1} - n_e n_r S_r - \frac{n_r}{\tau_c}, \tag{49}$$

where τ_c is the particle containment time. Measurements are based on absolute line intensities as function of time. These intensities are given by eqn. (23). Considering the first resonance line this equation reduces to

$$J_{21} = \frac{1}{4\pi} h\nu_{21} n_e n_r Q_{12} l, \tag{50}$$

where Q_{12} is the rate coefficient for excitation from the ground level (level 1) to the first excited level (level 2). Such a measurement needs an absolute calibration of the vacuum spectrograph.

A. Evaluation of excitation rate coefficients and oscillator strengths

A number of empirical or semi-empirical formulae have been proposed for excitation cross-sections. Among these let us take the following one which can be used for computing analytically the rate coefficients:

$$\sigma_{12} = 2.7 \times \pi a_0^2 f_{12} \left(\frac{\chi_H}{\chi_{12}}\right)^2 \frac{1}{u} \ln u, \tag{51}$$

where a_0 is the radius of the first Bohr orbit for hydrogen, f_{12} the oscillator strength, χ_H the ionization energy for hydrogen ($\chi_H = 13.6$ eV), χ_{12} the excitation energy for the first excited level, and u is the kinetic energy of incident electron in terms of the excitation energy: $u = E_e/\chi_{12}$. Therefore the rate coefficient Q_{12} is equal to the oscillator strength multiplied by a known function of electron temperature and excitation energy:

$$Q_{12} = f_{12} \Phi (T_e, \chi_{12}) \tag{52}$$

and

$$J_{21} = \frac{1}{4\pi} h\nu_{21} n_e n_r l f_{12} \Phi (T_e, \chi_{12}). \tag{53}$$

The parameters T_e and n_e are measured by independent methods. Knowing n_r, the measurement of J_{21} gives f_{12}. In practice, only the total amount of added impurity $\sum n_r$ is known from the initial conditions and the time history of the plasma parameters. The experimental procedure consists in comparing this value of $\sum n_r$, at a given time, with the value calculated from eqn. (53), using absolute line intensity measurements for every ion having an appreciable population density at that time, and adjusting the f values. Figure 17 shows results of such a procedure for a C-Stellarator discharge.[154] The problem may be complicated by other processes such as cascading, radiation trapping, and the existence of metastable levels.

B. Evaluation of ionization rate coefficients

Electron density n_e and confinement time τ_c, appearing in eqn. (49), are known from appropriate density measurements. Ion densities can be deduced from line intensities as above. Eqn. (49) can be solved for $S_r(T_e)$ as a function of $S_{r-1}(T_e)$:

$$S_r(T_e) = \frac{n_{r-1}}{n_r} S_{r-1}(T_e) - \frac{1}{n_e n_r} \left(\frac{dn_r}{dt} + \frac{n_r}{\tau_c}\right). \tag{54}$$

The experimental procedure consists in determining at a given time and temperature T_e, the slope of every ion density $n_r(t)$ which is obtained from absolute intensity measurements (see § A), and start with the known value of $S_0(T_e)$ for neutral atoms. The $S_r(T_e)$ values are successively deduced. Accurate determination of the slope of $n_r(t)$ is, of course, difficult. Nevertheless, a quantity can

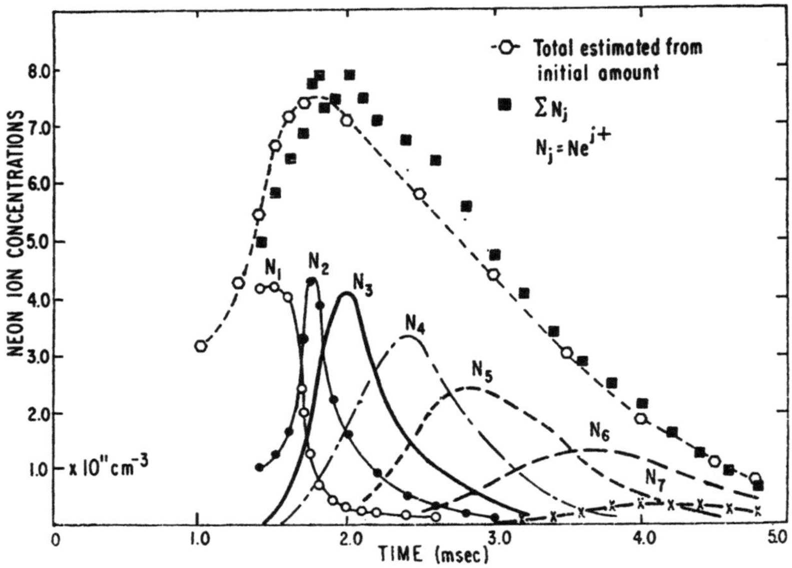

FIG. 17. Neon-ion concentration deduced from measured light intensities compared with total neon density estimated from the initial density and the temporal behaviour of the discharge.[154]

be measured more precisely, e.g. $S_1(T_e)$ at the maximum of $n_1(t)$ for which $dn_1/dt = 0$. This value can be utilized for calibrating a relative $S_1(T_e)$ curve which can be used, in turn, for measuring $S_2(T_e)$ at the maximum of $n_2(t)$ where $dn_2/dt = 0$ and so on.

References

1. WILSON, R., *J. Quant. Spectrosc. Radiative Transfer* **2**, 477 (1962).
2. MCWHIRTER, R. W. P., in *Plasma Diagnostic Techniques* (eds. HUDDLESTONE, R. H., and LEONARD, S. L.), Academic Press, New York and London 1965.
3. COOPER, J., *Rep. Prog. Phys.* **29**, 35 (1966).
4. GRIEM, H. R., *Plasma Spectroscopy*, McGraw-Hill, New York, 1964.
5. DRAWIN, H. W., and FELENBOK, P., *Data for Plasmas in Local Thermodynamic Equilibrium*, Gauthier-Villars, Paris, 1965.
6. OXENIUS, J., *Z. Naturforsch.* **25** A, 101 (1970); **25** A, 1302 (1970).
7. SHAW, J. F., MITCHNER, M., and KRUGER, C. H., *Phys. Fluids* **13**, 325 (1970); **13**, 339 (1970).

8. Drawin, H. W., *Ann. Phys.* **17**, 374 (1966).
9. Kieffer, L. J., and Dunn, G. H., *Rev. Mod. Phys.* **38**, 1 (1966).
10. Laborie, P., Rocard, J. M., and Rees, J. A., *Electronic Cross-sections, Macroscopic Coefficients*, vol. I, *Hydrogen and Rare Gases*, Dunod, Paris, 1968.
11. Moiseiwitsch, B. L., and Smith, S. J., *Electron Impact Excitation of Atoms*, NSRDS-NBS 25, National Bureau of Standards, Washington, DC, 1968.
12. Massey, H. S. W., Burhop, E. H. S., and Gilbody, H. B., *Electronic and Ionic Impact Phenomena*, Vol. I, *Collisions of Electrons with Atoms*, Clarendon Press, Oxford, 1969.
13. Seaton, M. J., in *Atomic and Molecular Processes* (ed. Bates, D. R.), Academic Press, London, 1962.
14. Mott, N. F., and Massey, H. S. W., *The Theory of Atomic Collisions*, Clarendon Press, Oxford, 1965.
15. Drawin, H. W., *Z. Physik* **164**, 513 (1961); **168**, 238 (1962); **172**, 429 (1963).
16. Schwob, J. L., *J. Phys.* **25**, 713 (1964).
17. Gryzinski, M., *Phys. Rev.* **138** A, 336 (1965).
18. Breton, C., thèse, Paris, 1967, Rapport CEA-R 3185.
19. Lotz, W., *Z. Phys.* **206**, 205 (1967); **216**, 241 (1968); **232**, 101 (1970).
20. Bates, D. R., and Dalgarno, A., in *Atomic and Molecular Processes* (ed. Bates, D. R.), Academic Press, London, 1962.
21. Burgess, A., *Astrophys. J.* **139**, 776 (1964); **141**, 1588 (1965).
22. Tucker, W. C., and Gould, R. J., *Astrophys. J.* **144**, 244 (1966).
23. Bates, D. R., and Damgaard, A., *Phil. Trans. Roy. Soc.* A, **242**, 101 (1949).
24. Allen, C. W., *Astrophysical Quantities*, Athlone Press, London 1963.
25. Wiese, W. L., Smith, M. W., and Glenon, B. M., *Atomic Transition Probabilities*, vol. 1, *Hydrogen through Neon*, NSDRS-NBS 4, National Bureau of Standards, Washington, DC, 1966.
26. Wiese, W. L., Smith, M. W., and Miles, B. M., *Atomic Transition Probabilities*, vol. 2, *Sodium through Calcium*, NSDRS-NBS 22, National Bureau of Standards, Washington, DC, 1969.
27. Chandrasekhar, S., *Radiative Transfer*, Dover Publications, New York, 1960.
28. Hearn, A. G., *Proc. Phys. Soc.* 81, 648 (1963); **84**, 11 (1964).
29. Mitchell, A. C. G., and Zemansky, M. W., *Resonance Radiation and Excited Atoms*, Cambridge University Press, 1961.
30. McWhirter, R. W. P., *Proc. Phys. Soc.* **75**, 520 (1960).
31. Hobbs, G. D., McWhirter, R. W. P., Griffin, W. G., and Jones, T. J. L., *Proceedings of the 5th Conference on Ion. Phen. Gases, Munich, 1961*, vol. 2, p. 1965.
32. Breton, C., *Proceedings of the 5th Conference on Ion. Phen. Gases, Munich, 1961*, vol. 2, p. 2247.
33. Bierman, L., *Naturwissenschaften* **34**, 87 (1947).
34. Wooley, R. V. D. R., and Allen, C. W., *Monthly Notice Roy. Astr. Soc.* **108**, 292 (1948).
35. Miyamoto, S., *Publ. Astr. Soc. Japan* **1**, 10 (1949).
36. Elwert, G., *Z. Naturforsch.* **7** A, 432 (1952).
37. Bates, D. R., Kingston, F. R. S., and McWhirter, R. W. P., *Proc. Phys. Soc.* A, **257**, 297 (1962); A, **270**, 155 (1962).
38. Bates, D. R., and Kingston, F. R. S., *Planetary Space Science* **11**, 1 (1963).
39. Hearn, A. G., and McWhirter, R. W. P., *Proc. Phys. Soc.* **82**, 641 (1963).
40. Hinnov, E., and Hirschberg, J. G., *Proceedings of the 5th Conference on Ion. Phen. Gases, Munich, 1961*, vol. 1, p. 638.
41. Hinnov, E., and Hirschberg, J. G., *Phys. Rev.* **125**, 795 (1962).
42. Breton, C., and Schwob, J. L., *Compt. Rend.* **260**, 461 (1965).
43. Breton, C. and Schwob, J. L., *Compt. Rend.* **261**, 1211 (1965).
44. Schwob, J. L., and Breton, C., *Compt. Rend.* **261**, 1476 (1965).
45. Schwob, J. L., thèse, Paris, 1967, Rapport CEA-R-3359 (1968).
46. Lochte-Holtgreven, W., *Plasma Diagnostics*, North-Holland, Amsterdam, 1968.

47. Platz, P., and Hirschberg, J. G., *Compt. Rend.* **261**, 1207 (1965).
48. Sawyer, G. A., Bearden, A. J., Henins, I., Jahoda, F. C., and Ribe, F. L., *Phys. Rev.* **131**, 1891 (1963).
49. Elton, R. C., and Griem, H. R., *Phys. Rev.* **135** A, 1550 (1964).
50. Bolt, G., and Cooper, W. S., *Z. Naturforsch.* **19** A, 968 (1964).
51. Elton, R. C., *J. Quant. Spectrosc. Radiative Transfer* **8**, 392 (1968).
52. Burgess, D. D., Fawcett, B. C., and Peacock, N. J., *Proc. Phys. Soc.* **92**, 805 (1967).
53. Eberhagen, A., and Wunderlich, R., *Z. Phys.* **232**, 1 (1970).
54. Burton, W. N., and Wilson, R., *Proc. Phys. Soc.* **78**, 1416 (1961).
55. Post, R. F., *Plasma Phys.* **6**, 217 (1961).
56. Bishop, A. S., Gibson, A., Hinnov, E., and Hofmann, F. W., *Phys. Fluids* **8**, 1541 (1965).
57. Kolb, A. C., and McWhirter, R. W. P., *Phys. Fluids* **7**, 519 (1964).
58. Samson, J. R., *Techniques of Vacuum Ultraviolet Spectroscopy*, Wiley, New York, 1967.
59. Jaeglé, P., *Compt. Rend.* **259**, 4556 (1964); Thèse, Paris, 1965.
60. Heroux, L. J., and Cohen, M., *Proceedings of the Conference on Calibration Methods UV and X-Rays, 1968*, ESRO SP **33**, 125.
61. Burton, W. M., Hatter, A. T., and Ridgeley, A., *Proceedings of the Conference on Calibration Methods UV and X-Rays, 1968*, ESRO SP **33**, 145.
62. Huber, M. C., Parkinson, W. H., and Reeves, E. M., *Proceedings of the Conference on Calibration Methods UV and X-Rays, 1968*, ESRO SP **33**, 231.
63. Schwinger, J., *Phys. Rev.* **75**, 1912 (1949).
64. Codling, K., and Madden, R. P., *J. Appl. Phys.* **36**, 380 (1965).
65. Pitz, E., *Appl. Opt.* **8**, 255 (1969).
66. Bogen, P., Rusbuldt, D., and Schluter, J., *Proceedings of the Conference on Calibration Methods UV and X-Rays, 1968*, ESRO SP **33**, 53.
67. Zaidel, A. N., Malyshev, G. M., and Shreider, E. Ya., *Zh. tek Fiz.* **14**, 815 (1961); [*Sov. Phys. Tech. Phys.* **6**, 93 (1961)].
68. Griffin, W. G., and McWhirter, R. W. P., *Conference Opt. Inst. Tech.*, Chapman & Hall, 1962.
69. Gladushchak, V. I., and Shreider, E. Ya., *Optika Spektrosk.* **14**, 815 (1963), and **17**, 144 (1964); [*Optics Spectrosc.* **14**, 432 (1963); **17**, 75 (1964)].
70. van Eck, J., de Heer, F. J., and Kistemaker, J., *Phys. Rev.* **130**, 656 (1963).
71. Hinnov, E., and Hofmann, F. W., *J. Opt. Soc. Am.* **53**, 1259 (1963).
72. Cristofori, F., Fenici, P., Frigerio, G. E., Molho, N., and Sona, P. G., *Phys. Letters* **6**, 171 (1963).
73. Breton, C., and Schwob, J. L., *Compt. Rend.* **262 B**, 667 (1966).
74. Breton, C., and Schwob, J. L., *J. Phys.*, suppl., **29**, C 3-114 (1968).
75. Goldman, L. M., and Kilb, R. W., *Plasma Phys.* **6**, 217 (1964).
76. Suckewer, S., *Plasma Phys.* **10**, 527 (1968).
77. Suckewer, S., *J. Phys.* B, **3**, 380 (1970); **3**, 390 (1970).
78. Akimov, E. M., and Malkov, I. P., *Optika Spektrosk.* **6**, 96 (1959); [*Optics Spectrosc.* **6**, 56 (1959)].
79. Balloffet, G., thèse, Paris 1960; *Ann. Phys. Paris* **5**, 1279 (1960).
80. Berthelot, Ch. A., *J. Quant. Spectrosc. Radiative Transfer* **5**, 495 (1965); *Spectrochim. Acta* **22**, 829 (1966).
81. Cunningham, S. P., *U.S. Energy Commission Rep.*, No. WASH-289, 279 (1955).
82. Sovie, R. J., *Phys. Fluids* **7**, 613 (1964).
83. Thonemann, P. C., *Optical Spectrometric Measurements of High Temperatures* (ed. Dickerman, P. J.), University Press, Chicago, 1961.
84. Strinnikov, V. M., *Optika Spektrosk.* **20**, 3 (1966); [*Optics Spectrosc.* **20**, 1 (1966)].
85. Mewe, R., *Br. J. Appl. Phys.* **17**, 1239 (1966).
86. de Vries, R. F., and Mewe, R., *Phys. Fluids* **9**, 414 (1966).
87. Drawin, H. W., *Z. Naturforsch.* **19** A, 1451 (1964).
88. Hinnov, E., *Phys. Fluids* **7**, 130 (1964).

89. HEROUX, L., *Proc. Phys. Soc.* **83**, 121 (1964).
90. BATES, D. R., and DAMGAARD, A., *Phil. Trans. Roy. Soc.* A, **242**, 101 (1949).
91. SCHWOB, J. L., *Compt. Rend.* **262** B, 1264 (1966).
92. VAN REGEMORTER, H., *Astrophys. J.* **136**, 906 (1962).
93. BELY, O., *Proc. Phys. Soc.* **88**, 587 (1966); *Annales d'Astrophys.* **29**, 131 (1966).
94. BURKE, P. G., TAIT, J. H., and LEWIS, B. A., *Proc. Phys. Soc.* **87**, 209 (1966).
95. SUCKEWER, S., *Phys. Rev.* **170**, 239 (1968).
96. KARZAS, W. J., and LATTER, R. L., *Astrophys. J.*, suppl., **6** (55), 167 (1961).
97. FEHRENBACH, C., in *Handbuch der Physik* **50**, 76 (1958).
98. COOPER, W. S., and KUNKEL, W., *Phys. Fluids* **8**, 482 (1965).
99. ECKERLE, K. L., and MCWHIRTER, R. W. P., *Phys. Fluids* **9**, 81 (1966).
100. BEARDEN, A. J., RIBE, F. L., SAWYER, G. A., and STRATTON, T. F., *Phys. Rev. Letters* **6**, 257 (1961).
101. SAWYER, G. A., JAHODA, F. C., RIBE, F. L., and STRATTON, T. F., *J. Quant. Spectrosc. Radiative Transfer* **2**, 467 (1962).
102. BOLAND, B. C., IRONS, F. E., and MCWHIRTER, R. W. P., *J. Phys. B (At. Mol. Phys.)* **1**, 1180 (1968).
103. PATOU, C., *J. Phys.* **31**, 339 (1970).
104. EHLER, A. W., *J. Appl. Phys.* **37**, 4962 (1966).
105. SCHWOB, J. L., BRETON, C., SEKA, W., and MINIER, C., *Plasma Phys.* **12**, 217 (1970).
106. SEKA, W., BRETON, C., SCHWOB, J. L., and MINIER, C., *Plasma Phys.* **12**, 73 (1970).
107. HENKE, B. L., ELGIN, R. L., LENT, R. E., and LEDINGHAM, R. B., *X-Ray Absorption in the 2-200 Å Region*, Norelco Reporter XIV No. 3-4, 1967.
108. JAHODA, F. C., LITTLE, E. M., QUINN, W. E., SAWYER, G. A., and STRATTON, T. F., *Phys. Rev.* **119**, 843 (1960).
109. ELTON, R. C., and ANDERSON, A. D., Naval Research Laboratory Report 6541, Washington, DC, 1967.
110. GRIEM, H. R., KOLB, A. C., LUPTON, W. H., and PHILIPS, D. T., *Nuclear Fusion*, suppl., **2**, 543 (1962).
111. BODIN, H. A. B., GREEN, T. S., NEWTON, A. A., NIBLETT, G. B. F., and REYNOLD, J. A., *Plasma Physics and Controlled Nuclear Research*, International Atomic Energy Agency, Vienna, 1966, vol. 1, p. 193.
112. KOLB, A. C., LUPTON, W. H., ELTON, R. C., MCLEAN, E. A., SWARTZ, M. S., and YOUNG, M. P., *Plasma Physics and Controlled Nuclear Research*, International Atomic Energy Agency, Vienna, 1966, vol. 1, p. 261.
113. KALECK, A., KONEN, L., NOLL, P., SUGITA, K., and WAELBROECK, F., *Plasma Physics and Controlled Nuclear Research*, International Atomic Energy, Agency, Vienna, 1966, vol. 2, p. 581.
114. MATHER, J. W., *Plasma Physics and Controlled Nuclear Research*, International Atomic Energy Agency, Vienna, 1966, vol. 2, p. 389.
115. PEACOCK, N. J., WILCOCK, P. D., SPEER, R. J., MORGAN, P. D., *Plasma Physics and Controlled Nuclear Research*, International Atomic Energy Agency, Vienna, 1966, vol. 2, p. 51.
116. BECKNER, E. A., *J. Appl. Phys.* **37**, 4944 (1966).
117. BOBIN, J. L., FLOUX, F., LANGER, P., and PIGNEROL, H., *Phys. Letters* **28** A, 398 (1968).
118. MANDEL'SHTAM, S. L., PASHININ, P., PROKHOROV, A. M., RAIZER, YU. P., and SUKHODREV, N., *Zh. éksp. teor. Fiz. Prisma* **49**, 127 (1965); [*Soviet Phys. JETP* **22**, 91 (1966)].
119. VANYUKOV, M. P., VENCHIROV, V. A., ISAENKO, V. I., PASHININ, P. P., and PROKHOROV, A. M., *Zh. éksp. teor. Fiz.* **7**, 321 (1968); [*JETP Letters* **7**, 251 (1968)].
120. ALCOCK, A. J., PASHININ, P. P., and RAMSDEN, S. A., *Phys. Rev. Letters* **17**, 528 (1966).
121. GLASSTONE, S., and LOVBERG, R. H., *Controlled Thermonuclear Reactions*, van Nostrand, New York, 1960.
122. EDLÉN, B., *Nova Acta R. Soc. Scient. upsal.* **9**, 62 (1934).
123. EDLÉN, B., *Rep. Prog. Phys.* **26**, 181 (1963).

124a. Balloffet, G., *Ann. Phys. Paris* **5**, 1243 (1960).
124b. Schwob, J. L., and Fraenkel, B. S., *Phys. Letters* **40A**, 81 (1972); **40A**, 83 (1972).
125. Alexander, E., Feldman, U., Fraénkel, B. S., and Hoory, S., *Br. J. Appl. Phys.* **17**, 265 (1966).
126. Fawcett, B. C., Gabriel, A. H., and Saunders, P. A. H., *Proc. Phys. Soc.* **90**, 863 (1967).
127. Fawcett, B. C., Jones, B. B., and Wilson, R., *Proc. Phys. Soc.* **78**, 1223 (1961).
128. Kaufman, A. S., Hughes, T. P., and Williams, R. V., *Proc. Phys. Soc.* **76**, 17 (1960).
129. Bockasten, K., Hallin, R., and Nilsson, N. R., *Ark. Fys.* **32**, 437 (1966).
130. Bockasten, K., Hallin, R., Johansson, K. B., and Tsui, P., *Phys. Letters* **8**, 181 (1964).
131. Goldsmith, S., and Kaufman, A. S., *Proc. Phys. Soc.* **81**, 544 (1963).
132. Hallin, R., and Hughes, T. P., *Proc. Phys. Soc.* **78**, 201 (1961).
133. Bockasten, K., Hallin, R., and Hughes, T. P., *Proc. Phys. Soc.* **81**, 522 (1963).
134. Fawcett, B. C., Gabriel, A. H., Griffin, W. G., Jones, B. B., and Wilson, R., *Nature* **200**, 1303 (1963).
135. Fawcett, B. C., Gabriel, A. H., Jones, B. B., and Peacock, N. J., *Proc. Phys. Soc.* **84**, 257 (1964).
136. Gabriel, A. H., Fawcett, B. C., and Jordan, C., *Proc. Phys. Soc.* **87**, 825 (1966).
137. Jahoda, F. C., Ribe, F. L., Sawyer, G. A., and McWhirter, R. W. P., *Proceedings of the 6th International Conference on Ion. Phen. Gases, Paris, 1963*, vol. 3, p. 347.
138. Peacock, N. J., Cooper, J., and Greig, J. R., *Proc. Phys. Soc.* **84**, 803 (1964).
139. Fawcett, B. C., and Gabriel, A. H., *Proc. Phys. Soc.* **84**, 1038 (1964).
140. Fawcett, B. C., and Irons, F. E., *Proc. Phys. Soc.* **89**, 1063 (1966).
141. Hallin, R., *Ark. Fys.* **31**, 511 (1966).
142. Hallin, R., *Ark. Fys.* **32**, 201 (1966).
143. Peacock, N. J., Speer, R. J., and Hobby, M. G., *J. Phys. B. (At. Mol. Phys.)* **2**, 798 (1969).
144. Basov, N. G., Boiko, V. A., Voinov, Yu. P., Kononov, E. Ya, Mandel'shtam, S. L., and Sklizkov, G. V., *Zh. éksp. teor. Fiz. Prisma* **5**, 177 (1967); [*JETP Letters* **5**, 141 (1967)].
145. Seka, W., Schwob, J. L., and Breton, C., *J. Appl. Phys.* **42**, 315 (1971).
146. Valero, F. P. J., Goorvitch, D., Ragent, B., and Fraenkel, B. S., Nasa Technical Note, NASA TN. D-6048, Washington, DC, 1970.
147. Fawcett, B. C., Gabriel, A. H., Irons, F. E., Peacock, N. J., and Saunders, P. A. H., *Proc. Phys. Soc.* **88**, 1051 (1966).
148. Ehler, A. W., and Weissler, G. L., *Appl. Phys. Letters* **8**, 89 (1966).
149. Fawcett, B. C., Burgess, D. D., and Peacock, N. J., *Proc. Phys. Soc.* **91**, 970 (1967).
150. Fawcett, B. C., Peacock, N. J., *Proc. Phys. Soc.* **91**, 973 (1967).
151. Archbold, E., Harper, D. W., and Hughes, T. P., *Br. J. Appl. Phys.* **15**, 1321 (1964).
152. Esteva, J. M., and Romand, J., *Compt. Rend.* **264**, 142 (1967).
153. Carillon, A., Jaeglé, P., and Dhez, P., *Phys. Rev. Letters* **25**, 140 (1970).
154. Hinnov, E., *J. Opt. Soc. Am.* **56**, 1179 (1966).
155. Johnson, L. C., *Phys. Rev.* **155**, 64 (1967).
156. Elton, R. C., and Koppendorfer, W. W., *Phys. Rev.* **160**, 194 (1967).
157. Kunze, H. J., Gabriel, A. H., and Griem, H. R., *Phys. Rev.* **165**, 267 (1968).
158. Boland, B. C., Jahoda, F. C., Jones, T. J. L., and McWhirter, R. W. P., *J. Phys. B. (At. Mol. Phys.)* **3**, 1134 (1970).

PART IV

PRINCIPLES OF VACUUM ULTRAVIOLET INSTRUMENTAL OPTICS

CHAPTER 9

Principles of vacuum ultraviolet instrumental optics

M. Pouey

Laboratoire des Interactions Moléculaires et des Hautes Pressions du CNRS, Bellevue, 92190 Meudon, France

Contents

1. Introduction

This chapter represents an attempt to describe the evolution of the design of vacuum ultraviolet (v.u.v.) mountings. In this spectral range the lack of highly transparent materials practically precludes the prism as a dispersing element. In addition, here, the reflectance of surfaces is much less than the 90-99% easily attainable in the visible, and it is essential to minimize the number of reflecting surfaces in order to attain the maximum light flux. Therefore, the concave grating which combines the principle of the diffraction grating with the focusing property of a concave mirror is generally employed. There are, nevertheless, many problems such as optical imaging, reflectance, and polarizance of the grating, and also reflective scattering from surfaces. The main part of this chapter will be concerned with the optical properties of concave gratings, the analysis being restricted to the study of the image evaluation and to the design of concave grating mountings.

The first-order theory, already well treated several times (see in particular Samson[1]), will be briefly reviewed. The focusing properties will then be considered from the point of view of both geometrical and diffraction theories. Indeed, in this spectral range it is essential to include their combined effects along with higher order terms in each case, the diffraction pattern depending on the values of the aberration. If the deviation Δ of the effective wave front over the exit pupil from a reference sphere centred in the middle of the Gaussian image is low, the diffraction pattern is only slightly modified. The image evaluation criterion could then be deduced from the concept of Strehl intensity familiar to lens designers. However, as the value of Δ increases, the pattern observed in the image plane tends towards a blurred figure formed by the intersection of geometrical rays with the image plane. In the x.u.v. region, where it is common to employ grazing incidence, the high values of Δ invalidate the Strehl criterion, and the image evaluation must be based on the association of a ray density with the energy flux.

To conclude this part, a comparison between different v.u.v. spectrometers will be made and the design of corrected mountings will be described along with a few examples.

Next, the scattering properties of optical surfaces will be treated. This question is very important in the v.u.v. range since σ, the r.m.s. height of the irregularities, cannot be made negligible with respect to wavelength. In particular, the dependence of the non-specular scattered flux in the direction α' on wavelength must be rederived since the law $\lambda^{-4}\mathscr{F}(\alpha'/\lambda)$, where $\mathscr{F}(\alpha'/\lambda)$ is the Fourier transform of the autocovariance function, can be applied only if $\sigma/\lambda \ll 1$.

2. Optical properties of the concave grating: first-order theory

After the earlier simplified theories of the concave grating by Rowland[2] and by Mack *et al.*[3,4] for grazing incidence, the first full analysis was given by Beutler;[5] Namioka[6,7] made this more rigorous and complete and studied the influence of the area of the ruled surface on the resolving power for several mountings. More recently, Welford,[8] Spencer and Murty,[9] and others[10,11] have given formulae for ray-tracing with the aid of electronic computers. A more detailed study of the aberrations and the principles for correcting them have been treated by Welford[12,13] and by Pouey.[14] A generalization to all types of ruled surfaces, and valid also for off-plane mountings, is presented by Werner.[15]

2.1. FIRST-ORDER PROPERTIES

We take a rectangular coordinate system (Fig. 1) with the origin O at the centre of the grating, the Z axis parallel to the rulings and the X axis normal to the grating surface at O. Let AP be the incident ray, the coordinates of A and of a point P on the grating surface being respectively (x, y, z) and (u, w, l). The surface of the ruled area is taken to be rectangular having the dimensions $W = 2b$ and $L = 2a$. Let the projection of A on the XY plane be A_0, with $OA_0 = r$, and the angle of incidence be α. The constant of the grating and its radius of curvature are respectively d and R. The projection of the grooves on the YZ plane are equally spaced parallel straight lines, conforming to the method by

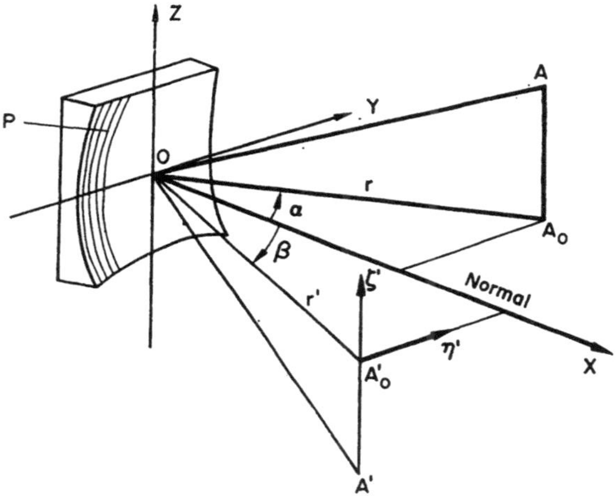

FIG. 1. Image formation by the concave grating.

which gratings are commonly ruled. Let the diffracted ray PB be associated with a point B (x', y', z') in the image space, a length r'_P, and an angle of diffraction β_P. The signs of α and β_P are opposite if A and B lie on different sides of the XZ plane. Then the optical path $L = AP + PB$ can be expanded as a power series in the variables w and l (for details see refs. 1, 5, and 6). To ensure that PB is a ray of wavelength diffracted at the point B, it is necessary, using Fermat's principle, to verify that:

L is stationary with respect to a small displacement of P in the Z direction: $\partial L/\partial l = 0$;

for a displacement d of P along the grating in the Y direction, L increases by an amount $m\lambda$, where m is the order of diffraction: $\partial L/\partial w = m\lambda/d$.

The directions of the pupil coordinates for which the conditions for image formation have to be fulfilled must define the collection of light rays all lying in the same plane instead of crossing each other. Because, generally, the principal rays of the incident and diffracted beams and the grating normal are not situated in the same plane, one cannot define a meridional plane in the usual way. Thus it is by no means clear, as pointed out by Werner,[15] along which direction in the pupil-plane, located on the grating, the partial derivatives of the path L must be taken in order to find the equations describing image formation. For off-plane mountings it appeared desirable to take the orientation of the coordinates around the normal to the grating surface as a variable.(15) For more simplicity only in-plane mountings will be henceforth considered.

As the above equations cannot be satisfied simultaneously, the horizontal focus is chosen so as to reduce the width of the spectral line. As this width decreases with w and l, the central ray can be considered. Then, in paraxial optics, to a point $A(r, \alpha, z)$ there corresponds a Gaussian image $A'(r'_t, \beta, z')$ in cylindrical coordinates such that

$$z/r + z'/r'_t = 0, \tag{1}$$

$$T + T' = 0, \tag{2}$$

where

$$T = \frac{1}{R}\left(\frac{\cos^2\alpha}{e} - \cos\alpha\right); \quad r = Re, \tag{3}$$

$$T' = \frac{1}{R}\left(\frac{\cos\beta}{e'_t} - \cos\beta\right); \quad r'_t = Re'_t. \tag{3'}$$

Since, in practice, z^2/r^2 and $z'^2/r'^2 \ll 1$, the grating equation can be taken as

$$d(\sin\alpha + \sin\beta) = \pm m\lambda, \tag{4}$$

where the negative sign applies when the spectrum lies between the central image ($\alpha = \beta$) and the tangent to the grating. The angular dispersion defined as $d\beta/d\lambda$ is easily determined by differentiating (4) for a fixed angle of incidence. The linear dispersion for a point in the XY plane is expressed by

$$dl/d\lambda = r'_t\,\frac{d\beta}{d\lambda} = \frac{mr'_t}{d\cos\beta}. \tag{5}$$

The coefficient of l in the equation $\partial L/\partial l = 0$ gives the equation for the vertical focus:

$$S + S' = 0, \tag{6}$$

where

$$S = \frac{1}{R}\left(\frac{1}{e} - \cos\alpha\right); \quad r = Re, \tag{7}$$

$$S' = \frac{1}{R}\left(\frac{1}{e'_s} - \cos\beta\right); \quad r'_s = Re'_s. \tag{7'}$$

Finally, for each wavelength an image plane can be defined by the axis $A_0'\zeta'$ parallel to the axis OZ and by the axis $A_0'\eta'$ perpendicular to $A_0'\zeta'$ and making an angle V'* with the direction OA_0' ($A_0'(r_t', \beta, 0)$ denoted $A_0'(r_t', \beta)$, is the projection of $A'(r_t', \beta, z')$ on to the XY plane).

2.2. VACUUM ULTRAVIOLET MOUNTINGS

Schematically the evolution of the designs[1] may be characterized by three stages. All the mountings were first designed to satisfy particular mathematical solutions of the equation $T + T' = 0$. In the second stage with the aim of making more compact mountings and of avoiding complicated mechanical linkages, further conditions were added to the requirement of eqn. (2). Finally, in order to take into account higher order terms in the equations derived from Fermat's principle, the basic equation of the new types is

$$T + T' = \varepsilon, \tag{8}$$

where ε is deduced from the equation giving aberration tolerances.

Spectrometers based on particular solutions of $T + T' = 0$

(i) *Wadsworth mounting.*[1] The point A is located at infinity by associating a collimating mirror with the concave grating, and the image distance is given by $r' = R\cos^2\beta/(\cos\alpha + \cos\beta)$. In addition the diffraction angle β is set equal to zero, so that the exit slit is at a distance

$$r' = R/(1 + \cos\alpha) = r_s' \quad (\beta = 0), \tag{9}$$

which is also the vertical-focus distance. The astigmatism is thus zero and remains low in the vicinity of the grating normal. In spite of this advantage, the Wadsworth mounting is not widely used in the laboratory because an extra mechanism is needed for scanning (the image distance r' being a function of α, the angle of rotation of the grating) and because of the additional optical defects introduced by the collimating system.[16] However, this mounting, which is half the size of the equivalent Rowland mounting, is employed in crossed-dispersion arrangements[17–19] and in space applications[20] the first surface serving as a filter for higher orders.[21] A recent interesting off-plane version is described by Werner.[15]

(ii) *Rowland circle mountings.* A second solution is based on the well-known Rowland circle[1–5] of diameter R, on which the horizontal focus is located provided that the entrance slit and the grating centre also lie on it and the grating normal is along a diameter. In this case we have

$$r = R\cos\alpha; \quad r' = R\cos\beta. \tag{10}$$

* As $A_0'\eta'$ is tangent to the focal curve given in polar coordinates by eq. (3'), it is possible to determine the value of the angle V' between OA_0' and this tangent by

$$\tan V' = \tan\beta\left(\frac{e_t'}{\cos\beta} - 2\right). \tag{3''}$$

Many variations of this arrangement, originally stated by Rowland in 1883, can be found in the literature, and the Paschen-Runge[22] mounting may be considered as a prototype of all normal incidence versions; in this case the exit slit is moved along the Rowland circle. Among others, the well-known Eagle mounting[23] in which α and β remain equal in order to minimize astigmatism, has the merit of compactness.

As the normal incidence reflectance of gratings becomes too low below 400 Å, a grazing incidence arrangement, such as the Paschen-Runge type, is employed in this region.[24, 25] However only the arrangement of Romand and Vodar for the x.u.v.[26] provides fixed direction for the exit beam. In this, the angle of incidence and the distance between the entrance slit and the grating are kept constant. The grating moves along the fixed direction of the exit beam while the entrance slit moves along a direction making an angle of $\theta = 90° - \alpha$ with the direction of the exit beam.

The various Rowland mountings offer optical advantages since the coma terms are zero (see. p. 296). However, mechanical linkages of considerable complexity are needed, and, in addition, the requirements of fixed slits and fixed exit beam are hard to satisfy. This is due to the fact that r' is a function of β (i.e. of λ) and from eqns. (2), (3), and (3′) is expressed by

$$r' = \cos^2 \beta \Bigg/ \left(\frac{\cos \alpha + \cos \beta}{R} - \frac{\cos^2 \alpha}{r} \right). \tag{11}$$

For scanning the spectrum it is necessary either to move the slit with the grating fixed or, keeping the slits fixed, to rotate and simultaneously translate the grating in order to keep the image in focus. Attention has thus been devoted to the design of spectrometers which do not depend on particular solutions of the condition $T + T' = 0$.

Spectrometers derived from the general solution of $T + T' = 0$

A mechanical condition is imposed to resolve eqn. (2). The most useful one is that the slits are fixed in position. A relation characterizing the motion of the grating is then obtained; generally, however, the mechanical linkages proposed do not satisfy it.

(i) *Modified Eagle mountings.* The focusing does not take place on the Rowland circle,[27–30] for one slit is inside and the other outside. The grating is constrained to move along the bisector of the angle subtended by the slits S and S' at the centre O of the grating and to rotate about an axis tangent at O and perpendicular to the plane of incidence. Translation and rotation can be made by a rod attaching the grating to a cam, or by two arms,[27–28] one of which is normal to the grating and the other is fixed to a point. In this case the two arm-lengths are equal to $R/2$ and linearization of the wavelength scale can easily be achieved.[30] It is also possible to have the entrance and exit slits symmetrically above and below the Rowland plane. However, in this off-plane Eagle design it is necessary to rotate the entrance slit with respect to the ruling.[7, 15]

(ii) *Miscellaneous mountings.*[31–36] Seya[31] sought the optimum condition for suppressing the translation of the grating. Putting $\alpha = \theta + \gamma$ and $\beta = \theta - \gamma$, where γ is the angle through which the grating is rotated, and setting eqn. (2) and its three successive derivatives with respect to γ equal to zero, he found for $\gamma = 0$ a solution characterized by

$$\theta = 35°15'; \quad r = r' = R \cos\theta. \tag{12}$$

A more detailed discussion based on diffraction theory has been made by Namioka;[7,32] for the same value of θ, the optimum value of object and image distance are slightly different as follows:

$$e = r/R = 0.8181; \quad e_t' = r'/R = 0.8176. \tag{13}$$

The simplicity of the focusing mechanism and the possibility of using a sine drive to obtain a linear wavelength scale is highly attractive provided that a resolution of a few angstroms is acceptable. The astigmatism is serious and a curved exit slit greatly improves the resolution.[1]

To minimize the defocusing in the above design, Johnson[33] has suggested rotating the grating about an axis displaced from its centre. A more detailed analysis by Onaka[34,35] gives the position of the axis of rotation which minimizes the angular shift in the principal rays.

More recently, Carlson[36] has designed a double monochromator for the x.u.v. The system employs fixed extrance and exit slits and two concave gratings which are rotated to vary the wavelength at the exit slit. Adjustment of the optical parameters and use of symmetry produce an approximate focusing condition comparable with that of the Seya mounting.

Spectrographs

Almost all spectrographs[1] are based on the Rowland circle on which the component parts—slit, grating, and plate are disposed. The short wavelength limit depends largely on the incidence angle[24,25] which determines the reflectance of the grating. However, occasionally strict adherence to the Rowland circle has been abandoned, and the entrance slit and the plate placed respectively outside and inside the circle.

2.3. LIMITATION OF THE FIRST-ORDER THEORY

In the classical mountings, the influence of the high-order terms in the equations of the Fermat principle is reduced by suitably limiting the ruled area, but the proposed values[1,3,6] are not correct with regard to the theory of aberration tolerances. On the other hand, when the equation $T + T' = 0$ is not strictly satisfied, the possibility of balancing different aberration terms has been ignored. By taking into account this possibility, first, the optimum ruled area, and thus the luminosity, must increase, and, second, it is possible to consider new corrected mountings without regard to peculiar mathematical solutions of the equation $T + T' = 0$. Before proceeding to the case $T + T' = \varepsilon$, we will describe the image evaluation in the case of both large and small phase errors.

3. Geometrical theory of concave grating imaging

As a first approximation, the geometrical theory of optical imaging neglects diffraction phenomena and hence is independent of wavelength. It may be regarded as the limiting case of the diffraction theory when $\lambda \to 0$. The image is then identified with the blurred figure formed by the points of intersection of the geometrical rays with the image plane. Its evaluation consists in determining the path of the light rays and the intensity distribution in the blurred figure as a function of the energy associated with each geometrical ray. Generally one introduces either the function Δ, which describes the deviation of the actual wave front over the exit pupil from a reference sphere, or the corresponding phase error $2\pi\Delta/\lambda$. The reference sphere, which is centred on the Gaussian image point, is characterized by the path L_0 and then

$$\Delta = L_0 - L = \Sigma_{ij} C_{ij} l^i w^j .$$

The transverse aberrations (i.e. the coordinates of the intersection of the diffracted rays PA' with the image plane) are

$$\eta' \simeq r' \frac{\partial \Delta}{\partial w}; \quad \zeta' = r' \frac{\partial \Delta}{\partial l}. \tag{14}$$

It is interesting to point out that these quantities are independent of the choice of the reference sphere and that the image evaluation can be made from the wave-front error over the plane of the grating which is the pupil.

The simplicity of this geometrical model arises essentially from the fact that the field behaves locally as a plane wave, and at optical wavelengths this model remains valid if $\Delta/\lambda \gg 1$. This condition is fulfilled in two cases, the first when there is a large bandpass and a high luminosity (large pupil area), and the second in x.u.v. grazing incidence mountings (large values of the coefficients C_{ij}).

3.1. ABERRATION TERMS

The wave-front error can be expressed by

$$\Delta'(w, l) = \Delta(w, l) + C_2(w^2 + l^2) + C_1 w \tag{15}$$

in order to choose the reference sphere centred on the brightest part of the image, even if this is not the centre of the image. The term $C_2(w^2 + l^2)$ is equivalent to a change of focus, and the last term $C_1 w = \eta' w/r'$ corresponds to a lateral displacement of the centre which occurs with coma type of aberration. The function

$\Delta'(w, l)$ is a sum of terms which characterize the different types of aberration. They have been studied by Beutler,[37] Dieke,[38] Namioka,[6] and in detail by Welford[12, 13, 39] and Pouey.[14]

As a first step, following Beutler, it is well to consider each type of aberration separately, neglecting the others. However, in the evaluation of the values of the coefficients C_1 and C_2 and of the ruled area dimensions, it is necessary to take the aggregate function $\Delta'(w, l)$. (The method suggested by Namioka[6] does not allow the possibility of balancing aberrations by a displacement of the reference sphere.)

Astigmatism and spectrum line curvature

The largest aberration is astigmatism, generally expressed by the ratio of the length of the astigmatic line to that of the ruling. This quantity

$$\frac{z'_{ast}}{2a} = \frac{r'_t}{r} \sin^2 \alpha + \sin^2 \beta \tag{16}$$

is equal to $\cos \beta \,(\sin^2 \alpha / \cos \alpha + \sin^2 \beta / \cos \beta)$ for Rowland circle mountings. In the grazing incidence case, the astigmatic distance D expressed by

$$D = r'_s - r' = r'_s \left(\frac{r'}{r} \sin^2 \alpha + \sin^2 \beta \right) \tag{17}$$

can reach several centimetres. Dieke[38] pointed out that for a given wavelength range the Eagle mounting has the least astigmatism of all Rowland circle designs. This is true for its modified versions.

There are two kinds of curvature, one C_1 due to the finite length of the entrance slit, and the other C_2 due to astigmatism. The sign of the RC_1 radius of curvature, given by

$$RC_1 = \frac{Re' \cos \beta}{\sin \alpha + \sin \beta}, \tag{18}$$

is such that the lines are concave towards longer wavelengths. On the other hand, even for a point in the XY plane, the focal lines are curved if considerable astigmatism is present. The effect of C_2 curvature, giving a radius RC_2 equal to

$$RC_2 = R^2 e' \cos \beta \,(S + S')^2 \times \left[\frac{S}{e} \sin \alpha + S' \frac{\sin \beta}{e'} - \frac{2 \sin \beta \,(S + S')}{e'} + R \sin \beta \,(S + S')^2 \right]^{-1} \tag{19}$$

combines with C_1 to give a resultant image as the envelope of one of the curves displaced along the other.

Defect of focus

The defect of focus includes two terms the first of which is zero for a point source in the XY plane and negligible for lengths of slit commonly employed.

The second characterizes the displacement of the reference sphere and takes the form

$$\Delta'_d = C_2 w^2 = -\frac{\varepsilon}{2} w^2 \tag{20}$$

if $T + T' = \varepsilon$.

Spherical aberration

The wave-front error $\Delta'_s(w, l)$ expressed by

$$\Delta'_s(w, l) = C_{22} w^2 l^2 + C_{04} w^4 + C_4 (w^2 + l^2)^2 \tag{21}$$

may be regarded as a generalized type of spherical aberration, but the terms involving l change their character in the presence of large astigmatism (see Welford[12]). The different coefficients take the form:

$$C_{22} = \frac{T}{4R}\left(\frac{S}{e} - \frac{S'}{e'}\right) - \frac{1}{2R^2}\left(\sin^2\alpha\,\frac{S}{e^2} + \sin^2\beta\,\frac{S'}{e'^2}\right), \tag{22}$$

$$C_{04} = \frac{T}{2R^2}\left(\frac{\sin^2\alpha}{e^2} - \frac{\sin^2\beta}{e'^2}\right) + \frac{T^2}{8R}\left(\frac{1}{e} + \frac{1}{e'}\right), \tag{23}$$

$$C_4 = \frac{-1}{8R^3}\left(\frac{\sin^2\alpha}{e} + \frac{\sin^2\beta}{e'}\right). \tag{24}$$

In fact only the term $C_4(w^2 + l^2)^2$ remains a true Seidel spherical type aberration, generally superior to the other terms for normal incidence mountings; as it is well known (see, for example, refs. 54 and 55), it must be balanced by a defect of focus.

Coma aberration

If the astigmatism is low, the coma is of the Seidel type, expressed by

$$\Delta'_c(w, l) = C_1 w + C_3(l^2 w + w^3) \tag{25}$$

with

$$C_3 = \frac{-1}{2R}\left(T\,\frac{\sin\alpha}{e} + T'\,\frac{\sin\beta}{e'}\right). \tag{26}$$

In the opposite case the l^2w term produces a curvature of the horizontal focus (C_2). Finally, even for a point object in the XY plane, the coma aberration is different from zero except for Rowland circle mountings for which $T = T' = 0$.

3.2. IMAGE EVALUATION IN THE PRESENCE OF LARGE PHASE ERRORS

In order to assign aberration tolerances it is necessary to define a quality factor. As diffraction phenomena have been neglected so far, this must be based on the association of ray density with energy flux. The method proposed by Herzberger[40] yields a pattern of points formed in the image plane for a system of rays emanating from a single point and distributed uniformly over the entrance pupil. This uniformity of distribution is necessary if each ray is to represent the same amount of energy. In the limit an energy conservation may be associated with each ray (Arakengy[41]), and recently de Velis[42] has justified the approximation made in the Herzberger formalism. As a result of the introduction of the ray-density method, certain averages over the image plane can be performed to obtain the quality factor, taken to be ω, the radius of gyration of the image pattern. Taking into account the fact that the geometrical intensity distribution in the image plane is exactly that given by Miyamoto,[43] and since there are more rays at the outer edges of the exit pupil than about the centre, a weighting factor like the radius of the exit pupil is generally used. Hence it can be shown[42] that averaging over the exit pupil is equivalent to averaging in the image plane. In the case of a rectangular aperture[14] the quality factor is taken to be $\langle\delta\lambda\rangle$, the tolerated bandpass at the exit slit, expressed in angstroms, and it is given by

$$\frac{1}{4ab}\int_{-a}^{a}\int_{-b}^{b}\left(\frac{\partial\Delta'}{\partial w}+\frac{\partial\Delta'}{\partial l}\right)^2 dw\,dl \leqq \frac{\langle\omega^2\rangle}{r'^2}=\frac{\langle\delta\lambda^2\rangle}{d^2\cos^2\beta}. \tag{27}$$

The value of $\langle\delta\lambda\rangle$ will be a minimum when

$$\partial\langle\delta\lambda\rangle/\partial C_1=\frac{\partial\langle\delta\lambda\rangle}{\partial C_2}=0.$$

From this there result two equations within variables a, b and the aberration coefficients which give the optimum focal position and the optimum ruled area. The tolerated value of $\langle\delta\lambda\rangle$ which depends on the ratio $a/b=\varrho$ is obtained for each aberration by putting the value of C_1 and C_2 in eqn. (27) (see Table 1, where $h=w/b$ and $K=l/a$). In the general case when several aberrations are present together in comparable amounts, the tilt or defocus coefficient must be deduced from eqn. (27). The optimum ruled area W_0 is then given for high values of the astigmatism by

$$W_0^6\left\{\frac{1}{175}(C_{04}+C_4)^2+\frac{1}{240}(C_{22}+2C_4)^2\varrho^2+\left[\frac{C_4}{60}(C_{22}+2C_4)+\frac{(C_{22}+2C_4)^2}{540}\right]\varrho^4\right.$$
$$\left.+\frac{1}{28}C_{4}^2\varrho^6\right\}+\frac{W_0^4}{20}C_{03}^2\leqq\frac{\langle\delta\lambda^2\rangle}{d^2\cos^2\beta}=Q^2. \tag{27'}$$

TABLE 1. ABERRATION TOLERANCES FOR A RECTANGULAR APERTURE IN THE CASE OF LARGE PHASE-ERRORS

Aberration type Aberration function	Corrected aberration function Δ'_G	Tilt or defocus coefficient	Tolerance
Fourth power aberration $C_2w^2 + C_{22}w^2l^2$	$C_{22}a^2b^2\left(h^2k^2 - \frac{1}{3}h^2\right)$	$C_2 = -\frac{1}{3}C_{22}a^2$	$\lvert C_{22}b_3\rvert \leqq \frac{\sqrt{(15)}\,Q}{2\varrho\left(1+\frac{4}{9}\varrho^2\right)^{1/2}}$
Fourth power aberration $C_2w^2 + C_{04}w^4$	$C_{04}b^4\left(h^4 - \frac{6}{5}h^2\right)$	$C_2 = -\frac{6}{5}C_{04}b^2$	$\lvert C_{04}b^3\rvert \leqq \frac{5\sqrt{(7)}}{8}Q$
Seidel spherical aberration $C_2w^2 + C_4(w^2 + l^2)^2$	$C_4b^4\left[(h^2+\varrho^2k^2)^2 - \frac{6}{5}\left(1+\frac{5}{9}\varrho^2\right)h^2\right]$	$C_2 = -\frac{6}{5}C_4b^2\left(1+\frac{5}{9}\varrho^2\right)$	$\lvert C_4b^3\rvert \leqq \frac{5\sqrt{(7)}\,Q}{8\left(1+\frac{35}{12}\varrho^2+\frac{385}{54}\varrho^4+\frac{175}{28}\varrho^6\right)^{1/2}}$
Third power aberration $C_1w + C_{03}w^3$	$C_{03}b_3\,(h^3 - h)$	$C_1 = -C_{03}b^2$	$\lvert C_{03}b^2\rvert \leqq \frac{\sqrt{5}}{2}Q$
Third power aberration $C_1w + C_{21}l^2w$	$C_{21}a^2b\left(k^2h - \frac{1}{3}h\right)$	$C_1 = -\frac{1}{3}C_{21}a^2$	$\lvert C_{21}b^2\rvert \leqq \frac{3Q}{2\left(\varrho^2+\frac{\varrho^4}{5}\right)^{1/2}}$
Seidel coma $C_1w + C_3(w^3 + l^2w)$	$C_3b^3\left[h^3 + hk^2\varrho^2 - \left(1+\frac{\varrho^2}{3}\right)h\right]$	$C_1 = -C_3b^2\left(1+\frac{\varrho^2}{3}\right)$	$\lvert C_3b^2\rvert \leqq \frac{\sqrt{(5)}\,Q}{2\left(1+\frac{5}{9}\varrho^2+\frac{\varrho^4}{9}\right)^{1/2}}$

with

$$C_1 = -\frac{C_{03}}{4} W_0^2, \tag{28}$$

$$C_2 = -\frac{W_0^2}{4}\left[\frac{6}{5}(C_{04} + C_4) + \frac{1}{3}(C_{22} + 2C_4)\varrho^2\right]. \tag{29}$$

If the astigmatism is low, the coma term is like Seidel coma, and the coefficient C_{03} has to be changed in $C_3\left(1 + \frac{1}{3}\varrho^2\right)$.

Astigmatic correction

If the astigmatic image is parallel to the rulings, this aberration has no effect on resolution (for crossed mountings see refs. 13-15), but it introduces a loss of luminosity. Unfortunately, as seen from eqn. (27), a correction for it is impossible without introducing another optical surface (its reflectance being less than 1 reduces the final gain in intensity). It is necessary to create an astigmatic object whose horizontal focus coincides with the entrance slit and the vertical focus in front of or behind this slit. For Rowland circle mountings, since the vertical image is to be located on the Rowland circle $r_s' = R\cos\beta$, the vertical focus must be located at a distance

$$r_s = R(\cos\alpha - \sin\beta\tan\beta)^{-1} \tag{30}$$

as deduced from eqn. (7).

Strictly, this correction is only valid at one wavelength λ^* corresponding to β^*, which may be suitably chosen to minimize the resulting astigmatism in the useful spectral range

$$\frac{z'_{\text{ast}}}{2a} = \sin^2\beta - \cos\beta\sin\beta^*\tan\beta^*. \tag{31}$$

The optimum value of β^* for an interval $\beta_m \leqq \beta \leqq \beta_M$ is given by ref. 16:

$$\sin\beta^* = A\left[-1 + \left(1 + \frac{2}{A^2}\right)^{1/2}\right]^{-1/2} \tag{32}$$

with

$$A = 2\surd(2)(\sin^3\beta_M - \sin^3\beta_m)[3(2\beta_M - 2\beta_m + \sin^2\beta_M - \sin^2\beta_m)]^{-1}. \tag{32'}$$

In grazing incidence Rowland mountings r_s is negative and toroidal mirrors are commonly employed for this correction. β_M is given by the mechanical condition of the instrument and the lowest wavelength limit, and hence β_m depends on the total reflection phenomenon at the grating. This method was first utilized by Rense and Violett[44] for a solar rocket spectrograph. The toroidal mirror M is located at a distance p from the entrance slit, for the object is at infinity. The ratio K between the radii of curvature R_1 and R_2 in the horizontal and vertical planes respectively is given by[14]

$$K^{-1} = R_2/R_1 = p^{-1} \cos^2 i(p + R \cos \alpha + |s|). \tag{33}$$

The only limitation for K is that imposed by the difficulty of manufacturing mirrors with large K (> 10) and p is immediately determined.

In the general case, the astigmatic distance D_t of the toroidal mirror must be positive; the value of p is given by

$$p = R_1 \cos i = \frac{D_t}{2} (2K \cos^2 i - 1) (1 - K \cos^2 i)^{-1}. \tag{34}$$

This is positive if

$$(2K)^{-1/2} \leq \cos i \leq K^{-1/2}. \tag{35}$$

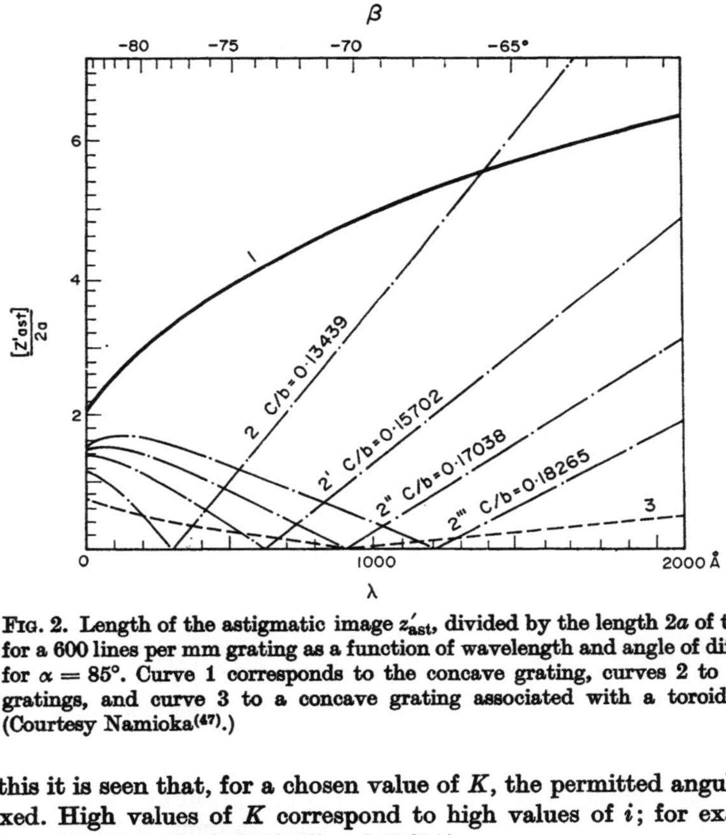

FIG. 2. Length of the astigmatic image z'_{ast}, divided by the length $2a$ of the rulings for a 600 lines per mm grating as a function of wavelength and angle of diffraction β for $\alpha = 85°$. Curve 1 corresponds to the concave grating, curves 2 to ellipsoidal gratings, and curve 3 to a concave grating associated with a toroidal mirror. (Courtesy Namioka[47].)

From this it is seen that, for a chosen value of K, the permitted angular range of i is fixed. High values of K correspond to high values of i; for example, if $K = 8$, i must lie between 69°17′43″ and 67°31′.

Some authors[45–47] have suggested aspherical gratings to compensate for the difference in the radii of curvature in the two planes of the wave front. Apart from the practical difficulties of manufacturing such gratings, the preceding method is better than the theoretical solution of an ellipsoidal grating (Fig. 2). Spherical gratings have also been bent into a toroidal shape to produce stigmatic solar images.[48] Naturally the recent possibility of making stigmatic gratings of high apertures by the holographic technique is of great interest for the future.[49]

4. Diffraction theory of concave grating imaging

The approximate geometrical model gradually loses validity as the aberrations become small. In particular, important deviations become manifest in the immediate neighbourhood of the boundaries of the shadow and in regions where a large number of rays meet.

4.1. DIFFRACTION PATTERN AND RESOLVING POWER

The amplitude distribution of the Fraunhofer diffraction pattern in the Gaussian image plane may be estimated from the Fresnel-Kirchhoff integral. Let g_n be the contribution of each groove; the diffracted amplitude in the image plane takes the form

$$U(r') = \frac{-i}{\lambda}\frac{U(A)}{rr'}\left(\frac{\cos\alpha + \cos\beta}{2}\right)\sum_{-b}^{b} dl \sum_{-a}^{a} g_n \exp i\Phi(w, l)\, dw, \qquad (36)$$

where $U(A)$ is the amplitude in the object plane and $\Phi(w, l)$ the phase error calculated from the geometrical approach. Neglecting the length of the rulings and assuming that g_n has a constant value (this is not true, in particular, for blazed concave gratings), the diffraction pattern for a two-dimensional grating[7, 50] can be deduced from eqn. (36). It has been customary to assume that the diffraction minima do not reach the value zero and to introduce a modified Rayleigh criterion for calculating the resolving power. However recent computations show that the symmetrical image profile has always a zero minimum in the direction of shorter wavelengths and often a zero minimum also in the opposite direction (Fig. 3). Then the Rayleigh criterion may be applied; two lines of equal intensity will be just resolved when their wavelength difference $\delta\lambda$ is such that the minimum total intensity is $8/\pi^2$ times that of the maxima. This value of $\delta\lambda$ determines the resolving power $R = \lambda/\delta\lambda$.

The resolving power cannot exceed a certain maximum R_t. For a grating with n $(= 1/d)$ lines per mm and of width W, with no aberrations present, $R_t = mnW = mN$ where m is the order of diffraction and N the total number of rulings exposed to incident radiation. In the case of low phase-errors, the maximum intensity in the diffraction image is lower than that at the Gaussian focus in an aberration-free system of the same aperture and focal length. The optimum width W_0 is determined by the aberration tolerance condition and the corresponding resolving power is limited to mnW_0. Finally, the practical resolving power R_p

depends on the slit width, i.e. on the dispersion of the instrument and on the slit illumination which can be generally considered as partially coherent, the degree of coherence being a function of the slit width.[51]

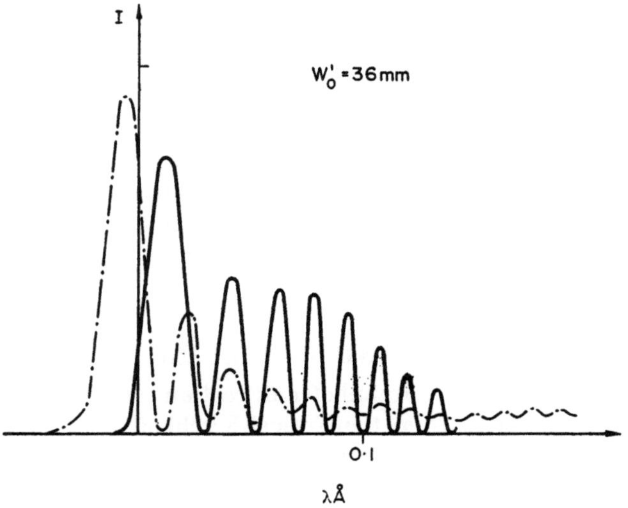

FIG. 3. Diffraction pattern, with (—·—·—) and without optical correction, for a symmetrical mounting. W_0' is the optimum ruled area for $L = 0$ (see pp. 308, 315).

4.2. IMAGE EVALUATION IN THE PRESENCE OF LOW PHASE ERRORS

The tolerance limit of $\lambda/4$ for the phase error was originally proposed by Rayleigh in the particular case of the Seidel spherical aberration due to a circular pupil, and this has been generally considered valid for all cases.[1] In addition, prior to the publications of Welford[13] and Pouey,[14] the possibility of correcting for concave grating aberrations by a displacement of the reference sphere had been practically ignored although attempts were made by Bowen[52] and by Kozlenkov[53] to study geometrically the surface focal in the neighbourhood of the Rowland circle.

In fact, the Strehl criterion[54, 55] applied in the case of a rectangular aperture, and a line image is appropriate. It defines the amount of aberration or combination of aberrations, and then the value W_0 of W, which lowers the maximum light intensity in the image pattern to a certain fraction of its value in the aberration-free image; if the fraction is taken as 0.8, it is commonly assumed that the image pattern remains unchanged, and that the r.m.s. error of the wave front is equal to $\lambda/13.5$.[55] The phase error being smaller than the wavelength, the image evaluating criterion for an incoherently illuminated slit is given by[13]

$$\frac{1}{4ab}\int_{-a}^{a}\left\{\int_{-b}^{b}\Delta'^2 dw\right\} dl - \frac{1}{8ab^2}\int_{-a}^{a}\left\{\int_{-b}^{b}\Delta' dw\right\}^2 dl \leq \frac{0.2\lambda^2}{4\pi^2}. \quad (37)$$

TABLE 2. ABERRATION TOLERANCES FOR A RECTANGULAR APERTURE IN THE CASE OF AN INCOHERENTLY ILLUMINATED LINE OBJECT

Aberration type Aberration function	Corrected aberration function Δ'_p	Tilt or defocus coefficient	Tolerance
Fourth power aberration $C_2w^2 + C_{22}w^2l^2$	$C_{22}a^2b^2\left(h^2k^2 - \frac{1}{3}h^2\right)$	$C_2 = -\frac{1}{3}C_{22}a^2$	$\lvert C_{22}b^4\rvert \leqq \frac{0.80\,\lambda}{\varrho^2}$
Fourth power aberration $C_2w^2 + C_{04}w^4$	$C_{04}b^4\left(h^4 - \frac{6}{7}h^2\right)$	$C_2 = -\frac{6}{7}C_{04}b^2$	$\lvert C_{04}b^4\rvert \leqq 0.94\,\lambda$
Seidel spherical aberration $C_2w^2 + C_4(w^2 + l^2)^2$	$C_4b_4\left[(h^2+\varrho^2)^2 - \frac{6}{7}\left(1 + \frac{7}{9}\varrho^2\right)h^2\right]$	$C_2 = -\frac{6}{7}C_4b^2\left(1 + \frac{7}{9}\varrho^2\right)$	$\lvert C_4b_4\rvert \leqq \frac{0.94\,\lambda}{\left(1 + \frac{49}{9}\varrho^4\right)^{1/2}}$
Third power aberration $C_1w + C_{03}w^3$	$C_{03}b^3\left(h^3 - \frac{3}{5}h\right)$	$C_1 = -\frac{3}{5}C_{03}b^2$	$\lvert C_{03}b^3\rvert \leqq 0.47\,\lambda$
Third power aberration $C_1w + C_{21}l^2w$	$C_{21}a^2b\left(hk^2 - \frac{1}{3}h\right)$	$C_1 = -\frac{1}{3}C_{21}a^2$	$\lvert C_{21}b^3\rvert \leqq \frac{0.58\,\lambda}{\varrho^2}$
Seidel coma $C_1w + C_3(w^3 + l^3w)$	$C^3b^3\left[h^3 + hk^2\varrho^2 - \frac{3}{5}\left(1 + \frac{5}{9}\varrho^2\right)h\right]$	$C_1 = -\frac{3}{5}C_3b^2\left(1 + \frac{5}{9}\varrho^2\right)$	$\lvert C_3b^3\rvert \leqq \frac{0.47\,\lambda}{\left(1 + \frac{35}{27}\varrho^4\right)^{1/2}}$

By applying the condition (37) the value of the coefficients C_1 and C_2 of eqn. (15) have to be determined, so that the left-hand side should be a minimum. This procedure is equivalent to taking the brightest part of the spectral line at the best plane of focus. Otherwise, in evaluating the aberration tolerance and eventually the optimum ruled area, it is assumed that the centre of the reference sphere does coincide with the brightest part of the spectrum line at the best focus. For example, if only Seidel coma and Seidel spherical aberration are present, by introducing the function

$$\Delta'(w, l) = C_1 w + C_3(l^2 w + w^3) + C_2 w^2 + C_4(w^2 + l^2)^2 \tag{38}$$

in eqn. (37) there results an equation in a, b and the aberration coefficients. By minimizing the resultant function with respect to C_1 and C_2, the optimum focal position is determined:

$$C_1 = -\frac{3}{5} C_3 b^2 \left(1 + \frac{5}{9} \varrho^2\right), \tag{39}$$

$$C_2 = -\frac{6}{7} C_4 b^2 \left(1 + \frac{7}{9} \varrho^2\right). \tag{40}$$

Introducing these values in eqn. (37), the joint condition in the two aberration coefficients C_3 and C_4 takes the form

$$\frac{1}{140}\left(1 + \frac{35}{27} \varrho^4\right)^{1/2} C_3^2 W^6 + \frac{1}{2205}\left(1 + \frac{49}{9} \varrho^4\right) C_4^2 W^8 \leqq \frac{\lambda^2}{\pi^2}. \tag{41}$$

It is seen that the left-hand side is the sum of the squares of the contributions from the coma and spherical aberration terms when each aberration type is considered separately—assuming the others negligible. The expressions deduced from criterion (37) are summarized in Table 2. It is to be remarked that any terms in $\Delta'(w, l)$ not involving w can be neglected, as the astigmatic focal line along the spectrum line does not alter the resolution. However, in the case of crossed and off-plane mountings, a tolerance for oblique astigmatism must be calculated.[13, 15]

4.3. COMPARISON BETWEEN GEOMETRICAL AND DIFFRACTION IMAGING

While it seems very difficult to express mathematically the progressive transformation of the diffracted pattern into the blurred figure as the aberrations increase, it is possible to handle the two limiting cases: the geometrical case by the ray density method and the other one by the Strehl criterion.

For example, in the case of the fourth power aberration

$$\Delta = C_{04} w^4 = C_{04} b^4 h^4 \tag{42}$$

in the presence of a defocusing $C_2 w^2$, the best focus is given by $C_2 = -6/7 C_{04} b^2$ for low phase-errors and by $C_2 = -6/5 C_{04} b^2$ for large phase-errors.

The corrected aberration functions are, respectively,

$$\Delta_p' = C_{04}b^4\left(h^4 - \frac{6}{7}h^2\right), \tag{43}$$

$$\Delta_g' = C_{04}b^4\left(h^4 - \frac{6}{5}h^2\right). \tag{44}$$

This type of aberration must be over-corrected in the geometrical case and under-corrected in the diffraction case (Fig. 4).

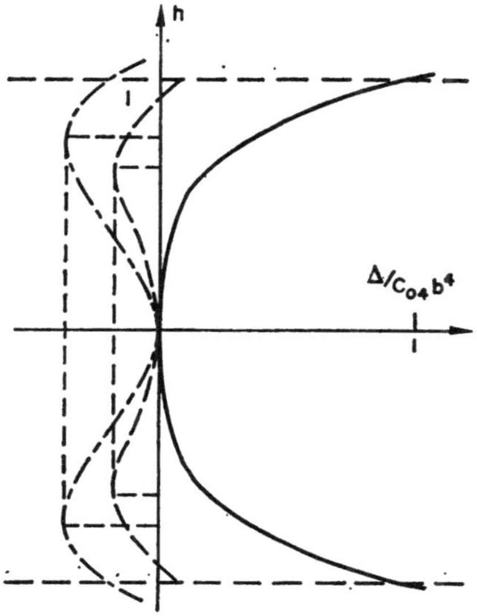

FIG. 4. Aberration functions in units $C_{04}b^4$ for the fourth power aberration $\Delta = h^4$ (———), for the corrected function $\Delta_p' = h^4 - 6/7\,h^2$ (— — —), and $\Delta_G' = h^4 - 6/5\,h^2$ (—·—·—).

This example illustrates the necessity of considering carefully the different aberrations and their relative proportions in determining the value of the optimum ruled area and of the defocusing coefficient ε.

5. Comparison of different spectrometers

The luminosity $\mathscr{L}$, which is equal to the flux at the exit slit for a source of unit luminance, depends on the slit width, and the practical resolving power R_p is inversely proportional to this width. In this case of partially coherent illumination,

$$\mathscr{L}R_p = \left(\tau WL\frac{z'}{r'}\frac{m\lambda}{d}\right)V, \tag{45}$$

where z' is the height of the illuminated exit slit and τ the efficiency of the mounting. The coefficient V depends on the degree of coherence at the entrance slit.

As pointed out by Mielenz,[56] the maximum of the above product is always reached when the resolving power of the spectrometer is equal to about $0.8R_0 = 0.8\,m/dW_0$, no matter how the slit is illuminated; the highest possible luminosity associated with a given resolving power is obtained when the entrance pupil is fully illuminated. Then eqn. (45) becomes

$$\mathscr{L}R_0 = \tau - \frac{z'}{r'}\frac{\lambda}{0.8}\frac{m}{d}\varrho W_0^2 = A\frac{m}{d}\varrho W_0^2 \tag{46}$$

in which the values of W_0 and L_0 are obtained from eqn. (37)—the tolerance associated with incoherent illumination nearly always applying (Welford[13]). In order to compare different mountings without introducing grating characteristics, one can consider[57, 58] the value W_0' of W_0 in the extreme case $L_0 = 0$ as a reference. From eqn. (27) or (37) the influence of the height of the rulings on the value of W_0 can be deduced and, finally, luminosities can be compared for a given resolving power.

5.1. LUMINOSITY OF ROWLAND CIRCLE MOUNTINGS

In these mountings a change of focus is not possible ($C_2 = 0$), and the main aberration term[12–14] which involves w is the Seidel spherical aberration

$$\Delta(w, l) = C_4(w^2 + l^2)^2. \tag{47}$$

If the height of the rulings is neglected,[6] the optimum value W_0' of W_0 takes the form[57]

$$W_0' = 1.438\left(\frac{\lambda}{C_4}\right)^{1/4} = 2.417\left[\lambda R^3 \frac{\cos\alpha\cos\beta}{\sin^2\alpha\cos\beta + \sin^2\beta\cos\alpha}\right]^{1/4}. \tag{48}$$

Apart from the numerical value of 2.417 for the constant c, this is essentially the same as the currently used result[1] derived from geometrical optics ($c = 2.38$) or physical optics ($c = 2.51$). The difference arises from the fact that generally only one Fresnel zone is taken as effective for the phase contribution to the image, and a general tolerance of $\lambda/4$ is usually assumed. If the height of the rulings is taken into account, W_0 is expressed in terms of W_0' by

$$W_0\left(1 + \varrho^4 + \frac{10}{7}\varrho^2\right)^{1/8} = W_0'. \tag{49}$$

This equation holds for all normal incidence Rowland mountings. In the case of the Abney mounting for which $\beta = 0$,

$$W_0 = 2.417\left[\frac{R^3}{nm}\right]^{1/4}\tan\alpha^{-1/4}. \tag{50}$$

The values of W_0 are plotted in Fig. 5 as a function of λ for two different gratings *a* and *b*.

The luminosity resolution product given by ref. 57

$$\mathscr{L}R_0 = A\,\frac{m}{d}\,(W_0')^2\varrho\left(1 + \varrho^4 + \frac{10}{7}\,\varrho^2\right)^{-1/4} \tag{51}$$

increases with L_0, but W_0 and hence R_0 will decrease. The variations of $\varrho = L_0/W_0$ and of the luminosity

$$\mathscr{L} = AW_0'\,\frac{\varrho}{U} = A\,\frac{W_0'}{U}\left\{\frac{-5 + \sqrt{[25 + 49(U^8 - 1)]}}{7}\right\}^{1/2} \tag{51'}$$

in units AW_0' are shown on Fig. 6 as a function of the parameter $U = W_0'/W_0$.

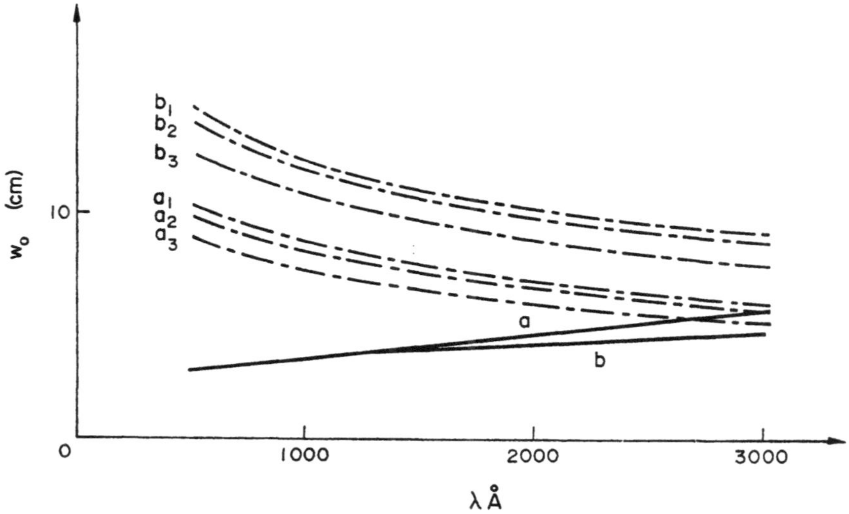

Fig. 5. Graphs of the optimum ruled area for the Abney (—·—·—) and the Wadsworth mountings fitted with the grating (a) ($R = 1000$ mm; $n = 2400$ lines per mm), and (b) ($R = 1000$ mm; $n = 1200$ lines per mm). Indices 1, 2, 3, correspond to the case of $\varrho = L_0/w_0 = 0$, $\varrho = 0, 5$, and $\varrho = 1$.

For large phase errors (grazing incidence mountings) W_0' takes the form[58]

$$W_{0t} = 1.743\left(\frac{Q}{C_4}\right)^{1/3}, \tag{52}$$

and eqn. (49) the form

$$W_{0t}\left(1 + \frac{\varrho^2}{105} + \frac{\varrho^4}{105} + \varrho^6\right)^{1/6} = W_{0t}'. \tag{53}$$

Figure 7 shows the variations of $\varrho_t = L_{0t}/W_{0t}$ and of the corresponding luminosity

$$\mathscr{L}_t = \frac{\tau z'}{5.29 R} C_4 W'^5_{0t} \varrho U_t^{-2} = A_t W'^5_{0t} \varrho U_t^{-2} \tag{54}$$

in units $A_t W'^5_{0t}$, as a function of the parameter $U_t = W'_{0t}/W_{0t}$.

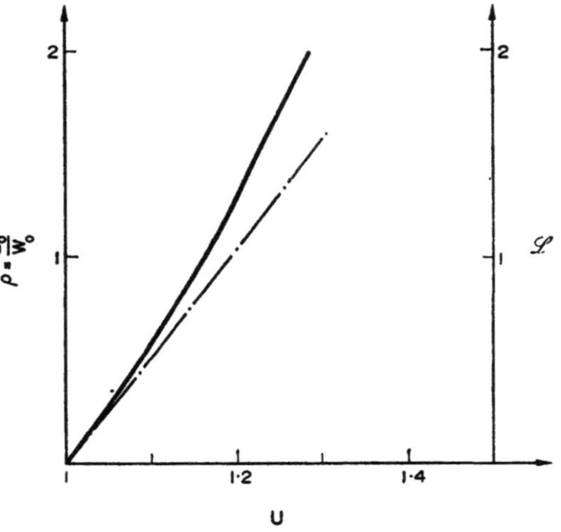

FIG. 6. Luminosity $\mathscr{L}$ (—·—·—) in units AW'_0 and $\varrho = L_0/W_0$ values as a function of U:W'_0/W_0 for normal incidence Rowland mountings.

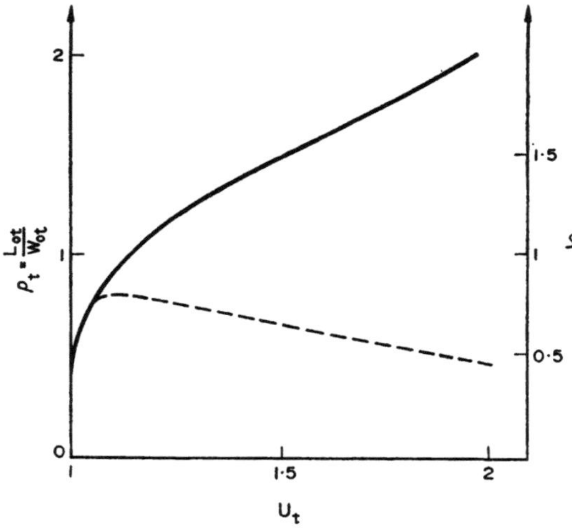

FIG. 7. Luminosity $\mathscr{L}_t$ (— — —) in units $A_t W'^5_{0t}$ and $\varrho_t = L_{0t}/W_{0t}$ values (———), as functions of $U_t = W'_{0t}/W_{0t}$ for grazing incidence Rowland mountings.

5.2. COMPARISON BETWEEN ABNEY AND WADSWORTH MOUNTINGS[58]

For the latter the aberration terms in the direction $\beta = 0$ reduce to

$$\Delta(w, l) = -\frac{\cos^2\alpha\,(1 + \cos\alpha)}{8R^3} \tag{55}$$

and

$$W_0 = 2.417 \left[\frac{R^3}{mn}\right] \left[\frac{\sin\alpha}{\cos^2\alpha\,(1 + \cos\alpha)}\right]^{1/4}. \tag{56}$$

It is seen from Fig. 5 that at $\lambda = 2650$ Å both the Abney and Wadsworth mountings, with gratings of the same optimum ruled area ($W_0 = L_0 = 54$ mm), have equal resolving power, while the luminosity of the latter is $\tau_c(1 + \cos\alpha)$ times (about 1.45) that of the former, assuming a reflectance $\tau_c = 85\%$ for the collimating mirror; the factor $(1 + \cos\alpha)$ arises from the ratio of the foci. Below this wavelength the resolving power is higher for the Abney than for Wadsworth mounting. Assuming τ_c values in use for this spectral range, the two mountings have the same luminosities at about 1650 Å, but the former has higher resolving power by a factor 1.65, and below this wavelength the increase of its optimum ruled area more than compensates the term $\tau_c(1 + \cos\alpha)$.

6. New corrected mountings

6.1. CORRECTED FAR ULTRAVIOLET SPECTROMETER DESIGN

As every spectrometer is characterized by luminosity and resolving power, the design of such a far u.v. instrument is mainly concerned with the choice of the most luminous mounting providing the effective resolution needed. The latter is given by the width of the triangular instrumental profile, which corresponds to the convolution product of the rectangular image of the entrance slit by the rectangular profile of the exit slit. If the mounting is not corrected, the image profile is broadened and the effective resolution decreases; in order to maintain the optimum effective resolution the entrance slit has to be narrowed, but the total incident flux on the grating will decrease.

Initially, from the condition $T + T' = 0$ the approximate values of the object and image distances are defined for a given mounting. It is then possible to calculate the aberration coefficients C_{ij}. For normal incidence these coefficients have generally small values, and $\Delta = \Sigma C_{ij} w^j l^i$ can remain inferior to λ even for high values of w and l. For a given height of the ruling the optimum ruled width can be deduced from the Strehl criterion [eqn. (37)], and the limiting resolution $\delta\lambda_{\text{lim}}$ is determined by

$$\delta\lambda_{\text{lim}} = \frac{\lambda d}{0.8\, m W_{0c}}. \tag{56'}$$

This value is obtained only if the condition $T + T' = \varepsilon$ is fulfilled, and for an exit slit width equal to $r'\lambda/0.8 \cos\beta\, W_{0c}$. However, generally the r'_{inst} distance between the centre of the grating and the exit slit determined by the linkage parameters is not strictly equal to the corresponding theoretical value; there results a residual instrumental defocusing $\delta\lambda_{\text{inst}}$ for which a tolerance can be estimated assuming approximatively that an aberration-free image is formed at the distance r' and observed at the distance r'_{inst}. In the case of small phase errors $\delta\lambda_{\text{inst}}$ must remain inferior to $t = 1.92\lambda d/mW_{0c} \simeq 1.54\delta\lambda_{\text{lim}}$. For high values of W_{0c} and L_{0c} (low resolution at normal incidence), or for high values of the C_{ij} coefficients (grazing incidence), the limiting resolution is given by the $\langle\delta\lambda\rangle$ value from eqn. (27), and the residual defocusing must remain inferior to

$$t' = \frac{2\sqrt{(3)}\,\langle\delta\lambda\rangle}{\cos\beta}.$$

6.2. CORRECTED SYMMETRICAL MONOCHROMATORS[14, 59]

For these mountings (Fig. 8) the object and image distances are equal and the grating is constrained to move along the bissector XX' of the angle 2θ subtended by the slits E and S at the centre of the grating and to rotate about an axis

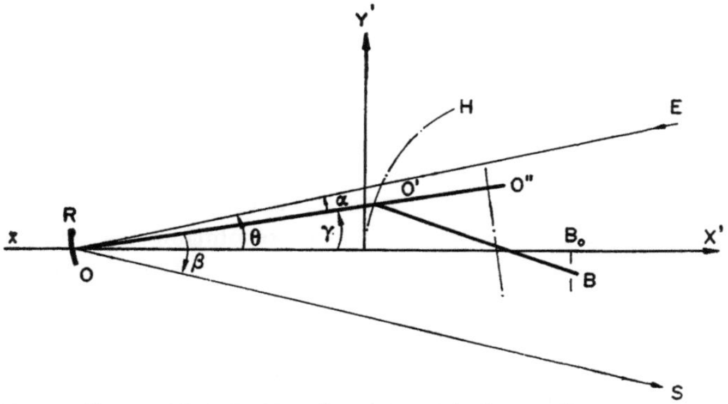

FIG. 8. Scanning mechanism for symmetrical mountings.

tangent to its centre and perpendicular to the plane $XX'ES$. Translation and rotation can be realized by two arms one of which $OO' = \varrho_1$ is normal to the grating, and the other $O'B = \varrho_2$ can rotate around a fixed point located at the distance $B_0B = \delta$ from the XX' axis.

The parameters $\varrho_1, \varrho_2, \delta$ are first computed in order to satisfy the first-order focusing conditions, which can be expressed by

$$U^3 - R\cos\gamma\, U^2 + f^2 U - R\cos\gamma\tan^2\gamma = 0, \tag{57}$$

where $U = r\cos\theta = f\cot\theta$ characterizes the distance between the grating and the intersection of the XX' axis with the segment ES of length $2f$, and γ is the angle of rotation of the grating. The difference between the theoretical and the mechanical displacement has to be minimized by a proper choice of the parameters ϱ_1, ϱ_2, and δ. In the next step this difference is determined in order to compensate for the high-order terms by solving the following equation:

$$\frac{1}{e}(\cos^2\alpha + \cos^2\beta) = \cos\alpha + \cos\beta + R\frac{W_{0c}^2}{2}\left[\frac{6}{7}(C_4 + C_{04}) + \frac{1}{3}(C_{22} + 2C_4)\varrho^2\right]. \tag{58}$$

In the case of a "Jobin et Yvon" grating having 1831.8 lines per mm and a radius of curvature of 997.9 mm, the optimum ruled width is about 71 mm for $L_{0c} = 54$ mm. Indeed, for the distance $ES = 2f = 200$ mm the coefficients C_{04} and C_{22} can be neglected in comparison to C_4 and then the optimum ruled area W_{0c} is given by[57–59]

$$W_{0c}\left(1 + \frac{49}{9}\varrho^4\right)^{1/8} = W'_{0c} = 1.367 W'_0, \tag{59}$$

where W'_{0c} is the optimum ruled area for $\varrho = 0$ in the corrected case and where W'_0 is given by eqn. (48). Then eqn. (51) becomes

$$\mathscr{L} = 0.655 A W'_0 U^{-1}(12.25U^8 - 1)^{1/4}, \tag{60}$$

where U is now equal to W'_0/W_{0c}. Figure 9 shows the ϱ values and the luminosity of this mounting, along with the luminosity of the corresponding Rowland type, versus U. At equal luminosities $\mathscr{L} = 0.75 A W'_0$ the resolving power of the corrected mounting is 1.46 times that of the Rowland mounting.

Computed results[59] show that the residual instrumental defocusing for the actual ruled area (54 by 54 mm^2) remains inferior to t; with the following linkage parameters $\varrho_1 = 453$ mm, $\delta = 0$ and $387.5 < \varrho_2 < 388.5$ mm, the limiting resolving power remains equal to 79,100.

This value is about 41,300 in the case of a grating having 2400 lines per mm and a radius of curvature of 400.7 mm, if the ruled area is limited to $W_{0c} = 21.5$ mm and $L_{0c} = 25$ mm (Fig. 10). In order to increase the luminosity in the case of lower resolution, the optimum values of the mechanical linkage parameters have been computed according to eqn. (27) in which W_{0c} and L_{0c} have respectively the value 40 and 30 mm corresponding to the actual size of this particular grating and from eqn. (58) in which, however, the coefficient of $C_4 + C_{04}$ is now $\frac{6}{5}$

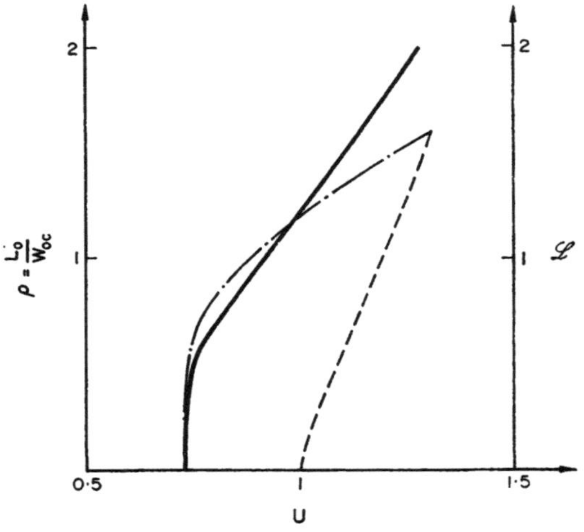

Fig. 9. $\varrho = L_{0c}/W_{0c}$ values (———) and luminosity $\mathscr{L}$ in units AW'_0 as functions of $U = W'_0/W_{0c}$ for the symmetrical mountings (—·—·—), and luminosity for the corresponding Rowland mounting (— — —).

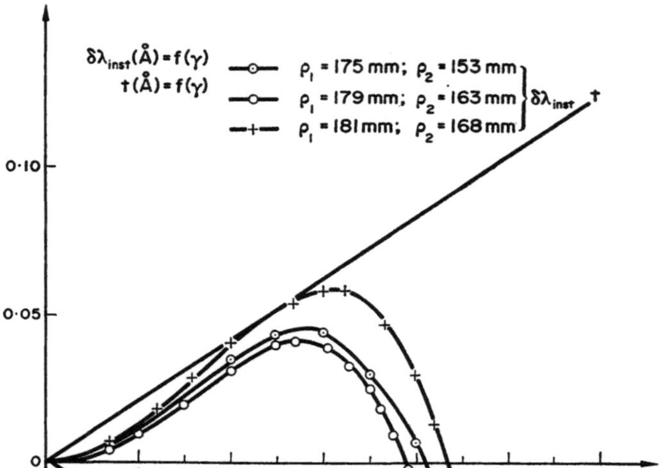

Fig. 10. Instrumental defocusing $\delta\lambda_{\text{inst}}$ and tolerance value t (———) as functions of γ (degrees) for a grating having 2400 lines per mm and a radius of curvature of 400.7 mm. The ruled area is limited to $W_{0c} = 21.5$ mm and $L_{0c} = 25$ mm. $\delta = 0$ mm.

instead of $\frac{8}{7}$. The results show that in this case (Fig. 11) the resolving power R_p remains about 18,300 at 1000 Å.

This example illustrates the gain in luminosity which arises from optical corrections and the possibility of engineering new mountings not based on the Rowland circle.

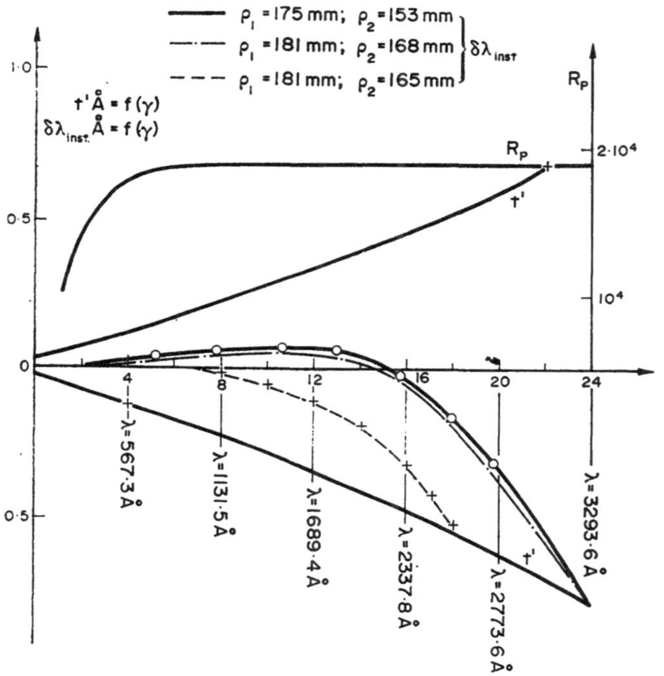

Fig. 11. Resolving power R_p (——), instrumental defocusing $\delta\lambda_{inst}$, and tolerance value t' (——) as functions of γ (degrees) for a grating having 2400 lines per mm and a radius of curvature of 400.7 mm. The ruled area is limited to $W_0 = 40$ mm and $L_0 = 30$ mm. $\delta = 0$ mm.

6.3. SPECTROPHOTOMETERS

There is at present a growing need to extend the range of spectrophotometers towards shorter wavelengths (below 2000 Å). While a high level of perfection has been attained in the technique of double-beam spectrophotometry in the infrared, visible, and near u.v. regions, the situation is not so favourable in the far u.v. This is due to the well-known practical difficulties encountered at short wavelengths. One may mention the problems of the beam splitter, of controlling the shape of the beam emerging from the exit slit of the monochromator, and of achieving high luminosity.

In a double-beam automatic ratio system the problem is to cancel out the wavelength- and time-dependence of the light source intensity, the transmittance and reflectance of the optical materials, the transmittance of the sample cell, and possibly of the solvent, and the sensitivity of the detector. This is done by setting up two equivalent paths, except for the sample absorption, and obtaining the absorption spectrum of the sample by an optical or electronic null method.

The first mountings[60, 61] circumvented the beam-splitter problem by inserting into the optical path a moving[60] blade coated with sodium salicylate in order to produce a reference beam. While this system compensates for source fluctuations in time and is usable with wavelength selectivity, it precludes the use of a reference cell because the characteristic radiation is not present, as such, in the reference beam.

To avoid tedious single-beam measurements and double-beam technical difficulties, Waggoner *et al.*[62] have proposed a mounting which employs two equivalent Hopfield continuum light sources with an intervening sample region. Direct-current discharges sources and electronic device are used for obtaining two time-separated signals instead of two separated beams. This arrangement is, in fact, a direct recording absorption spectrometer.

Some double-beam devices have been proposed in which the comparison is also made between two time-separated signals; they are limited to use of a steady source with a frequency much greater than the modulating frequency of the beam-splitter. Schmitt and Brehm[63] have proposed a beam-splitter device consisting of an oscillating toroidal mirror, at 11.3 c/s and arranged so that only a single reflection in each beam is required. Similarly, the McPherson model 225 double-beam attachment splits the light beam from the exit slit by means of a vibrating mirror, operated at grazing incidence. To avoid the numerous problems associated with beam splitting, Nihei *et al.*[64] alternatively introduce the sample and the reference cell in a single beam by rotating the cells at a suitable frequency.

In order to allow the use of any type of source, particularly the spark-source emitting a continuum in a wide spectral range,[65] a fixed beam-splitter is necessary. An attempt has been made in this direction by Enard and Pouleau[66] by using two lithium fluoride prisms having a small vertex angle to split the monochromatic beam, but this technique is limited in wavelength by the LiF cut-off. In the instrument described by Castex *et al.*[67] (Fig. 12) the beam-splitter consists of a plane reflecting mesh tilted with respect to the light beam emerging from the exit slit of the monochromator. Both the mesh and a mirror, which reflects the light transmitted through the mesh, are made of aluminium-coated nickel and are adjusted to have the same angle of incidence (70°). Under these conditions, the intensity ratio of the two beams can be kept constant to less than 5% in the range from 2500 to 1200 Å.

The monochromator employed is the corrected symmetrical mounting designed by Pouey[14] (see p. 313). The two separated beams are focused by toroidal mirrors M_1 and M_2 in order to traverse the corresponding cells. Finally, the beams are converted into visible light by two plates coated with sodium salicylate and detected by two photomultipliers. The photomultiplier dark currents are compensated by conventional circuits, and the two output signals are handled by an appropriate electronic device and sent to a ratio recorder.

An attempt has been made with this instrument to test the possibility of the optical balancing method, which is commonly used in conventional spectrophotometers. The balance was obtained by a comb attenuator (having the same transmission in every spectral region) placed in the path of the reference beam at the tangential image of the exit slit given by the focusing mirror M_2. It has been found

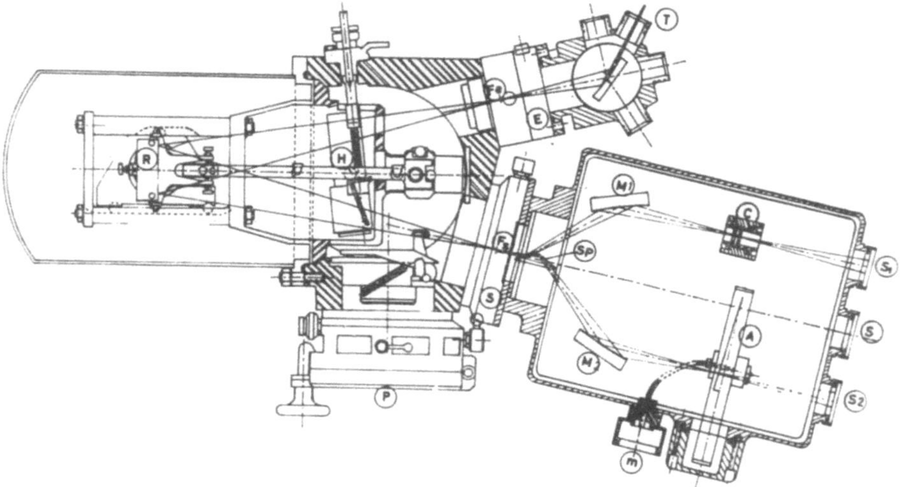

Fig. 12. Schematic of the spectrophotometer: *T*, source unit with toroidal mirror; F_e, F_s entrance and exit slits; *P*, photographic plate; *SP*, spatial beam splitter; M_1 and M_2, focusing mirrors; *C*, sample cell; *A*, comb attenuator; *M*, servomotor; S_1 and S_2, sodium salicylate plates.[67]

experimentally that a considerable improvement of the performance of the instrument is achieved by the removal of the comb and, thus of the very delicate optical adjustment of the astigmatism responsible for luminosity losses (factor of 4) and by the possibility of checking the scale of the intensity ratio. In the ratio recording method, the ratio of the two electric signals is plotted by the recorder on a linear scale to give the transmission. In this case there is no longer any need to produce a constant astigmatism at the comb position. Consequently it is possible to eliminate the astigmatism of the focusing system in the photometer by employing toroidal mirrors and thereby increasing the intensity.

As shown earlier by Lombos and Sauvageau,[68] even in the region below 1600 Å the hydrogen emission spectrum can be used as a light source with a reasonably good resolution (0.3 Å). The spectral range as well as the measurable optical density may be extended by using a windowless cell and the three electrode vacuum spark.[65] The use of a similar beam-splitter proved convenient for adapting a pulsed light source to photoelectric detection and to the scanning of a monochromator. The intensities of the two beams are compared for each spark, with no need to integrate over several sparks. The detection device comprises conventional nuclear electronics, including amplitude convertors and scaler counters.[69]

7. Vacuum ultraviolet reflective scattering distributions

How much does a roughened or coated surface reduce and redistribute reflected v.u.v. radiation? What is the meaning of the term "polished surface" in a spectral range where the r.m.s. height of the irregularities has the same order of magnitude as the wavelength λ? Although much attention has been given to scattering in other wavelength regions, only very little work[70–76] has been done at this time in the far u.v. to answer these questions.

Technically, a detector having a relatively small area is scanned at different angles around the sample and the detected energy integrated over the scattering angle. Let Φ_0 be the incident flux on a random surface A, and $\langle|\Phi_d|\rangle$ be the mean value of the scattered flux collected by the detector; let E_0^2 be the incident power density and $\langle|E_p|^2\rangle$ the mean value of the power density at one point P, where the active detector dimensions S subtend a solid angle $\Delta\Omega$. We have

$$\langle|\Phi_d|\rangle = \Phi_0 \frac{|\langle E_p|^2\rangle}{E_0^2} \frac{S}{A} = \Phi_0 \frac{A \cos^2 \alpha}{\lambda^2} \langle qq^*\rangle \Delta\Omega, \tag{61}$$

where α is the incident angle, and $\langle qq^*\rangle$ is essentially the mean square of E_p/E_0. The intensity distribution, in the direction α' is given by[74]

$$g(\alpha, \alpha') = \frac{\langle|\Phi_d|\rangle}{\Phi_0 \Delta\Omega} = \frac{A \cos^2 \alpha}{\lambda^2} \langle qq^*\rangle. \tag{62}$$

The measured ratio of the scattered to the incident flux characterizes the intensity distributions. In the case of a plane surface the luminance distribution is given by

$$f(\alpha\alpha') = \frac{g(\alpha, \alpha')}{\cos \alpha'} = \frac{A \cos^2 \alpha}{\lambda^2 \cos \alpha'} \langle qq^*\rangle. \tag{63}$$

According to the theory of Beckman and Spizzichino[77] $\langle q\rangle$ can be estimated by the Kirchhoff integral expressed in terms of the characteristic function associated with the distribution of the height of the irregularities. However, the rough surface is not only described by this distribution but also by the density of the irregularities, i.e. by the autocorrelation function. Then, with the restriction that both the height and the density of the irregularities are described by normal distributions, we can find $f(\alpha, \alpha')$ by determining the value of $\langle qq^*\rangle$, which is related to the variance of q, which in turn can be expressed in terms of the characteristic function of the height of the irregularities. A general series expansion can be made[77] for $\langle qq^*\rangle$ in which the relative importance of the various terms depends on the particular case involved.

For "polished" surfaces, if light of sufficiently long wavelength is used (i.e. $\sigma/\lambda \ll 1$)[78-81] the decrease in measured specular reflectance due to surface roughness is a function only of the r.m.s. height σ and, in the specular direction, $\langle|\Phi_d|\rangle$ is proportional to $1/\lambda^2$. Then, it has been shown mathematically that in other directions the scattered flux is proportional to $\lambda^{-4}\mathscr{F}(\alpha'/\lambda)$ where $\mathscr{F}(\alpha'/\lambda)$ is the Fourier transform of the autocorrelation function; but the experimental verification is very difficult because of the structure of the modified diffraction patterns.

As σ/λ increases, i.e. as λ decreases, the amount of incoherently reflected light tends to increase and may even dominate over the coherently reflected light. Waves producing incoherently reflected light are uncorrelated in phase unless they arise from the same localized region of the surface; the light intensity is then diffusely reflected in all directions, and the corresponding scattered flux is proportional to $1/\lambda^2$.

It is very important to consider this evolution in attempting to explain the characteristics of the far u.v. reflected profiles from polished surfaces. Indeed, measurements made at 1216 and 584 Å on several glass and fused silica substrates with various degrees of surface roughness, show[72] effectively that the profiles are broadened. According to recent experiments on machined metallic surfaces[75] and on ground or polished MgF_2 surfaces, the amount of the total scattered flux is proportional to $1/\lambda^2$, but depends also on the autocorrelation distance. Indeed, if the autocorrelation distance decreases, i.e. if the slope of the irregularities increases, the amount of diffusely reflected flux increases, then two surfaces having the same σ value can exhibit completely different luminance functions. An example of this is given by the values of a quantity, not too dependent on the reflectance, the scattering ratio (ratio of the scattered flux to the total reflected flux, the latter including the scattered and the specular reflected flux). For fused silica and several glasses[72] it is of the order of 0.20% at 1216 Å (particle size of the polishing agent 1.5-2 μ). For machined-metal surfaces[75] ($\sigma \simeq 0.15\ \mu$) the ratio is about 16% at 1608 Å.

Figure 13 shows the influence of substrate roughness and film thickness on the scattering from evaporated films at 1216 Å. Some values of the hemispherical reflectance at this wavelength have also been measured by Johnson[73] for specular and intermediate reflectors; for example he found 28.4% for evaporated aluminium. In the case of the no. 304 stainless steel, the hemispherical reflectance is about 10% for bulk polished samples and 15% for sandblasted.

However, as pointed out by Pouey,[82] even for "polished" surfaces the second law (total scattered flux proportional to $1/\lambda^2$) cannot be systematically used in the v.u.v. for all materials—in particular, if they are very hard to polish, like beryllium. In this case, σ may be greater than λ, and the "polished" surface must be considered to be rough. For such random rough surfaces the profiles are almost independent of the wavelength over a wide range of the v.u.v. This results from the general expression for the term $\langle qq^*\rangle$, or can be directly obtained by use of Fourier analysis, taking into account the fact that the phase variations are very large over the pupil. Then it can be shown[82] that, in the case of $\sigma/\lambda \gg 1$, the

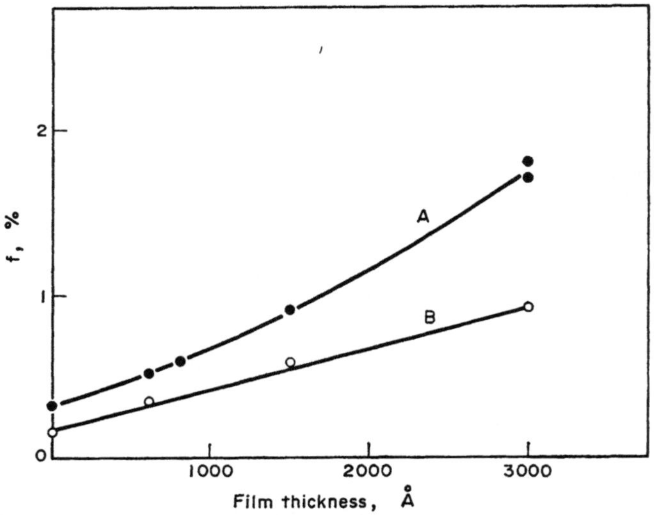

Fig. 13. Influence of substrate roughness and film thickness on the scattering from evaporated aluminium films at 1216 Å. Substrate *A* was commercial plate glass (as purchased). Substrate *B* was the same glass given a high optical polish before deposition of the aluminium. (Courtesy Johnston *et al.*[72])

important parameter is the characteristic function of the slope distribution $P(\tan \beta)$ and from eqn. (63) that the intensity distribution is given by

$$g(\alpha, \alpha') = \frac{\cot \beta_0}{\sqrt{(\pi)} \cos^2 \beta} P(\tan \beta). \tag{64}$$

It follows that, at normal incidence, and for low values of $\tan \beta_0$, the mean square value of the slopes, the scattering distribution resembles a Lambert distribution to a good approximation even for large values of α'. Effectively

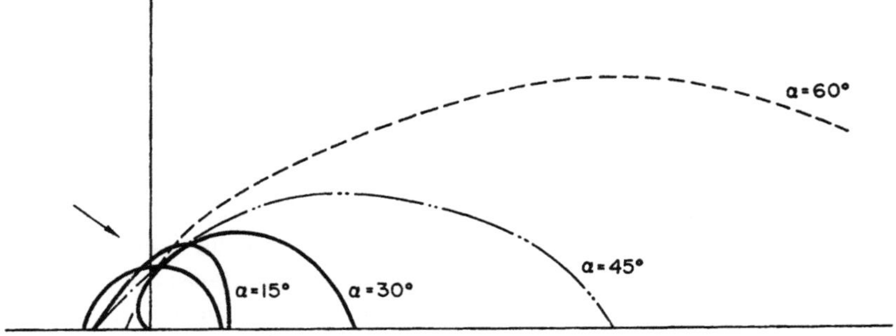

Fig. 14. Luminance distribution $f(\alpha, \alpha')$ for a ground glass surface at 1608 Å. At normal incidence $\alpha = 0$ the luminance distribution is like a Lambert distribution with an albedo of 5.6×10^{-2}.

Lambert distributions have been observed at normal incidence in the case of ground-glass surfaces[74] (Fig. 14) and of gold black on gold substrate[71] (Fig. 15) (hemispherical reflectance of 0.76% after[79]); this is also the case for a thick layer of sodium salicylate (fluorescence or v.u.v. distribution)[70, 74, 76] which can act as a perfect diffuser in this spectral range. Moreover, Johnson[78] and Pouey[74] have also observed that the reflectivity of surfaces which appear "black" does not

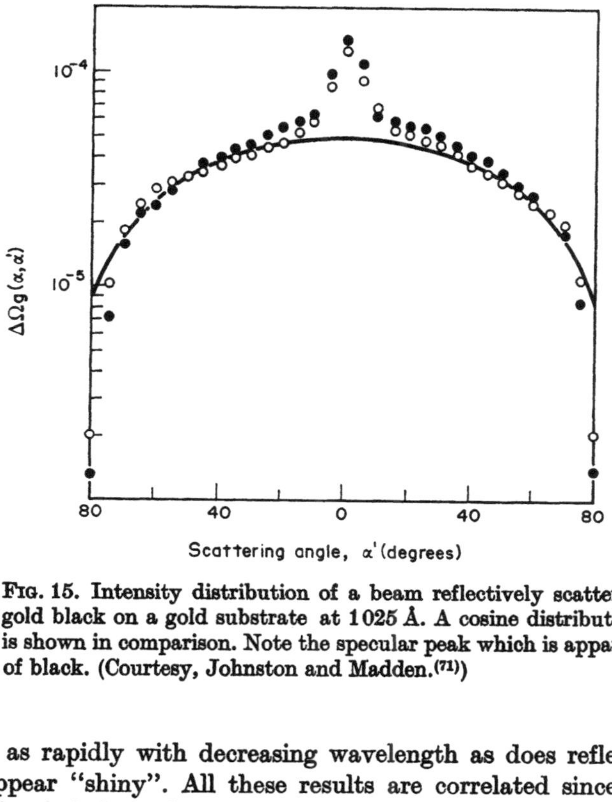

FIG. 15. Intensity distribution of a beam reflectively scattered from a sample of gold black on a gold substrate at 1025 Å. A cosine distribution (perfect scatterer) is shown in comparison. Note the specular peak which is apparent for this thickness of black. (Courtesy, Johnston and Madden.[71])

decrease as rapidly with decreasing wavelength as does reflectivity of surfaces which appear "shiny". All these results are correlated since, if the scattering distribution is independent of λ, then the variation of the reflectivity of diffuse reflectors from visible to far u.v. depends only on the specific reflectance. At large angles of incidence, off-specular peaks have been observed for both metallic and non-metallic surfaces.[83] Specular reflection from randomly disposed mirror-like facets, plus a diffuse component due to multiple reflections and/or internal scattering, has been postulated as the basic mechanism explaining the off-specular peak phenomenon.[82–83] This latter can be used to determine the r.m.s. slope of the irregularities and must be taken into account in the determination of optical constants from reflectance measurements.

References

1. Samson, J. A. R., *Techniques of Vacuum Ultraviolet Spectroscopy*, Wiley, New York 1967.
2. Rowland, H. A., *Phil. Mag.* **16**, 197 (1883).
3. Mack, J. E., Stehn, J. R. S., and Edlén, B., *J. Opt. Soc. Am.* **22**, 245 (1932).
4. Mack, J. E., Stehn, J. R. S., and Edlén, B., *J. Opt. Soc. Am.* **23**, 184 (1933).
5. Beutler, H. G., *J. Opt. Soc. Am.* **35**, 311 (1945).
6. Namioka, T., *J. Opt. Soc. Am.* **49**, 446 (1959).
7. Namioka, T., *J. Opt. Soc. Am.* **49**, 460 (1959).
8. Welford, W. T., *Optica Acta* **9**, 389 (1962).
9. Spencer, G. M., and Murty, M. V. R. K., *J. Opt. Soc. Am.* **52**, 672 (1962).
10. Rosendahl, G. R., *J. Opt. Soc. Am.* **51**, 1 (1961).
11. Neupert, K., *J. Opt. Soc. Am.* **53**, 10 (1963).
12. Welford, W. T., in *Progress in Optics*, North-Holland, Amsterdam, 1965, vol. 4.
13. Welford, W. T., *Optica Acta* **10**, 121 (1963).
14. Pouey, M., thèse, Paris 1965; Pouey, M., and Romand, J., *Rev. Opt.* **44**, 445 (1965).
15. Werner, W., *Appl. Opt.* **6**, 1691 (1967).
16. Namioka, T., in *Space Astrophysics*, McGraw Hill, New York, 1961.
17. Harrison, G. R., and Stroke, G. W., *J. Opt. Soc. Am.* **50**, 1153 (1960).
18. Stroke, G. W., and Stroke, H. H., *J. Opt. Soc. Am.* **53**, 333 (1963).
19. Tousey, R., Purcell, J. D., and Garett, D. L., *Appl. Opt.* **6**, 365 (1967).
20. Tousey, R., *Space Sci. Rev.* **2**, 3 (1963).
21. Kunz, C., Haensel, R., and Sonntag, B., *J. Opt. Soc. Am.* **58**, 1415 (1968).
22. Weinard, J., *Z. angew. Physik* **7**, 584 (1955).
23. Eagle, A., *Astrophys. J.* **31**, 120 (1910).
24. Sprague, G., Tomboulian, D. H., and Bedo, D. E., *J. Opt. Soc. Am.* **45**, 736 (1955).
25. Mendlowicz, M., *Phys. Rev. Letters* **9**, 149 (1962).
26. Romand, J., and Vodar, B., *Optica Acta* **9**, 371 (1962).
27. Vodar, B., *Rev. Opt.* **21**, 97 (1942).
28. Robin, S., *J. Phys. Radium* **14**, 551 (1953).
29. Tousey, R., thesis, Harvard University, 1933.
30. Romand, J., and Vodar, B., *Rev. Opt.* **39**, 167 (1960).
31. Seya, M., *Sci. Light* **2**, 8 (1952).
32. Namioka, T., *J. Opt. Soc. Am.* **49**, 951 (1959).
33. Johnson, P. D., *Rev. Scient. Instrum.* **28**, 833 (1957).
34. Onaka, R., *Sci. Light* **7**, 23 (1958).
35. Onaka, R., and Fujita, Y., *Sci. Light* **9**, 31 (1960).
36. Carlson, R. W., *Appl. Opt.* **8**, 1129 (1969).
37. Beutler, H. G., *J. Opt. Soc. Am.* **35**, 320 (1945).
38. Dieke, G. H., *J. Opt. Soc. Am.* **23**, 274 (1933).
39. Welford, W. T., *J. Opt. Soc. Am.* **53**, 766 (1963).
40. Herzberger, M., *J. Opt. Soc. Am.* **37**, 485 (1947).
41. Arakengy, A., *Am. J. Phys.* **25**, 519 (1957).
42. de Velis, J. B., *J. Opt. Soc. Am.* **55**, 165 (1965).
43. Miyamoto, K., *J. Appl. Phys. Japan* **26**, 421 (1947).
44. Rense, W. A., and Violett, T., *J. Opt. Soc. Am.* **35**, 311 (1945).
45. Haber, H., *J. Opt. Soc. Am.* **40**, 153 (1950).
46. Greneir, H., and Schaffer, E., *Optik* **16**, 288 (1959); **16**, 350 (1959).
47. Namioka, T., *J. Opt. Soc. Am.* **51**, 4 (1961); **51**, 13 (1961).
48. Purcell, J. D., and Tousey, R., *J. Opt. Soc. Am.* **47**, 1057 (1957).
49. Labeyrie, A., and Flamand, J., *Optics Commun.* **1**, 5 (1969).
50. Stephan, G., Lemonnier, J. C., and Robin, S., *Optica Acta* **12**, 345 (1965).

51. MIELENZ, K. D., *Optik* **13**, 437 (1956).
52. BOWEN, I. S., *J. Opt. Soc. Am.* **23**, 313 (1933).
53. KOZLENKOV, A. I., *Optika Spektrosk.* **8**, 692 (1960); [*Optics Spectrosc.* **8**, 365 (1960)].
54. STEEL, W. H., *J. Opt. Soc. Am.* **47**, 405 (1957).
55. MARECHAL, A., *Handbuch der Physik*, Springer-Verlag, Berlin, 1956, vol. 24.
56. MIELENZ, K. D., *J. Opt. Soc. Am.* **57**, 66 (1967).
57. POUEY, M., *Optics Commun.* **1**, 111 (1969).
58. POUEY, M., *Optics Commun* **2**, 339 (1970).
59. POUEY, M., *Optics Commun.* **3**, 201 (1971).
60. ONAKA, R., and EJIRI, A., *Appl. Opt.* **2**, 321 (1963).
61. SYS, A., ALEMANY, C., and JACOBS, G., *Rev. Scient. Instrum.* **36**, 385 (1965).
62. WAGGONER, M. G., CHOW, K. W., and SMITH, A. L., *Chem. Phys. Letters* **3**, 151 (1969).
63. SCHMITT, R. G., and BREHM, R. K., *Appl. Opt.* **5**, 1111 (1966).
64. NIHEI, Y., ODA, S., NIKI, E., and KAMADA, M., *Spectrochim. Acta* **24 A**, 457 (1968).
65. DAMANY, H., RONCIN, J.-Y., and DAMANY-ASTOIN, N., *Appl. Opt.* **5**, 297 (1966).
66. ENARD, D., and POULEAU, J., *Rev. Opt.* **44**, 557 (1965).
67. CASTEX, M. C., thèse, Paris 1967; CASTEX, M. C., MONTLOUIS, P., and ROMAND, J., *Rev. Phys. Appl.* **3**, 175 (1968); CASTEX, M. C., ROMAND, J., and VODAR, B., *Rev. Scient. Instrum.* **39**, 331 (1968).
68. LOMBOS, B. A., and SAUVAGEAU, P., *J. Molec. Spectrosc.* **24**, 253 (1967).
69. European Research Office of USA Army Contract nb DAJA 37-69 C 0756 and 37-70-C-0863.
70. ALLISON, R., BURNS, J., and TUZZOLINO, A. J., *J. Opt. Soc. Am.* **54**, 747 (1964).
71. JOHNSTON, R. G., and MADDEN, R. P., *Appl. Opt.* **4**, 1574 (1965).
72. JOHNSTON, R. G., CANFIELD, L. R., and MADDEN, R. P., *Appl. Opt.* **6**, 719 (1967).
73. JOHNSON, M. C., *Appl. Opt.* **7**, 879 (1968).
74. POUEY, M., FABRE, D., and ROMAND, J., *Optica Acta* **15**, 159 (1968).
75. POUEY, M., and FABRE, D., *Optica Acta* **16**, 471 (1969).
76. FABRE, D., and POUEY, M., *Optica Acta* **16**, 483 (1969).
77. BECKMANN, P., and SPIZZICHINO, A., *The Scattering of Electromagnetic Waves from Rough Surfaces*, Pergamon Press, Oxford, 1963.
78. BENNETT, H. E., and PORTEUS, J. O., *J. Opt. Soc. Am.* **51**, 123 (1961).
79. BENNETT, H. E., *J. Opt. Soc. Am.* **53**, 1389 (1963).
80. PORTEUS, J. O., *J. Opt. Soc. Am.* **53**, 1384 (1963).
81. MARECHAL, A., *Optica Acta* **5**, 70 (1958).
82. POUEY, M., *Optics Commun.* **1**, 105 (1969).
83. TORRENCE, K. E., and SPARROW, E. M., *J. Opt. Soc. Am.* **57**, 1105 (1967).

Note added in proofs

Different approaches have been made for solving the generalized focusing conditions, for symmetrical and Seya-Namioka mountings, in the cases of low and large phase errors. While symmetrical mountings afford a higher luminosity resolution product, new interesting solutions lead to simple rotation of the grating. In this case, the slits are fixed and the angle between incident and diffracted beams can be choosen between 10 and 170°. The pecularities of these new corrected mountings have lead to specific designs of monochromators. They may have several exit or entrance slits and they can match the geometry of the synchrotron radiation. Also, in connection with the rapid development of the so-called holographic concave gratings, the phase balancing method has been used to obtaining nearly stigmatic mountings in the v.u.v.

New relevant papers are listed in the next bibliography.

Additional bibliography

JOHANNIN-GILLES, A., Marche des rayons diffractés par un réseau sphérique concave dans un montage symétrique, *Compt. Rend.* **273B**, 952 (1971).

NAMIOKA, T., NODA, H., and SEYA, M., Possibility of using the holographic concave grating in vacuum monochromators, *Sci. Light* 22, 77 (1973).

POUEY, M., Corrected Seya-Namioka design, *Optics Commun.* **3**, 158 (1971).

POUEY, M., Nouveau procédé de focalisation des réseaux concaves, *Compt. Rend.* **272B**, 1200 (1971).

POUEY, M., Design of far ultraviolet corrected monochromators, *Third International Conference on Vacuum Ultraviolet Radiation Physics*, Tokyo, 1971, 31a A2-2.

POUEY, M., Design of new corrected v.u.v. mountings for synchrotron radiation, *First International Symposium for Synchrotron Radiation Users*, Daresbury, 1973, 88.

POUEY, M., Monochromateur à double entrée pour le test des composants optiques dans l'ultraviolet lointain, *Quatrième journées d'Optique Spatiale*, Marseille, 1973.

POUEY, M., Conditions de stigmatisme pour les montages de Rowland équipés de réseaux holographiques concaves, *Compt. Rend.* **276B**, 531 (1973).

POUEY, M., Conditions de stigmatisme pour les montages à simple rotation du réseau concave, *Compt. Rend.* **277B**, 459 (1973).

POUEY, M., PRIOL, M., QUÉMERAIS, A., and ROBIN, S., Commentaires sur le monochromateur à réseau concave en incidence normale, *Optica Acta* 18, 555 (1971).

PRIOL, N., QUÉMERAIS, A., and ROBIN, S., Monochromateur à réseau concave en incidence normale pour l'ultraviolet lointain, *Optica Acta* **17**, 603 (1970).

Index

Bei Fragen zur Produktsicherheit wenden Sie sich bitte an:
If you have any questions regarding product safety,
please contact:

Walter de Gruyter GmbH
Genthiner Straße 13
10785 Berlin
productsafety@degruyterbrill.com